D0622229

The Geology of the Mexican Republic

By

Dante Morán-Zenteno

Universidad Nacional Autonoma de México (U.N.A.M.)

Collaborators:
From U.N.A.M.:

J. Urrutia Fucugauchi
G. Silva Romo
C. Caballero Miranda
E. Cabral Cano
S. Alarcon Parra
G. Mora Alvarez
S. Campos

From Instituto Nacional de Estadistica, Geografia e Informatica (INEGI):

J. Alvaro Iruretagoyena
J. Luis Moreno
E. Campos Madrigal
J. Uribe Luna
J. Olivera

Translated
and with additional annotated bibliography by
James Lee Wilson
and
Luis Sanchez-Barreda

AAPG Studies in Geology #39

Published by
The American Association of Petroleum Geologists
1994
Printed in the U.S.A.

Translation of: ***Geología de la República Mexicana***
First Spanish edition, 1984, copublished by INEGI & UNAM
Second Spanish edition, 1985
First Spanish reprinting © 1990

ISBN: 0-89181-047-1

Association Editor: Kevin T. Biddle
Science Director: Richard Steinmetz
Publications Manager: Cathleen P. Williams
Special Projects Editor: Anne H. Thomas
Production: Custom Editorial Productions, Inc., Cincinnati, Ohio

On the cover: Canyon of Sumidero, incised in the Lower and middle Cretaceous Sierra Madre Limestone near Tuxtla Gutiérrez, State of Chiapas, southern Mexico. Photo by R. K. Goldhammer, Exxon E&P Research. Inset photo shows the Middle Cupido Formation in Cortinas Canyon, Sierra Madre Oriental.

This and other AAPG publications are available from:
The AAPG Bookstore
P.O. Box 979
Tulsa, OK 74101-0979
Telephone: (918) 584-2555; (800) 364-AAPG (USA—book orders only)
FAX: (918) 584-0469; (800) 898-2274 (USA—book orders only)

About the Author

Dante J. Morán-Zenteno has been a research scientist at the Geophysics Institute of the National University of Mexico (UNAM) since 1992. He spent eight years working in the geological mapping program of the National Institute of Statistics, Geography and Information Technology (INEGI). During this time he prepared most of the present book. Since 1984 he has participated in research projects of the Geophysics Institute dealing with the tectonic structure and paleogeographic evolution of southern Mexico. As part of these projects, he has applied paleomagnetic and isotope geochemical analyses to the study of the tectonic evolution of the southwestern continental margin of Mexico. Since 1983 he has been lecturing on geology at the Engineering Faculty of UNAM. From 1988 to 1990 he was at the University of Munich, in Germany, carrying out isotope analyses of Mexican rocks. In 1992 he received his Ph.D.

About the Translators

James Lee Wilson was born in Waxahatchie, Texas, and raised in Houston. He attended Rice University and the University of Texas at Austin, where he received his B.A. and M.A. degrees. He received his Ph.D. from Yale University in 1949.

Jim Wilson was a field geologist in the Rocky Mountains, Associate Professor at the University of Texas at Austin, and from 1953 to 1966 worked as a research geologist for Shell Development Company in Houston. During this period he spent three years in the Netherlands working on Mesozoic geology of the Middle East.

In 1966 Jim returned to academia as Professor of Geology at Rice University; he joined the University of Michigan in 1979. In 1975 he completed a book, *Carbonate Facies in Geologic History* (Springer-Verlag). Jim was President of SEPM in 1975–1976, became an Honorary member in 1980, and was elected an Honorary member of AAPG in 1987. In 1990 he received the Twenhofel Medal from SEPM. He has participated in carbonate field and lecture courses with the Laboratory of Comparative Sedimentation of Miami University, Florida; with ERICO of London; the University of Houston; AAPG; and MASERA Corp. of Tulsa, Oklahoma. His field experience includes work in Mexico, New Mexico, North Africa, the Rocky Mountains, the Austroalpine area, and the Middle East.

Jim is now Professor Emeritus at the University of Michigan and adjunct Professor at Rice University in Houston. He resides in New Braunfels, Texas. As a consultant, he is working on the geology of Mexico and is involved in a worldwide study of carbonate platforms.

Luis A. Sanchez-Barreda is currently senior consultant for Barreda and Associates, Navasota, Texas. He received his B.S. degree in Oceanography in 1972 from the University of Baja California and in 1976 an M.A. degree in geology from Rice University. He began his career as a field geologist in Libya and Spain. After receiving his Ph.D. from the University of Texas at Austin in 1981, he worked as an explorationist for Pecten International (Shell Oil Company). In 1987 he left Pecten to form his own consulting company. Luis has more than 20 years of geologic experience working in Mexico, and presently specializes in frontier exploration throughout Mexico, the Caribbean, and Central and South America. His main areas of interest focus on seismic/structural interpretation of sub-Andean, forearc, and passive margin basins of Latin America.

Table of Contents

Preface

PRESENTATION

The National Institute of Statistics, Geography, and Technical Information Technology (INEGI) and the Faculty of Engineering of the National Autonomous University of Mexico offer this work as a joint effort to contribute to knowledge of the geology of Mexico, employing the new concepts related to the dynamics of the earth and as a step toward the teaching and further development of professionals in Earth Science.

This volume collects and interprets a large part of the information gathered during more than 15 years of geological mapping by the General Directory of Geography and forms a compendium of scientific contributions related to the Geology of Mexico, many of which result from research investigations within the National Autonomous University of Mexico.

TRANSLATORS' NOTE

Even though ten years old, Dante Morán-Zenteno's summary of the Geology of the Mexican Republic remains the most complete report of this very large, structurally complex, and economically important area that forms the southwestern margin of the North American craton. Because Morán-Zenteno's work was in Spanish only, it has not become well known north of the border. Zoltan De Cserna's excellent 32-page Outline of the Geology of Mexico (1989, DNAG Volume A) was until this point the only other good general description of Mexican geology in English. The translators have attempted to render an accurate and readable English text while retaining some of the use of passive voice and indirect style of the elegant Spanish language.

The translators and the American Association of Petroleum Geologists are grateful to Dr. Dante Morán-Zenteno for his work and for his review of the translation. We are also grateful to the Instituto Nacional de Estadística, Geografía, e Informática (INEGI) for permission to publish this version, which is complete except for outcrop photos whose originals were lost after a move of INEGI headquarters from Mexico City to Aguascalientes.

The geological map at the scale of 1/1,000,000 (eight sheets) that accompanies the original text shows the general features of the geologic structure of the country, treating not only the different types of rocks that outcrop at the surface, but also the geologic times in which they were formed, i.e. their relative positions within the stratigraphic column.

However, the geological map at 1/1,000,000 scale described above has been recently superceded by a map of 1/2,000,000 scale that includes important changes, particularly in southern Mexico. This map is accompanied by text describing the various formations in the nation. For this reason the appendix of the original work, "Methodology of Formulation of the Geologic Map at a Scale of 1/1,000,000," has not been included in the English translation.

A fairly complete and briefly annotated bibliography of Mexican geology from 1983 to 1993 has been added to update the original text references and comprises Section II of this publication.

Dr. James Lee Wilson
Professor Emeritus, University of Michigan
Ann Arbor, Michigan, USA
and Adjunct Professor, Rice University,
Houston, Texas, USA

Dr. Luis A. Sanchez-Barreda
Barreda and Associates
Navasota, Texas, USA

Section I
The Geology of the Mexican Republic

Introduction

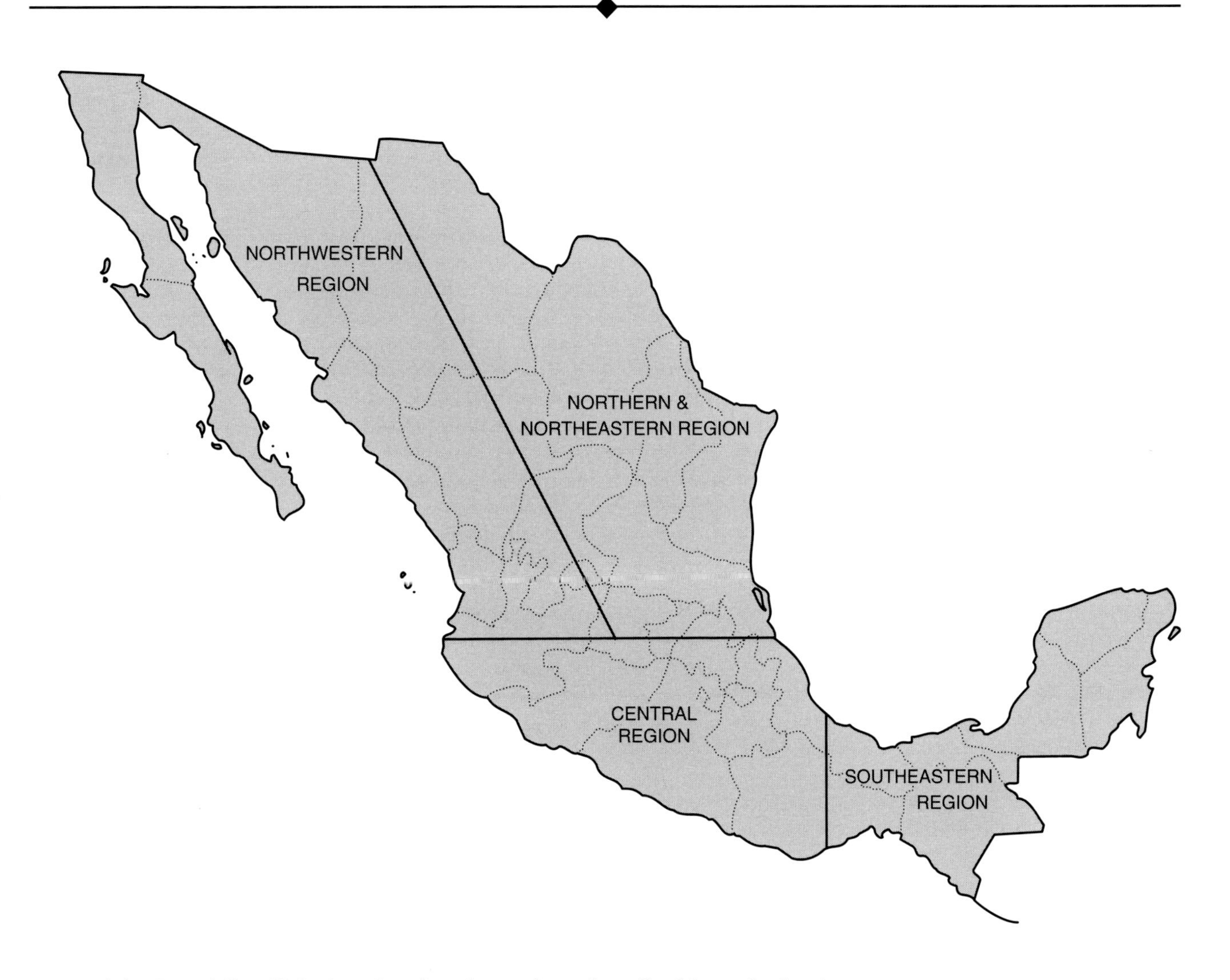

Map of the Republic of Mexico showing the regions described in each chapter.

Available knowledge concerning the origin and geologic structure of Mexico is still incomplete. Each day scientific discoveries, advances in mapping, and new techniques of exploration offer more information toward the development of our understanding. Nevertheless, it is still difficult to achieve a complete description of the geologic character of the territory of the nation, as well as to work out functional models to explain the origin of its geologic structure. Along with the development of geologic studies that science has made in Mexico, there have been a few attempts to formulate general works covering the many aspects of geology that the country presents. Nevertheless, one must recognize that the lack of information about certain periods in the geological history of the national territory, and the numerous unexplored areas, have constituted some principal obstacles toward achieving a finished work of this type.

In reviewing past information, it is worth indicating that in 1896 a somewhat unsettled state of knowledge resulted in a work in Spanish entitled *A Sketch of the Geology of Mexico* formulated by Jose Guadalupe Aguilera and Ezequiel Ordonez of the Geologic Institute of Mexico, a descriptive work that constituted an important complement to and summary for the *Geologic Map of Mexico,* which had been published earlier. Nevertheless, it was not until 1949 that V. Garfias and T.C. Chapin published the work entitled *Geology of Mexico,* in which reconstructions of the events that occurred during the geological history of the Republic are included.

A more recent work is *The Geology of Mexico,* whose author was Ing. Manuel Alvarez, Jr., and which the Faculty of Engineering of UNAM printed as notes of the subject matter of the *Geology of Mexico,* presented by the same author. Finally, in 1979, Ing. Ernesto López-Ramos published his work, *The Geology of Mexico,* in two volumes. That publication constitutes at present the most widely known text because it contains detailed descriptions of lithostratigraphic units and references to numerous unpublished works, principally from Petróleos Mexicanos.

The present book has the double objective of offering a geological synthesis of Mexico as a general reference work for all readers, and a presentation of themes in organized and didactic form so that it can be utilized in upper level courses related to the geology of Mexico.

The first edition of this work was the responsibility of the National Institute of Statistics, Geography, and Technical Information (INEGI) as a complement to the geological maps that the General Department of Geography had prepared. The preparation of the text was the responsibility of Ing. Dante J. Morán-Zenteno, then chief of Petrography and Paleontology of the same department.

This second edition is the result of combined forces of INEGI, an administrative unit decentralized between the Secretariat of Planning and Budget, and the Faculty of Engineering of UNAM. Ing. Dante J. Morán-Zenteno gives courses on the geology of Mexico and physical geology at UNAM and in addition is a researcher in the Institute of Geophysics at the same university.

To develop the present work it was necessary to divide the Republic into different regions, defined by natural limits, which are described in each of the chapters that form the work. This division does not correspond to that of the original eight maps of the Republic at the 1:1,000,000 scale, which were designed at a scale to show topography, culture, and use of the substrate. In this work, the geologic maps that pertain to each chapter are mentioned.

The information in this second edition can be used as a point of departure for regional projects of investigation and guidance. It offers, together with the geologic maps at 1:1,000,000 scale, a general key for localizing areas and objectives of economic or particular scientific interest. It should make clear, furthermore, key characteristics that are pertinent for geological interpretation of certain regions, and it comments on the most recent models concerning tectonic evolution and the geologic origin of significant economic deposits. The bibliography that accompanies each chapter makes it possible to organize a wide variety of consulting work aimed at studying in depth certain aspects or certain particular areas.

1. Geology of the Northwest Region of Mexico

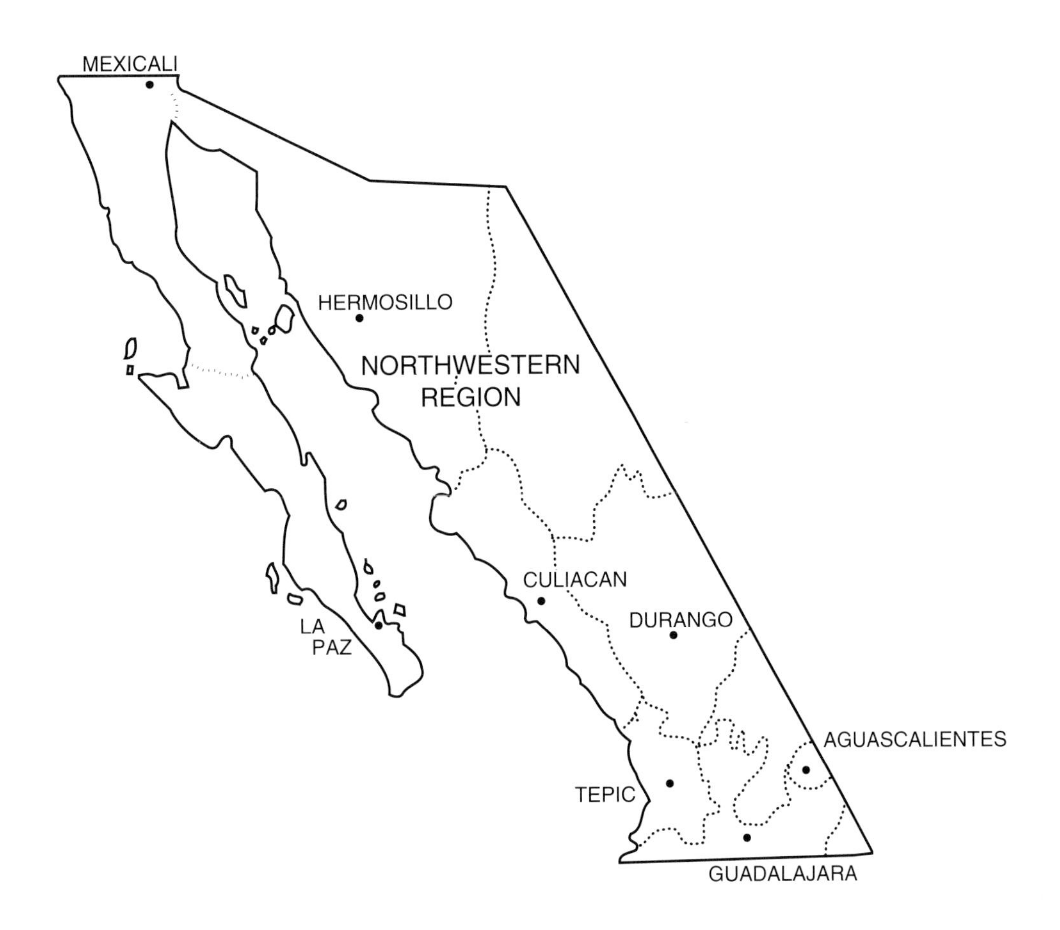

GENERAL CONSIDERATIONS

For the description of northwestern Mexico the following natural limits have been selected in this work: to the east, the volcanic sequence of the Western Sierra Madre; to the west, the Pacific coasts of Baja California and Sinaloa; and to the south, the northern edge of the Neovolcanic axis.

In accordance with the physiographic division of the General Department of Geography (see Figure 1.1), the provinces of Baja California, the Sonoran Desert, Sierra Madre Occidental, and the Pacific Coastal Plain are included within this region. The climate varies in general from dry in Baja California, Sonora, and northern Sinaloa, to subhumid in the higher parts of the Sierra Madre Occidental and south of Mazatlán. In almost all the region rainfall comes in the summer, except in the north of Baja California where the rain is in winter.

PENINSULA

The peninsula of Baja California as shown on the geologic map at 1:1,000,000 scale (General Department of Geography, DGG) offers a high structural complexity and rocky outcrops, which make it difficult to reconstruct a stratigraphic column for this region and to ascertain events that have occurred. Nevertheless a subdivision has been made, as rational as possible, that permits explanation, with a certain clarity, of the geologic concepts of this province and that coincides in large part with the physiographic divisions of the DGG and with the division into the geologic provinces of López-Ramos (1979).

Figure 1.1. Physiographic framework of the Republic of Mexico.

A Portion of Northern Baja California

In this zone exposures of a stratigraphic sequence whose geochronologic range varies from Paleozoic to Recent are encountered. The configuration of the different units forms three pre-Tertiary belts (Figure 1.2) that run the length of this part of the peninsula and that present clearly differentiated petrographic, structural, and stratigraphic characteristics. These belts are covered indiscriminately by volcanic bodies and Tertiary and Quaternary sedimentary deposits.

The first belt, located in the extreme western peninsula, is composed of a sequence of marine and continental Upper Cretaceous sediments that are poorly consolidated and lack appreciable tectonic deformation. This band of outcrops is of maximum width at the latitude of Punta San Antonio, a little less than lat. 30°N (Figure 1.3). The sequence was designated by Beal (1948) as the Rosario Formation and consists of subhorizontal strata of silty, shaly, and conglomeratic sandstone that contain both marine fossils and saurian bones. All this attests to the development of environments that vary from continental out to the platform and slope with fluctuating coastlines oriented more or less parallel to the line that divides this belt from the terranes located to the east. The latter constitute the source of supply for the sediments that comprise this sequence since in this time emergence occurred and formed mountainous masses exposed to erosion. Gastil et al. (1975) defined the outer limit of the cited belt as the "Santillan-Barrera Line" (Figure 1.2) and considered this feature to have controlled the depositional history of Baja California for long time periods. These authors cite numerous paleontological determinations that stratigraphically position the Rosario Formation in the Campanian and Maastrichtian Stages.

Mina (1957) correlated this formation with clastic sediments that crop out on the western border of the State of Southern Baja California and called it the Valle Formation.

The sequence that comprises this western portion of Baja California covers, in angular discordance, older volcanic and sedimentary rocks, as well as intrusives; it underlies Quaternary volcanics and continental and marine sediments of the Tertiary and Quaternary.

The next belt is located to the east of that described above and is made of sequences of volcanic rocks, volcaniclastics, and sedimentary rocks whose age is principally Lower Cretaceous (see Figure 1.4). The upper and more extensive part of the sequence was

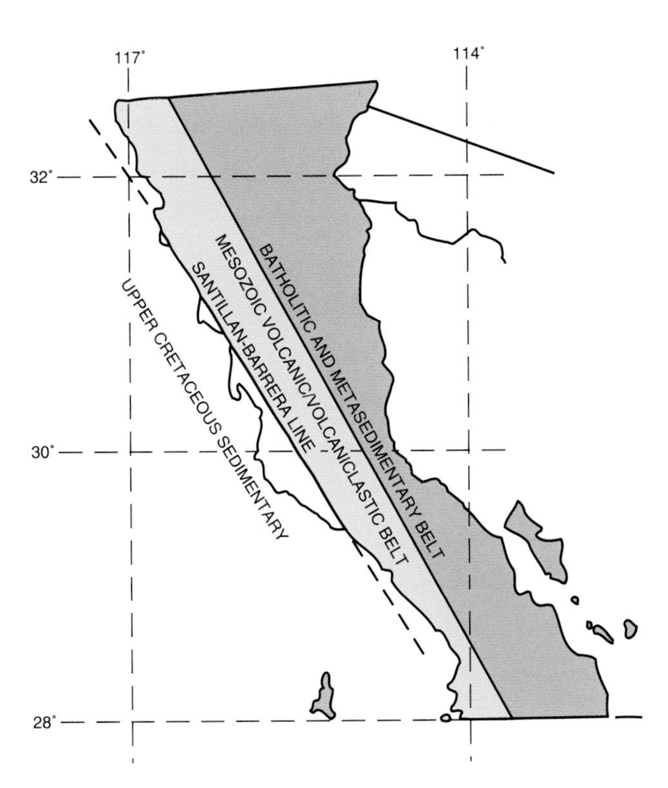

Figure 1.2. Pre-Tertiary terranes of Northern Baja California.

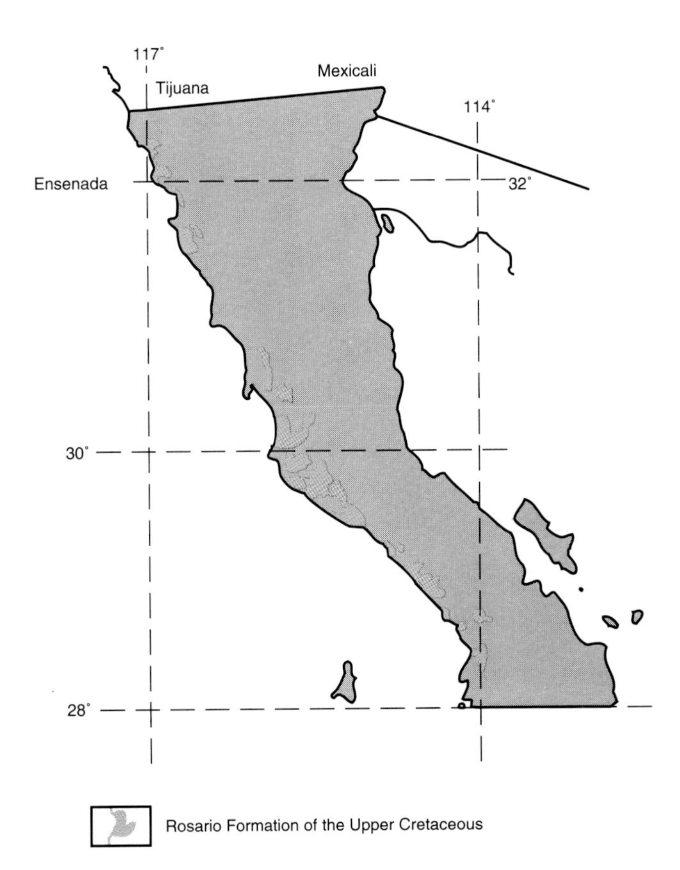

Figure 1.3. Distribution of outcrops of marine sedimentary rocks of the Upper Cretaceous.

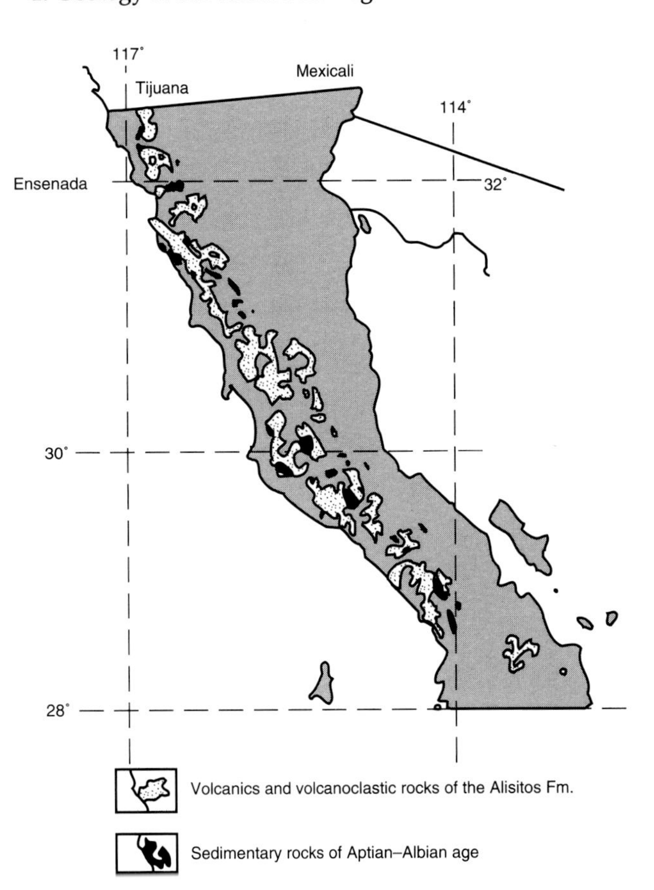

Figure 1.4. Distribution of volcanic, volcanoclastics, and sedimentary rocks of the Lower Cretaceous.

originally named by Santillán and Barrera (1930) the Alisitos Formation from exposures on the Rancho Alisitos located to the south of Ensenada. It is constituted chiefly of pyroclastic rocks and lavas of dacite-andesite composition, by bodies of reefal limestone with Aptian and Albian fossils, as well as clastic rocks derived from volcanics. This formation covers discordantly in some localities volcanic and sedimentary rocks of Triassic and Jurassic age. It is deformed and partly metamorphosed, and it is affected by numerous faults and by emplacement of bodies of intrusive granite of Cretaceous age. It underlies discordantly the Rosario Formation and extends persistently along all northern parts of the Baja California peninsula. Numerous outcrops of this type of sequence exist, correlative with the Alisitos Formation, mainly along the western border of Mexico.

Rangin (1978) has interpreted this sequence as one of the volcano-sedimentary belts that were developed in northwest Mexico during the Mesozoic, that formed in a similar manner to vulcanism in Sonora, and that evolved on continental crust. These belts have been related to subduction and partial fusion associated with one or more convergent borders (see Figure 1.5) developed in northwest Mexico. The convergent borders seem to be tectonic features common to all of western Mexico since there exist numerous volcano-sedimentary outcrops along this side of the country.

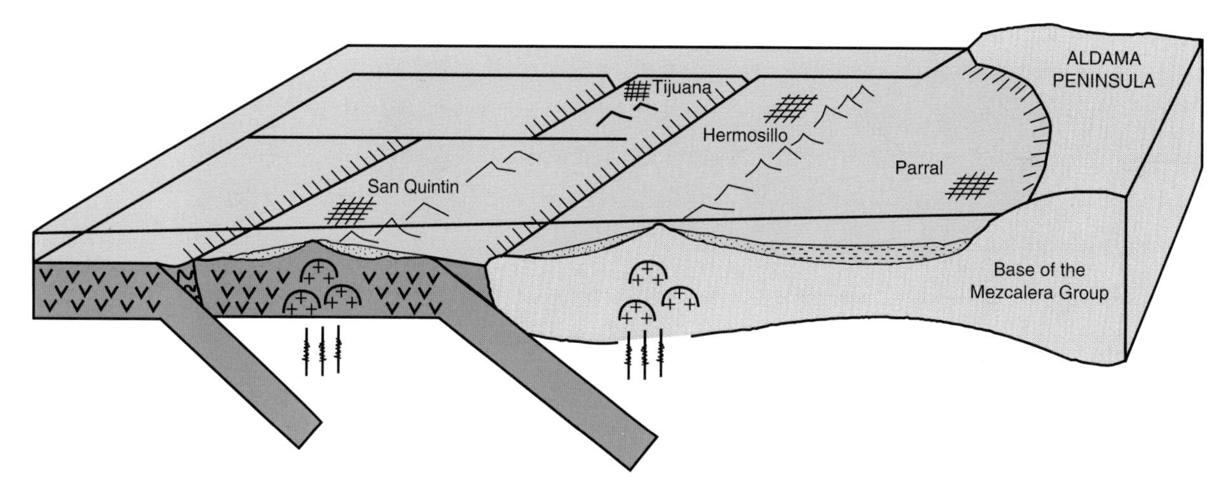

Figure 1.5. Idealized block diagram that shows the tectonic situation of northwest Mexico for the Late Jurassic. Based on ideas of Gastil et al. (1980), Márquez-Casteñeda (1984), and R. Garza (in Márquez-Castañeda, 1984).

This process developed during the opening of the Atlantic Ocean and the movement of North America toward the northwest. The sediments that form the Alisitos Formation were subjected to a period of compression at the beginning of the Upper Cretaceous. They were folded and partially metamorphosed. The terranes that form this second belt emerged during this period; and to the west of them sediments formed that were to become the Rosario Formation.

The third belt, located on the eastern border of the northern part of the peninsula of Baja California, is composed of complex outcrops of intrusive rocks and metamorphics derived principally from the regional metamorphism of sedimentary rocks. To this belt belong the Mesozoic batholiths (Figure 1.6) of the northern part of Baja California and the pre-batholithic metamorphic rocks formed before the Alisitos Formation. Their age has still not been well defined.

The plutonic rocks that comprise the batholiths vary in mineralogical composition from tonalites to granodiorites and granites. In contrast, in some localities small plutons of diorite and gabbro are mapped. Some authors (Gastil and Krummenacher, 1978; Silver and Anderson, 1978), citing radiometric studies, have postulated that in northwestern Mexico there occurred a migration in time and space of this type of plutonic emplacement from the Cretaceous in Baja California to the Cenozoic in Chihuahua. The major part of this batholithic emplacement occurred during and after the sedimentation and magmatic extrusions that originated the Alisitos Formation.

Pre-batholithic metamorphic sequences associated with this third belt present various metamorphic facies, but their ages have not been determined. However, McEldowney (1970) reported the presence of Paleozoic crinoids, corals, and bivalves in sedimentary rocks that crop out southeast of Ensenada. There also exist on the eastern edge of the peninsula some outcrops of metamorphic calcareous rocks that probably are related to the Paleozoic limestones which crop out in the State of Sonora.

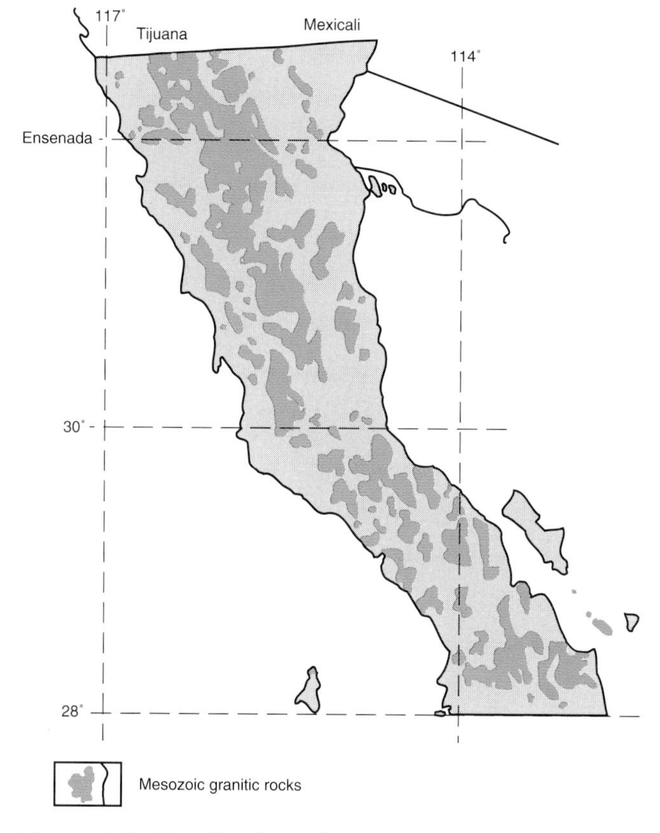

Figure 1.6. Distribution of Mesozoic plutonic outcrops in northern Baja California.

The Cenozoic history of the northern part of Baja California is characterized by the accumulation of great thicknesses of continental sediments that crop out in numerous localities; by the development of marine deposits, particularly on the western edge of the peninsula; and by important volcanic activity that partly covers the Mesozoic belt described above.

During the Paleocene and Eocene, sediments accumulated in nearshore and deltaic environments (Gastil et al., 1975) on the western border of the northern portion of the peninsula, following a coastline that is located slightly to the east of the present shore. These sediments came from emergent areas to the east where time-equivalent continental sediments are encountered.

Santillán and Barrera (1930) termed the marine Paleocene sediments that were encountered between Punta San Isidro and Mesa de San Carlos the Sepultura Formation. This formation can be correlated with the Santo Domingo, Tepetate, and Malarrimo formations described by Mina (1956) in the southern half of Baja California. The Pliocene and Miocene sediments correspond, it seems, to great thicknesses of fluviatile and eolian strata that crop out at lat. 31° in the area of San Augustín and some localities located about at the latitude of the bays of Las Animas and San Rafael. These sediments are found generally capped by lava extrusions of Miocene and Pliocene age. The Miocene contains outcrops of marine sediments that are the oldest Cenozoic strata to appear in the northeastern part of the Peninsula and that marked the earliest advances of the sea over the area that would become the Gulf of California.

According to Gastil et al. (1975), in the Eocene the Mesozoic mountains were completely denuded and formed only small isolated hills. These areas were drained by stream courses that flowed toward the Pacific and fed their sediments into the marine deposits on the western edge. Later in the Eocene, the eastern part of the region experienced some subsidence where the Gulf of California later developed. Some of the interior fluviatile stream courses were directed toward this area.

Coastal deposits that formed in the littoral of the Pacific during the Pleistocene are found above a series of terraces developed in that epoch. Some of these reach up to 500 m in altitude. These terraces have been related to glacial changes of sea level (Gastil et al., 1975) that were superimposed on a tectonic setting of a series of uplifts and downwarps in the coastal zone of the peninsula during the Pleistocene (Ortlieb, 1978). In contrast, in the interior of the peninsula during this time alluvial, eolian, and lacustrine deposits accumulated. Many of these sediments continue to develop today.

The Cenozoic vulcanism of the northern part of the peninsula can be referred principally to four zones where wide exposure of the volcanic rocks that originated in this area are encountered and that mark the Miocene as the major epoch of volcanic activity (see Figure 1.7). The first zone, located in the southern part of Sierra de Juárez between latitude parallels 31° and 32°, contains an important sequence of siliceous pyroclastic rocks of diverse types that are found capped in some localities by basaltic flows of Pliocene and Quaternary age (Figure 1.7). The second zone, located on the coast of the Gulf of California at the latitude of 30°, is represented by siliceous pyroclastic sequences that are seen to cover andesite flows in some localities and are capped in other places by basaltic flows of Pliocene–Quaternary age. The third zone corresponds to extensive flows of alkaline basalts of the upper Tertiary located in the central part of the peninsula at the latitude of Canoas Point. These flows are similar to those that are localized with minor distribution around the Mesa de San Carlos and San Quintín. The last zone is composed of outcrops of basaltic and pyroclastic rocks, principally rhyolites, that occur on the island Angel de la Guarda and to the south of the 29° parallel. These cover great thicknesses of continental and mixed sedimentary rocks. This zone can be considered as a northward extension of the Miocene sequence that constitutes the Sierra de la Giganta in southern Baja California.

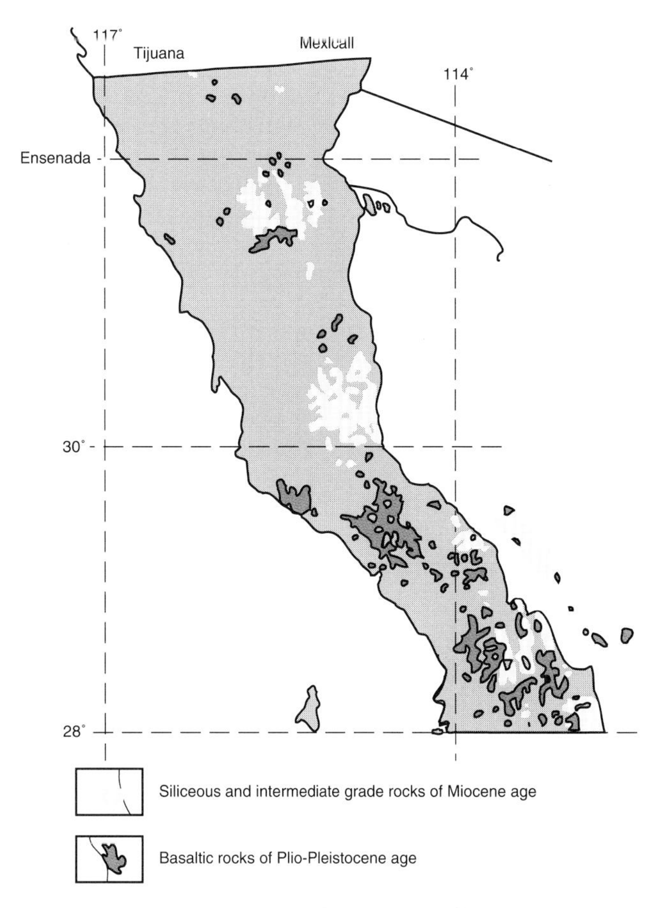

Figure 1.7. Distribution of outcrops of Cenozoic volcanics of northern Baja California.

Sierra de la Giganta

The Sierra de la Giganta, located in South Baja California, is composed of an impressive sequence of pyroclastic rocks, lava flows, and continental sandstones that together reach 1200 m in thickness. Outcrops of this sequence are persistent in most of the eastern half of the southern peninsula. Originally, Heim (1922) termed these deposits the Comondú Formation (see Figure 1.8) and assigned their stratigraphic position to upper Miocene. Later, Escandón (1977) indicated that the upper member of this formation belongs to the lower Pliocene. This sequence presents strong lateral variation and is composed principally of volcanic agglomerate, pumice-tuffs, ignimbrites, basalts, litharenite sandstones, and conglomerates. On the other hand, the strata cover discordantly the main sedimentary sequences of the Tertiary that crop out more widely in the basins of Purísima-Iray and Vizcaíno, and the plutonic rocks that are a southern continuation of the batholiths of northern Baja California. Mina (1956) considered that the source of supply of this great quantity of volcanic sediments should have been located in a volcanic belt to the east of the present coast of the Gulf of California.

The sequence that makes up the Comondú Formation does not show strong tectonic deformation. However, it reveals accentuated epeirogenic uplift and an inclination of its strata gently toward the west.

The Basins of Vizcaíno and Ballenas-Iray-Magdalena

The basins of Vizcaíno and Ballenas-Iray-Magdalena, which take in the western half of the larger part of the State of South Baja California, are represented by zones of low and smooth topography in which exposed sequences are encountered whose geochronologic ranges vary from Triassic to Recent. Structurally, these areas constitute two large synclinal depressions with general northwest–southeast orientation. They are composed of Cretaceous and Cenozoic rocks (see Figure 1.9). Lozano (1976) has interpreted the existence, at depth, of an uplifted block of ophiolitic rocks that separates these two structural depressions, based on geophysical data and wells drilled by Petróleos Mexicanos. This block might have an orientation perpendicular to the general structural tendency of the peninsula and would be located between parallels 27° and 28°. Above this structural high, the Cretaceous sequences wedge out, but these strata reach great thicknesses in the center of both depressions. The southwestern flank of these major structures is represented by outcrops of older rocks that form ophiolitic complexes and partly metamorphosed Triassic and Jurassic sequences (see Figure 1.10). In the axial portion of the structures, outcrops of the younger Cenozoic formations occur, while on the northeastern flank some bodies of the batholithic complex of Baja California are found, although generally these are covered by the Miocene and Pliocene sequence of the Comondú Formation.

The oldest sequence of this region is composed of partly metamorphosed volcanic and sedimentary rocks that crop out in Punta Prieta, Punta San Hipólito, and Cedros Island (Figure 1.10). Originally

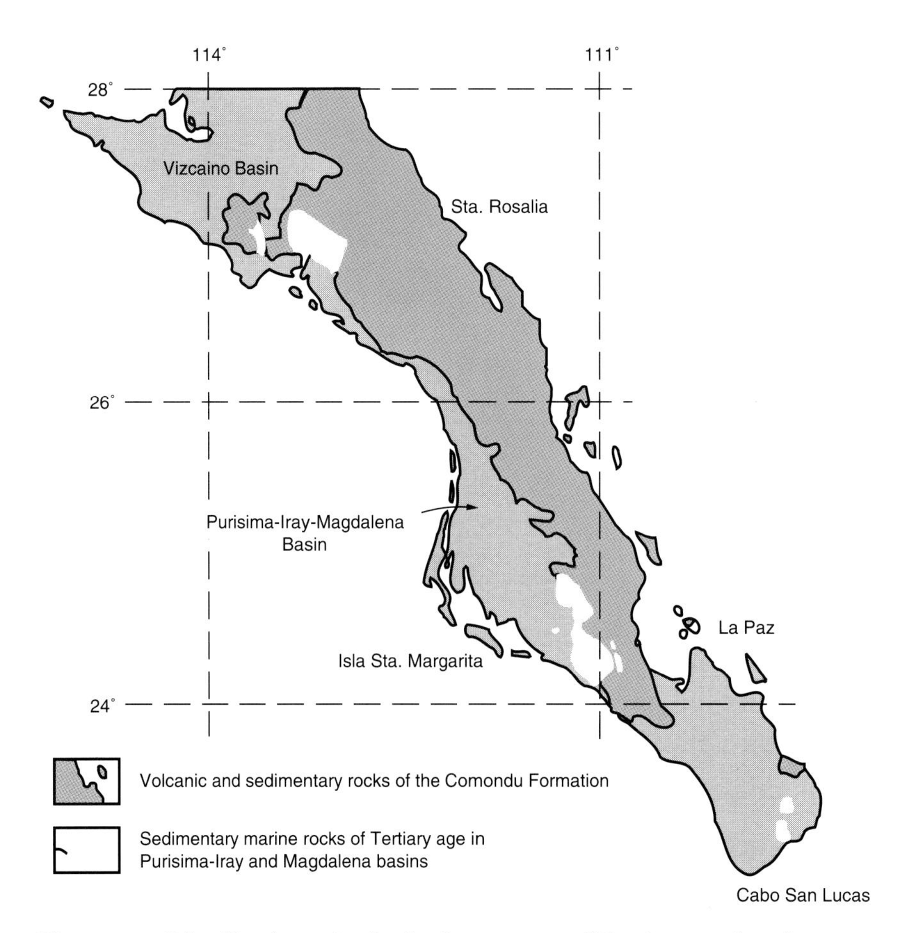

Figure 1.8. Distribution of principal outcrops of Tertiary rocks of southern Baja California.

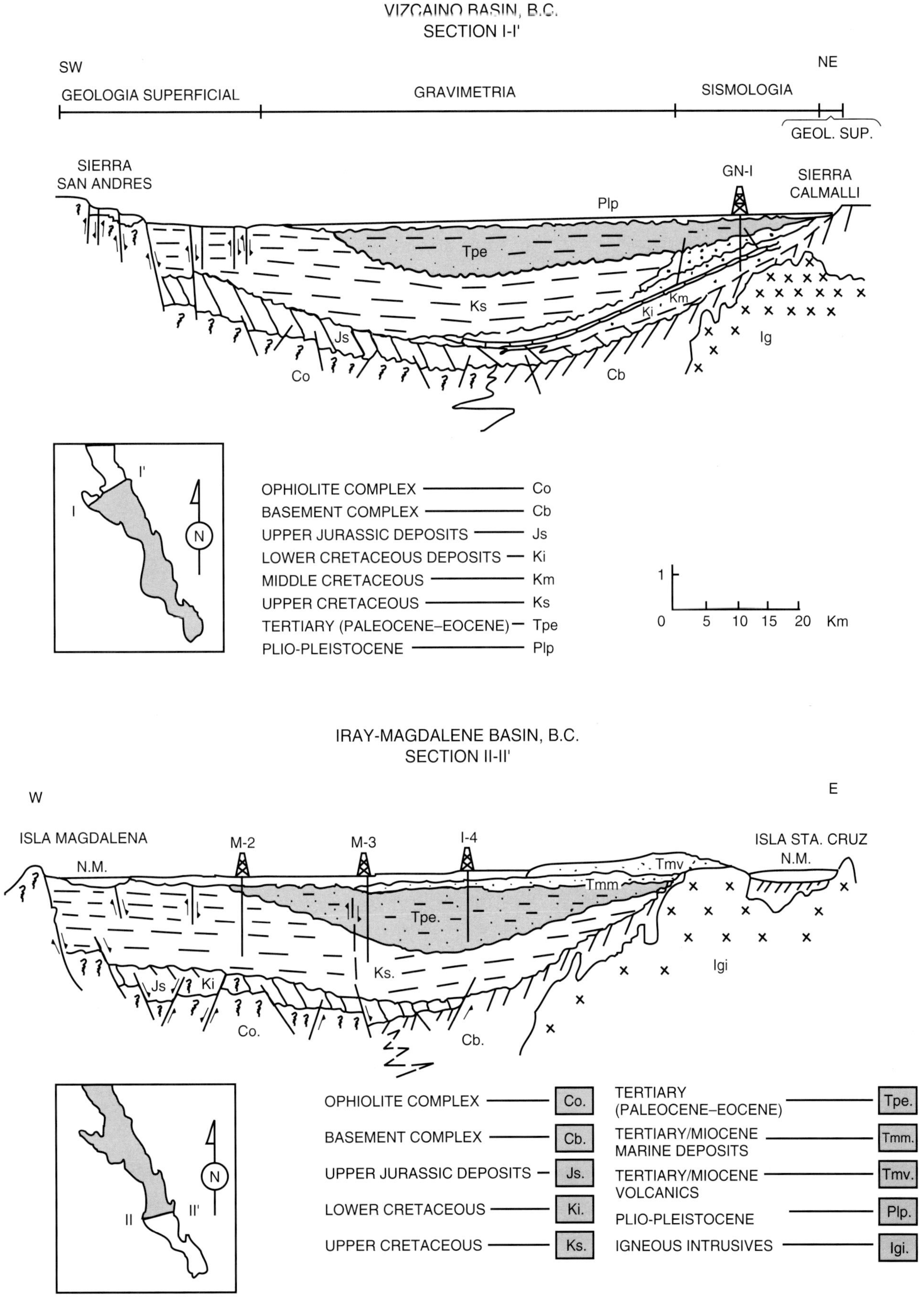

Figure 1.9. Vizcaíno Basin, B.C. sect. I–I′ and Iray-Magdalena Basin, B.C. sect. II–II′ taken from Petroleum Evaluation of the Peninsula of Baja California, by F. Lozano (1976).

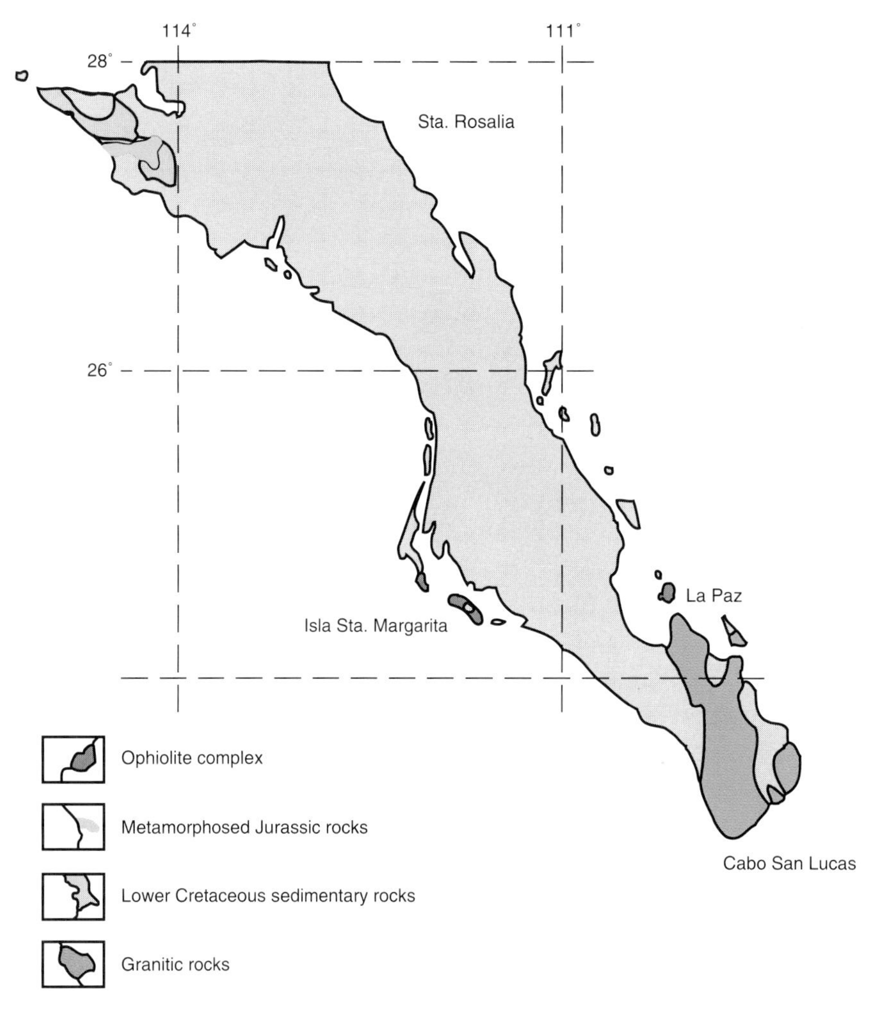

Figure 1.10. Distribution of the principal outcrops of Mesozoic rocks of southern Baja California.

Mina (1956) designated these rocks the San Hipólito Formation. They crop out in the type locality and correlate lithologically with the Franciscan Formation of California. For this reason he tentatively designated these rocks as Jurassic. Later, Lozano (1976) reported a fauna of the Late Triassic at the top of the sequence and for this reason considered the rocks to be of this age.

In Cedros Island, the Vizcaíno Peninsula, and the area of Magdalena Bay, rocks exist that petrographically resemble those of the Mesozoic and that form an intricate mosaic of terranes of both oceanic and volcanic arc affinity. Although Mina originally had postulated correlation with the Upper Jurassic Franciscan Formation, Finch and Abbott (1977) later placed these beds in the Upper Triassic, because of their content of macrofossils and radiolarians. The association of chert, litharenitic volcanics, and the inclusion of reefal limestone blocks forming a sequence underlain by pillow basalts, as well as the apparent absence of detritus derived from the craton, indicate that this unit was deposited in an oceanic basin associated with a volcanic island arc within a convergent tectonic framework (Finch et al., 1979; Gastil et al., 1981).

There also exist other outcrops of sequences with oceanic affinity that resulted in ophiolites and melanges; these also have been attributed to the Jurassic because of their radiolarian content (Rangin, 1978). These units crop out both on Cedros Island and on the Vizcaíno Peninsula. On the islands of Santa Margarita and Magdalena, partially serpentinized ultramafic rocks crop out and are apparently a fraction of an ophiolitic complex related to those of Vizcaíno and Cedros. There have been recognized in this region combined volcanics, volcaniclastics, and sedimentary rocks of Late Jurassic and Early Cretaceous age with an ophiolitic basement-forming sequence originally termed the Eugenia Formation by Mina (1956).

The Upper Cretaceous is represented in the region of southern Baja California by a detrital sequence of Cenomanian to Maastrichtian age that overlies earlier sequences with apparent angular discordance. This unit was designated the Valle Formation by Mina (1956) and includes turbiditic toe-of-slope fan deposition (Patterson, 1979). It has been recognized in outcrops of the Vizcaíno Peninsula and in the subsurface of the two Cenozoic basins of this region.

The Cenozoic sedimentary formations form the larger part of the basin-fill of Vizcaíno and Purísima-Iray-Magdalena downwarps and are principally characterized by little consolidation, subhorizontal position of the strata, and a marine clastic lithology.

Outcrops of Paleocene sediments are rare, notwithstanding that a thickness of more than 2000 m can be recognized in the subsurface, thanks to the wells drilled by Petróleos Mexicanos (Lozano, 1976). These wells encountered diverse lithologies with predominance of slope shale facies. To this epoch belong the Santo Domingo and Malarrimo formations (Mina, 1956), the latter of which rests in discordance on the Cretaceous formations. The outcrops of the Eocene are represented principally by sandy and shaly sequences that have been designated as the Bateque Formation in the area of Vizcaíno and as the Tepetate Formation in the La Purísima area where the lower part of the sequence belongs to the Paleocene. Sediments corresponding to this epoch have been recognized in the Pemex wells (Lozano, 1976), principally in the La Purísima area where they reach a thickness of up to 500 m. In this part of Baja California, Oligocene sedimentary rocks do not crop out, attesting to a period of emergence for that time.

The Miocene is found amply exposed in the regions of Vizcaíno and Purísima and consists of sedimentary and volcanic rocks. The lower Miocene is represented in the area of Vizcaíno by agglomerates, sands, and shales of the Zacarías, Santa Clara, La Zorra, and San Joaquín formations (Mina, 1956). In the area of Purísima it is composed of lutites with diatomite intercalations of the Monterrey Formation (Darton, 1921) and white sandstones of the San Gregorio Formation (Heim, 1922). The middle Miocene is formed by diverse sequences that show much lateral variation and are composed of tuffaceous sandstones, bentonitic shales, and sandstones of the Isidrio (Beal, 1948), San Ignacio, Tortugas, and San Raymundo formations (Mina, 1956) that represent the coastal and lagoonal platform environments.

The above formations underlie discordantly the continental sedimentary and volcanic deposits of the Comondú Formation that reach maximum development in the Sierra de la Giganta, located to the east of this described region.

During the Pliocene, sediments of coastal environments were deposited in the Vizcaíno and Purísima basins, with discordance over the Miocene formations. These are represented by the Almejas Formation in the area of Vizcaíno (Mina, 1956) and the Salada Formation in the Purísima area (Heim, 1922).

Region of the Cape (El Cabo)

The extreme south of the Baja California peninsula breaks abruptly across the general geologic aspect of the basins described above; the area is formed by a massive batholith that is expressed in the form of a mountainous complex and that is interrupted in the central part by a depression known as the Santiago Valley and in the northern part by the La Ventana Valley.

The batholith that composes this mountainous zone has characteristics similar to those masses that crop out in the north of Baja California and is made up of granodiorite and granite. The rectilinear borders of these mountains suggest faults of great displacement that juxtapose and elevate the batholithic region above the level of the areas of Cenozoic outcrops.

In the northern and western portions of the Sierra de la Laguna, one can recognize the existence of a prebatholithic metamorphic complex formed principally by metasedimentary rocks derived from lutites, sandstones, and limestones with some apparently metavolcanic bodies bearing epidote and amphibole. In the metasedimentary sequence, Ortega-Gutiérrez (1982) has identified a closely juxtaposed succession of isograds of biotite, andalucite, sillimanite, and cordierite. There exist also cataclastic belts and dioritic and gabbroic intrusions that form north–south-oriented lineations. In the neighborhood of the main batholithic body, inside the metamorphic complex, zones of migmatites and numerous intrusions of felsitic character are present apparently associated with the batholith.

The Santiago Valley is structurally a tectonic graben in which the principal sedimentary sequences of the region are developed. The base of these sequences is formed by conglomeratic deposits that are correlative with the Comondú Formation and that rest above the crystalline basement represented by intrusive Cretaceous rocks. The outcrops of these deposits are located chiefly in the extreme north of Santiago Valley. Over the above sequence, the sediments of the Pliocene Trinidad Formation rest with angular discordance (Pantoja-Alor and Carrillo-Bravo, 1966). These sediments form a sequence of sandy clays and silts with some diatomite horizons, all of which attest to a marine depositional environment. Above this unit rests in concordance a sequence of marine sandstones that represents the Salada Formation (Heim, 1922), which has isolated outcrops along the Santiago Valley. The Cenozoic sedimentary sequence that fills this tectonic graben is covered discordantly by a series of sandy-conglomeratic deposits of the Pleistocene that are seen in the form of ancient piedmont fans and belts.

Tectonic Summary

The principal tectonic elements of the Baja California Peninsula can be summed up as follows (see Figure 1.11):

1. It is possible to recognize on the western border at Cedros Island, the Vizcaíno Peninsula, and the islands in Magdalena Bay the presence of combined tectonically controlled rock types of oceanic affinity that include portions of ophiolite complexes and typical melange sequences that range from Triassic to Late Jurassic. These combinations of sediments have been interpreted to result from the structural evolution of a paleo-oceanic crust and to mark an ancient line of convergence. They have been related, furthermore, to similar lithologic

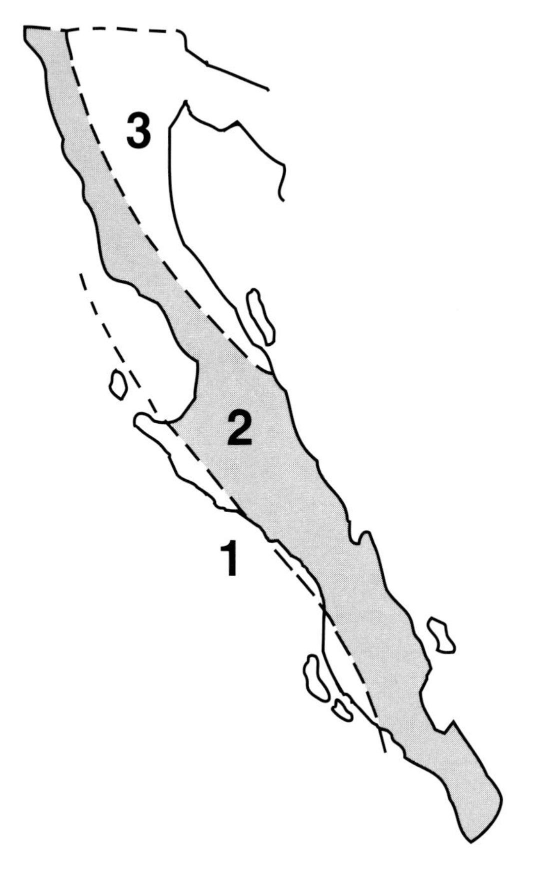

Figure 1.11. Tectonic belts of Baja California (1) dominantly of oceanic affinity; (2) dominantly volcanic, volcanoclastic, and sedimentary rocks of the Jurassic and Cretaceous; (3) dominantly metasedimentary rocks of Paleozoic?–Triassic age.

combinations in California that include the Franciscan Formation.

2. To the east of the above combinations of sediments, a volcanic-plutonic fringe of Jurassic to Early Cretaceous age is developed, at least partly, on oceanic crust and constitutes an ancient calcalkaline volcanic arc like those that typically evolve as fringes parallel to convergent borders. This volcanic-plutonic belt occurs in the western half of northern Baja California, is extended beneath the volcanic cover of the Sierra de la Giganta, and stretches probably to the El Cabo region.
3. To the east of the above domain, a belt of metasedimentary continental border clastic sequences appears (Gastil et al., 1981), overlapped partially by the combined volcanic-plutonic rocks. This belt is probably of Triassic age and forms the eastern half of northern Baja California. In the extreme east, some isolated outcrops of calcareous and very badly deformed detrital rocks occur that have been attributed to the Paleozoic.

The tectonic evolution of Baja California during the Paleozoic seems to have been related to the Cordilleran continental margin of the western border of North America, but nevertheless offers some distinctive details in its own evolution. There are sequences exposed in the east of northern Baja California, in addition to the calcareous and detrital sequences of Sonora, which reveal the presence during this era of a passive margin domain for northwest Mexico. This type of tectonic situation has been interpreted for a large part of the North American Cordillera. Two episodes of orogenic deformation have been identified for this region. The first of them occurred in the Devonian–Carboniferous (Antler Orogeny) and the second in the Permian–Triassic (Sonoma Orogeny). Both events have been interpreted recently as marking pathways of collisions of intraoceanic arcs against the passive margin of North America that induced the emplacement of allochthons of the Roberts and Galconda Mountains over the miogeoclinal sequences of the Cordillera (Dickinson, 1979).

In contrast, passive margin conditions are recognizable between these two events. In northern Baja California, neither episodes of collision nor allochthonous arc sequences have been identified. However, Gastil and his co-workers (1981) have suggested the possible existence of a trench or basin marginal to the cratonic edge.

In northern Baja California, passive margin conditions persisted during the Triassic, evidenced by an apparent tectonic stability in Sonora and by the absence of volcanic arc components in the metasedimentary sequence of the Peninsula. Only the Upper Triassic San Hipólito Formation in the Vizcaíno region reveals a probable boundary of convergence toward the ocean interior, later accreted landward (Gastil et al., 1981).

In the Jurassic, the development of an island arc dominion was initiated to the west of the cratonic border and its Triassic sedimentary wedge. This arc apparently evolved contemporaneously with the one reported above the continental crust in Sonora (Rangin, 1978). Gastil et al. (1981) consider that these two arcs were associated with different zones of subduction that evolved in parallel, one of them related to an intraoceanic trench and the other to a trench bordering the craton (see Figure 1.5).

The collision of the intraoceanic arc related to the Alisitos Formation against the cratonic margin apparently occurred in distinct episodes owing to the presence of transform faults between trenches that displaced distinct segments of the arc. The principal episodes of collision seem to have occurred in Baja California during the Cenomanian. This phenomenon generated a primary phase of deformation that folded, metamorphosed, and elevated the volcanic, volcanoclastic, and earlier sedimentary sequences at the same time of the principal emplacement of batholiths (Gastil et al., 1981).

In the Late Cretaceous and part of the Cenozoic, only a subduction zone persisted, located on the western margin of Baja California and marking the convergent boundary between the Farallon and North American plates. Generally the uplifts of Baja California and northwest Mexico during the Late Cretaceous made for an important contribution of detritus directed toward the east within a general framework of eastward marine regression. The vulcanism associated with the subduction on the western margin of Baja California during the Late Cretaceous and Paleogene has been recognized, chiefly in the continental part of Mexico, and as late as Miocene time it is expressed in the Peninsula by pyroclastic sequences in the Sierra de la Giganta and other eruptive centers in northern Baja California.

In Oligocene time the collision of the Pacific ridge with the North American plate was initiated. This ridge divided the Farallon plate, now extinct, from the Pacific plate and is apparently formed of segments displaced by numerous transform faults. According to a model of McKenzie and Morgan (1969) and Atwater (1970), the collision of the first segment of the ridge against the North American plate was initiated approximately 30 million years ago, at a point located in present-day Baja California. Starting from the first contact of the Pacific and North American plates, there began a right lateral movement along the growing border of both plates with a velocity of 6 cm per year (see Figure 1.12). This lateral movement could have occurred in its initial stages along the continental border of North America and later could have occupied the present belt of the San Andreas System and the Gulf of California (Atwater, 1970). The opening of the Gulf of California and the development of its ridge system was initiated about 4 million years ago. This system is the manifestation of the relative movement between the North American and Pacific plates and is the southern prolongation of the San Andreas System. The movement of Baja California toward the northwest is possibly

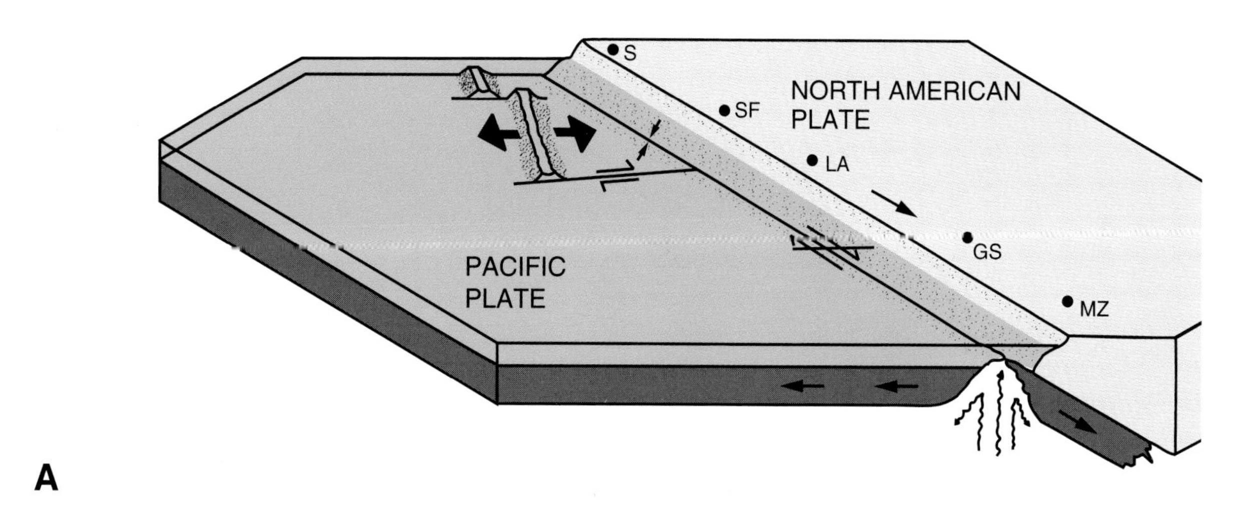

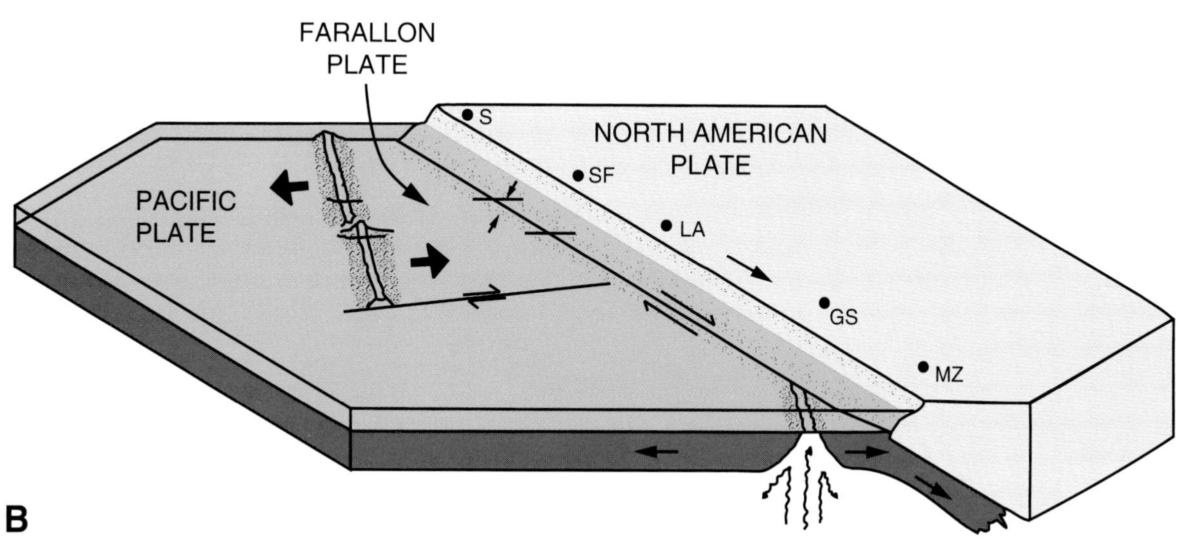

Figure 1.12. Tectonic evolution of northwest Mexico in the Tertiary. Different stages in the collision of the eastern Pacific oceanic crest, and the development of right lateral movement between the North American plate and the Pacific plate. S = Seattle, SF = San Francisco, LA = Los Angeles, GS =Guaymas, MZ =Mazatlán (after Atwater, 1970). (A) 10 million years before the present; (B) 20 million years before the present; (C) 30 million years before the present; (D) 40 million years before the present.

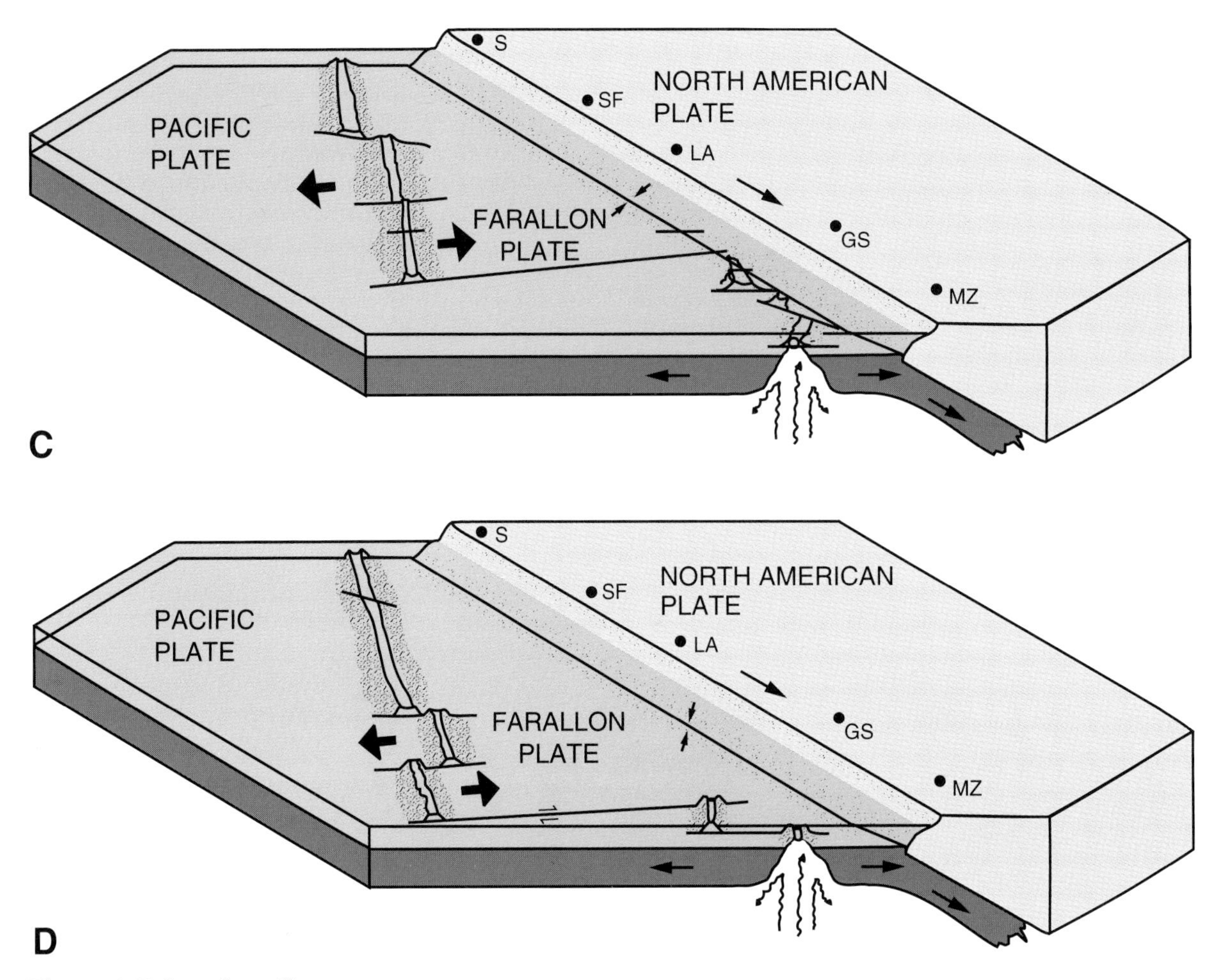

Figure 1.12 (continued).

related to the tectonic lineaments that cut the Peninsula diagonally and to the alkaline basaltic extrusions of the Pliocene–Pleistocene that are encountered in numerous localities.

Economic Deposits

According to Gastil et al. (1975), the northern portion of the Baja California Peninsula can be divided into five mineral provinces (Figure 1.13). The most western of these comprises deposits of mesothermal iron and copper sulfides as well as oxides of iron. These deposits are found emplaced in the partially metamorphosed volcanic sequence of Mesozoic age and have been attributed to hydrothermal origin related to the Cretaceous granitic intrusions. The principal known localities that manifest this type are: El Sueño mine (loc. 1), San Antonio (loc. 4), Mision San Vicente (loc. 11), San Isidro Point (loc. 10), Rancho Rosario (loc. 12), zones to the east of El Rosario (locs. 17, 18), and to the southeast of San Fernando (locs. 19–21). To this province also belong the deposits of the El Arco mine (loc. 29), which accounts for one of the most important copper reserves of the nation.

The second province comprises veins of gold contained in metasedimentary rocks that are distributed along the axis of the peninsula. Their occurrence, restricted to metasedimentary rocks, leads to the possibility that they have been reworked from ancient placers before their metamorphism. The principal localities known for this type of deposits are: Las Cruces (loc. 7), El Alamo (loc. 9), Socorro (loc. 13), Arroyo Calamajú (loc. 23), Cerro San Luis (loc. 24), Desengaño (loc. 25), León Grande (loc. 26), and Columbia Mine (loc. 27).

The third province comprises tungsten deposits related to contact metamorphism of pre-batholithic calcareous rocks where precious stones may be found. The intrusives that affect the calcareous sequences were emplaced chiefly in the Cretaceous. The localities known are: La Olivia (loc. 3), Los Gavilanes and El Fenomeno (loc. 6), as well as in the Sierra de los Cucapá, Sierra Mayor, and Sierra San Pedro Mártir.

The fourth province includes superficial deposits of travertine with sulfides of manganese, copper, silver, and lead as well as deposits of wulfenite, stibnite, and other minerals. These deposits have the peculiarity of having formed in the Cenozoic near the edge of the Gulf of California. The distribution of these deposits is very complex and the localities are numerous.

The final province includes deposits of placer gold developed in the Cenozoic. The principal localities

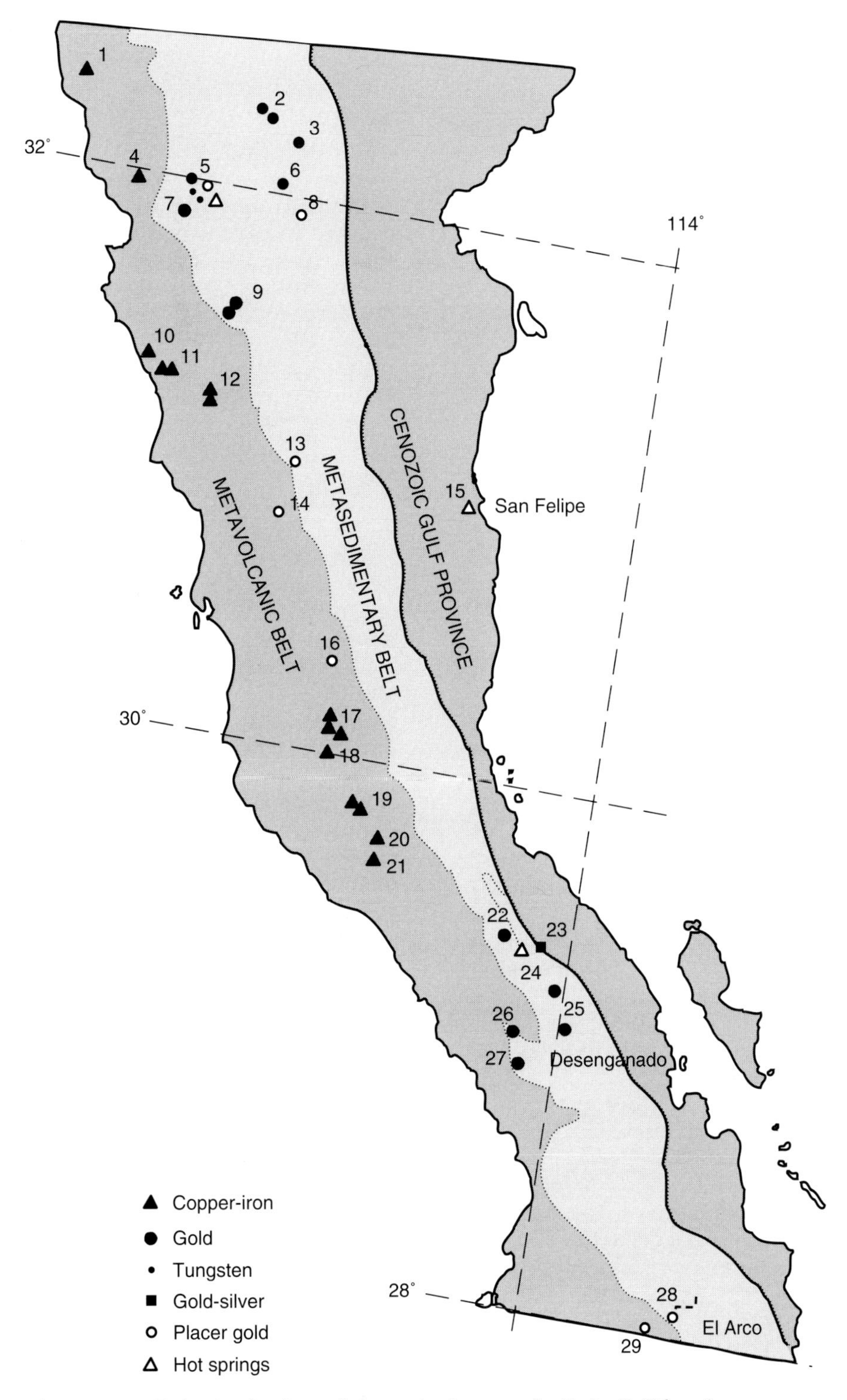

Figure 1.13. Principal mineral deposits known in Baja California: copper-iron, gold, tungsten, gold-silver, placer gold, hot springs.

are: Campo Juárez (loc. 2), Los Pinos and Campo Nacional (loc. 8), Socorro (loc. 12), Valledores (loc. 36), Los Enjambres (loc. 40), Real del Castillo (loc. 14) and Pozo Alemán (loc. 28).

In the southern part of the Peninsula of Baja California, occurrences of mineral deposits are rarer because Mesozoic rocks are less exposed (Figure 1.14). On the coast of the Gulf of California some deposits of manganese exist, but these are of little importance. They are in the form of oxides occurring as hydrothermal veins. The principal localities are: Lucifer (loc. 1), Mulegé (loc. 21), and Misión San Juan (loc. 3).

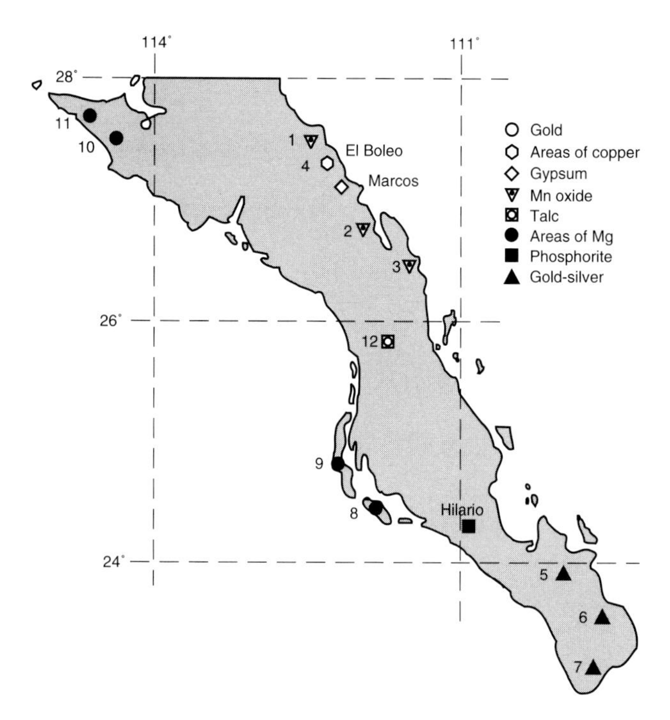

Figure 1.14. Principal mineral deposits known in southern Baja California (taken from the metalogenetic map of the Mexican Republic, Ing. Guillermo P. Salas, 1975). Gold, copper deposits, gypsum, manganese oxides, talc, magnesite deposits, phosphorite, gold-silver.

The most important copper deposits are in the form of the sulfides of El Boléo (loc. 4) developed in Mesozoic volcanic rocks of the Santa Rosalía area. Their metallic deposits are represented by hydrothermal deposits of gold and silver in the Cabo region (locs. 5–7), as well as in the Vizcaíno area.

Various occurrences of nonmetallic deposits exist including the magnesite deposits found in Magdalena Bay (locs. 8, 9) and Eugenia Point (locs. 10, 11), the talc deposits of Comondú (loc. 12), and the phosphorite deposits of San Hilario area. These last constitute the major reserves of phosphorite in Mexico.

The most important petroleum occurrences are localized in Paleocene sediments in the Purísima basin, as seen in exploratory wells drilled by Petróleos Mexicanos (Lozano, 1976) and in some oil seeps of this same region.

SONORA AND SINALOA

In the States of Sonora and Sinaloa, one observes as in Baja California a great complexity of rocky outcrops owing to similarly intricate structure and to the great lithologic heterogeneity of the various units; particularly those of the pre-Tertiary form highly variable stratigraphic columns in this region. This geologic terrane contrasts markedly with that observed to the east of the Sierra Madre Occidental where the structures are more regular and the stratigraphy more homogeneous.

To describe effectively the geologic characteristics of this region, the physiographic division of provinces defined by the General Department of Geography have been followed because they present suitable natural limits for better description.

Sonoran Desert

This zone is characterized by the presence of complex mountains, separated by alluvial valleys that become wider toward the northwest portion of the state, where they contain important eolian deposits. The complex mountains are found to conform with pre-Tertiary terranes that are covered toward the east by the piles of Cenozoic volcanics forming the Sierra Madre Occidental. Here they turn up in the form of isolated outcrops, under the ignimbrite cover.

In the State of Sonora, units of rock are exposed that have a geochronologic range varying from Precambrian to Recent.

The Precambrian is represented by two well-defined groups of rocks (see Figure 1.15): one older group composed of metamorphic rocks derived from igneous and sedimentary rocks, and a younger group composed of sedimentary sequences of quartzite and dolomite that overlie discordantly the earlier strata.

The metamorphic Precambrian occurs in northwest Mexico as an extension of the Precambrian shield that crops out widely in the United States and Canada. This Precambrian basement in North America shows a series of provinces that are older toward the nucleus of the craton, suggesting accretionary development of the continental crust of this region. In northern Sonora, two Precambrian metamorphic terranes exist that are of different ages and are structurally juxtaposed along one major strike-slip zone that originated in the Jurassic. It crosses the northern part of Sonora diagonally with a northwest-southeast orientation. This strike-slip zone has been proposed by Silver and Anderson (1974) as the "Mojave-Sonora Megashear," with a left-lateral movement that is extended toward the states of Arizona and California (see Figure 1.16).

The Precambrian block located to the southwest of the "megashear" is represented by outcrops of metamorphics in the area of Caborca, where the oldest known rocks of Mexico are located. This block has rectilinear limits, both to the south and to the west, marked by the sudden disappearance of Precambrian outcrops. As far away as Sinaloa, rocks this old appear again as outcrops, represented by the Sonobari Complex (Rodríguez and Córdoba, 1978); however, their age has not yet been confirmed.

The Precambrian metamorphic outcrops of the Caborca area are represented by igneous rocks and by metasedimentary rocks of greenschist and amphibolite facies (Anderson et al., 1978) formed during a period of from 1700 to 1800 million years (Silver and Anderson, 1979). These metamorphic units have been designated by Longoria et al. (1978) as the Bamori Complex. They have suggested that the existence of massive anorthositic rocks could correspond to the unification of two Precambrian continental areas.

In opposition to the above ideas of unification, there exist to the northeast of the zone of "megashear" Precambrian metamorphic rocks, such as those

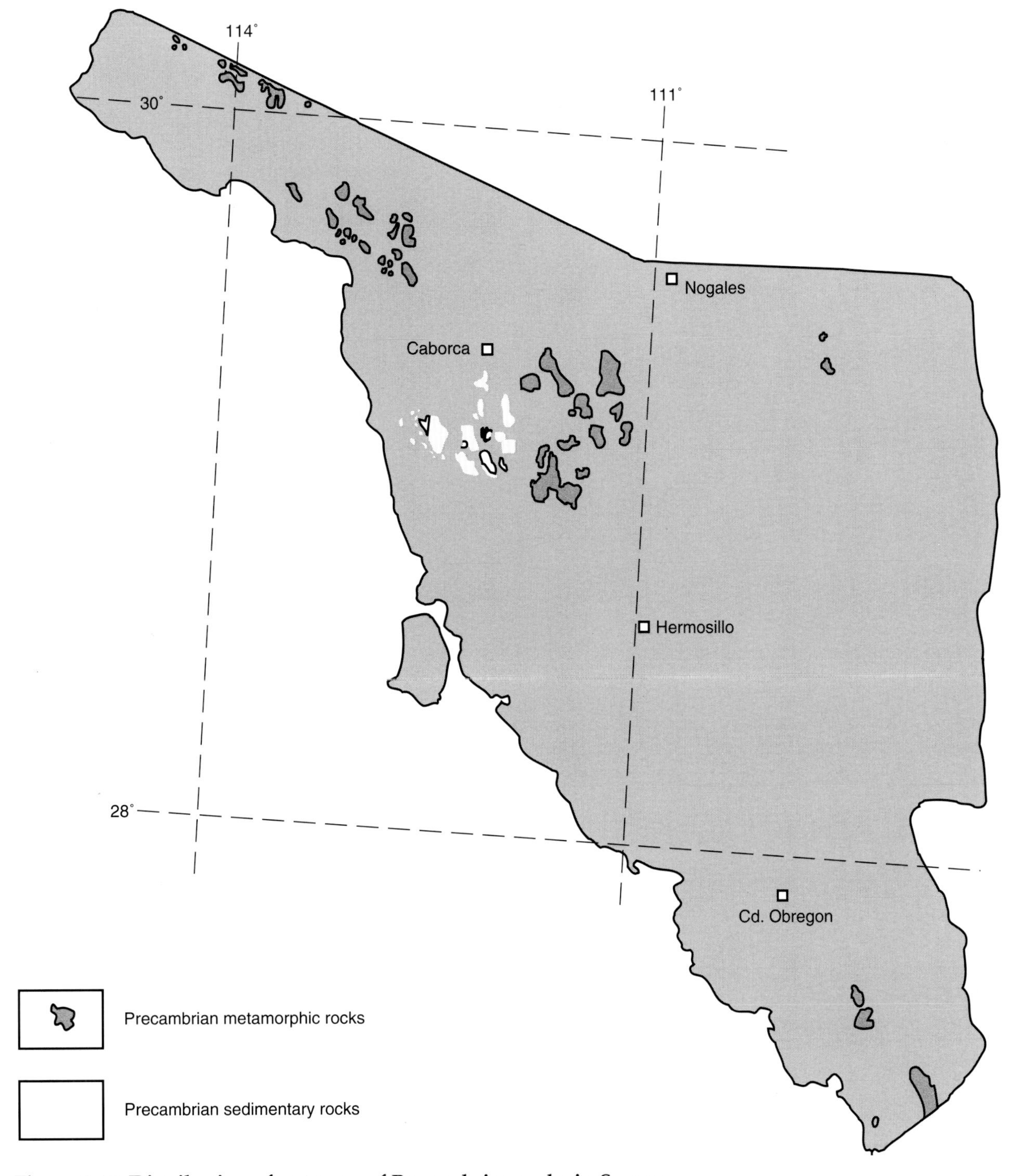

Figure 1.15. Distribution of outcrops of Precambrian rocks in Sonora.

cropping out in the Sierra de los Ajos, whose ages range between 1600 and 1700 million years and that have been correlated with the Pinal del Sur schists of southern Arizona.

A sedimentary group of late Precambrian age crops out in the Caborca area and covers, with tectonic discordance, the metamorphic Precambrian (Longoria et al., 1978). Originally this sequence was named by Keller and Wellings (1922) as the Gamuza beds, and later Stoyanow (1942), on the basis of Collenia algal reefs, placed it in the late Precambrian. The sequence includes the Pitiquito and Gamuza formations (Longoria and Pérez, 1978) and is composed principally of dolomite with stromatolites and with quartz sandstone and shale. The upper contact of the Gamuza Formation is discordant with the overlying Paleozoic sequence.

The Paleozoic sequence crops out in numerous localities in the State of Sonora and is composed principally of limestones and sandstones that were

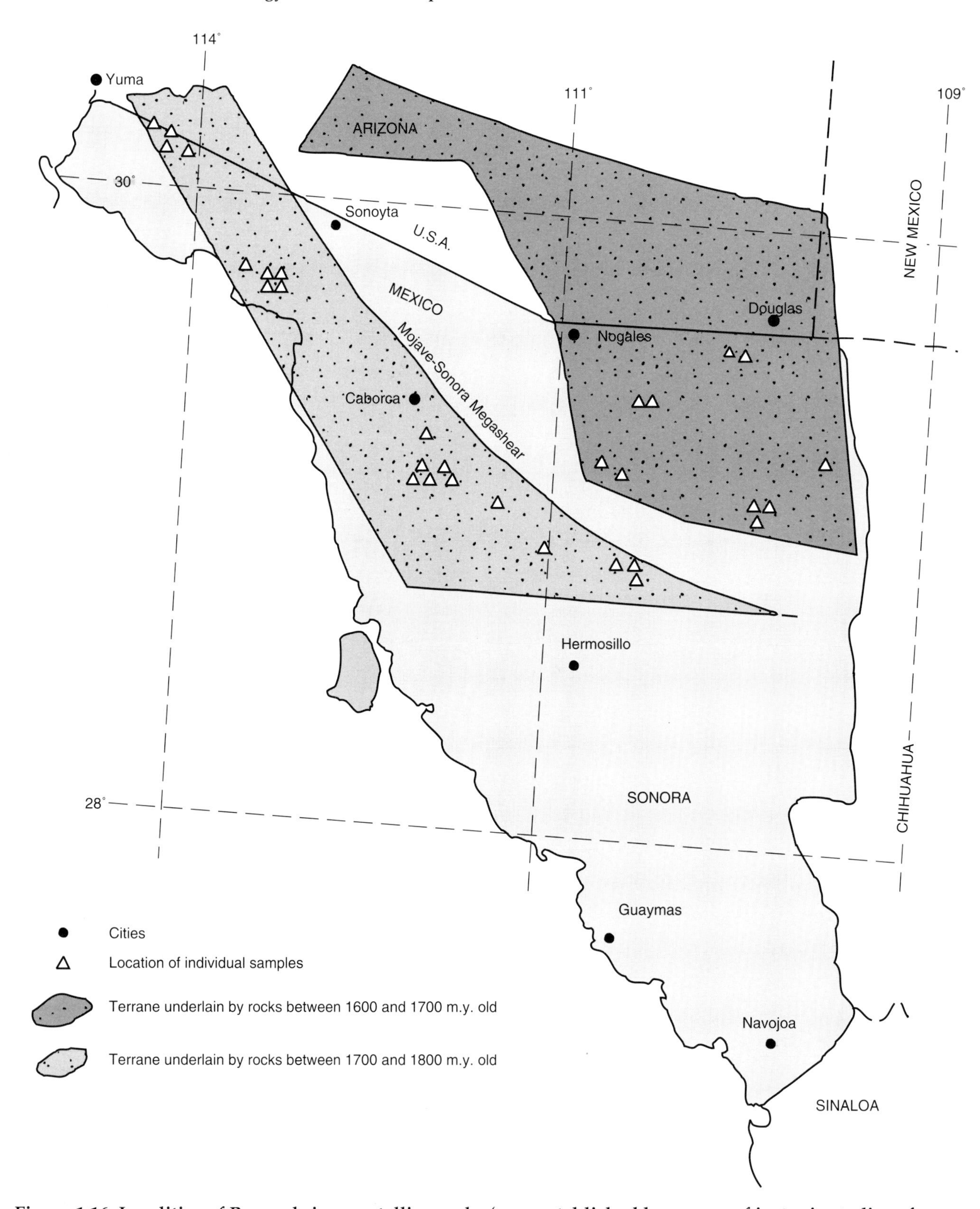

Figure 1.16. Localities of Precambrian crystalline rocks (ages established by means of isotopic studies of Anderson and Silver, 1979). Cities, localities of individual samples, terranes underlain by rocks of ages between 1600 and 1700 million years, terranes underlain by rocks with ages of 1700–1800 million years.

deposited in a platform environment (see Figure 1.17). This ancient continental platform would be a southward continuation of the miogeosynclinal Cordilleran belt. Fries (1962) proposed the name Sonoran Trough for this southern extension of the Cordilleran Geosyncline and indicated that during the whole

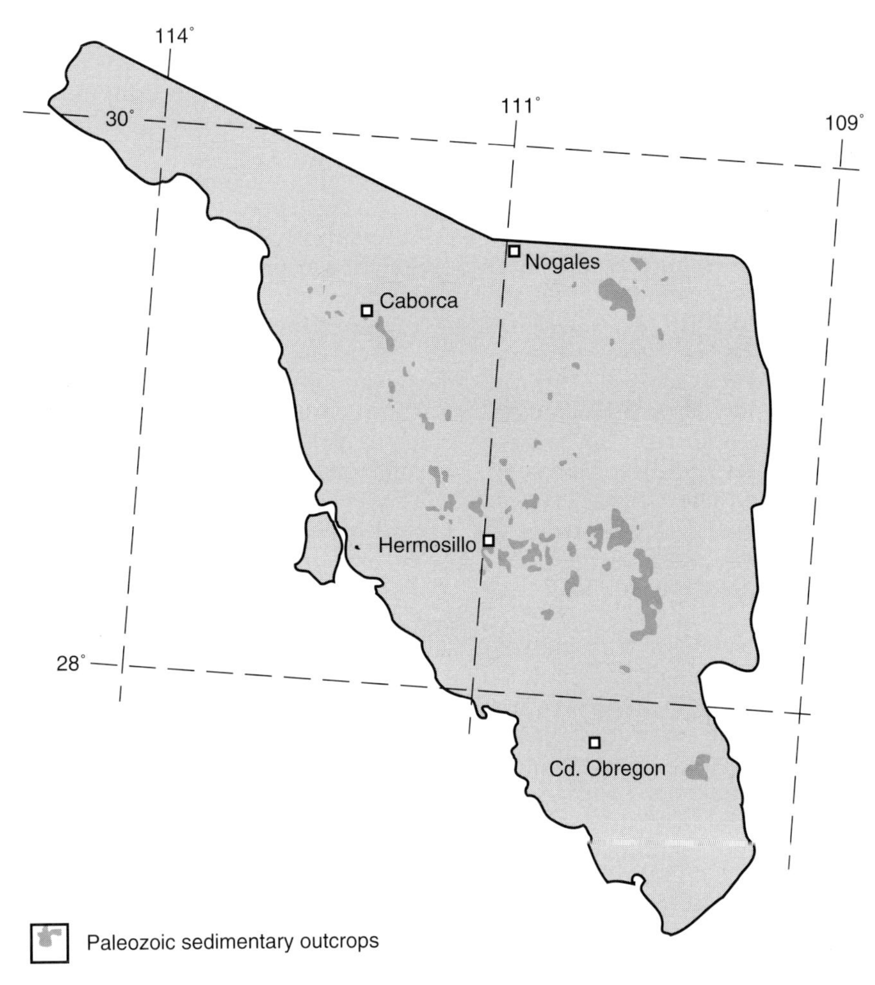

Figure 1.17. Distribution of outcrops of Paleozoic rocks in Sonora.

Paleozoic this area underwent a slow and uninterrupted subsidence. According to F. Rangin (1978), at the time of the Cambrian and Ordovician periods there existed a gradation from platform facies in the north of Sonora to a more internal facies toward the south of the state where the Paleozoic facies have a tectonic style of much more intense deformation. The calcareous facies of the Carboniferous and Permian constitute a more homogeneous facies over all the state. The two above major intervals are separated by a phase of major deformation that occurred in the Devonian.

The Paleozoic of the Caborca area is represented in ascending stratigraphic order by the following formations: Puerto Blanco, Provedora Quartzite, Buelna, Cerro Prieto, Arroyos, and Tren. All of these belong to the Cambrian (Cooper et al., 1952) and consist of sequences principally of calcareous-detrital lithology. Small isolated outcrops of calcareous strata that represent parts of the Ordovician, Silurian, Devonian, and Mississippian Systems (Cooper and Arellano, 1946) exist in the Bisani area. In the vicinity of Antimonio, a Permian sequence crops out that consists of beds of shale and sandstone with limestone lenses and that was termed the Monos Formation by Cooper and Arellano (1946). In the mineral district of Cananea there exists a Cambrian sequence of quartzite and limestone comprising the Capote Quartzite and Esperanza Limestone (Mulchay and Velasco, 1954; Valentine, 1936) as well as limestones of Devonian, Mississippian, Pennsylvanian, and Permian age. In northwest Sonora there exist calcareous outcrops of Paleozoic strata in Cabullona (Taliefferro, 1933), Sierra del Tigre, Nacozari, and the Sierra de Moctezuma (Imlay, 1939). In the Hermosillo area and in the region located more to the south, isolated outcrops of Ordovician and Permian exist (King, 1939).

The first deposits after the Paleozoic in the state of Sonora are continental sediments of the Upper Triassic and Lower Jurassic belonging to the Barranca Formation that crops out in the central and southern portions of the state. There are in addition marine deposits of sandstone, limestone, and shale in the areas of Antimonio and Santa Rosa in northwest Sonora. The deposits of the Antimonio area form a marine sequence some 300–400 m thick. They contain ammonites, belemnites, and bivalves whose age varies from Late Triassic to Early Jurassic, are exposed principally in the Sierra de El Alamo, and have been informally designated as the Antimonio Formation by González (1979). This sequence is correlative with the lower part of the sedimentary and volcanoclastic sequence of the Rajon Group (Longoria and Pérez, 1978) that crops out in the hill of the same name located to the southeast of Caborca. According to Alencaster (1961), the region that includes these localities constituted a former bay in which were deposited sediments coming from the northern and northeastern parts of the state. In the area of San Marcial, to the southeast of Hermosillo, a sequence crops out that consists of fine-grained clastics with coal beds and with limestone intercalations; this was deposited in a swampy basin contemporaneously with the marine deposits of the Sierra de El Alamo. This sequence was named by King (1939) the Barranca Formation, and Alencaster (1961) later elevated its deposits to the rank of Group. The absence of Lower Triassic deposits and the disconformable relationship observed within the Upper Triassic sequence above Permian rocks in the Sierra de El Alamo reveal important tectonic movements in the region during the close of the Paleozoic and the initiation of the Mesozoic.

The Jurassic in the State of Sonora is characterized by the development of an important volcanic-plutonic arc with a general northwest-southeast direction, evidenced by numerous outcrops of andesitic volcanic and volcanoclastic rocks (see Figure 1.18). The development of this arc has been related to episodes of subduction occurring on the Pacific margin of Mexico where an oceanic plate was being subducted under the continental crust of Mexico.

In the area of Cucurpe, to the southeast of Santa Ana, Rangin (1977a) reported a sedimentary sequence with volcanic intercalations that contain Jurassic ammonites. In the Sierras de la Gloria (Corona, 1979), El Alamo (González, 1979), and in various localities in northwest Sonora, Mesozoic volcanic and volcanoclastic rocks of probable Jurassic age have been reported but unconfirmed. In some localities these rocks are partly affected by dynamic metamorphism and are generally of andesitic composition. Anderson and Silver (1978) have reported U-Pb ages in various localities of volcanic and volcanoclastic rocks varying between 180 and 150 million years. According to these authors the volcanic-plutonic activity in the Jurassic originated owing to the presence of a zone of plate convergence to the west and was interrupted by the initiation of lateral displacement along the so-called Mojave-Sonora Megashear. The prevalence of the magmatism that remained in the convergent zone is also evidenced by volcanic and volcanoclastic Cretaceous rocks that also crop out in various localities in Sonora.

In the Cretaceous, two realms in Sonora that have clearly distinctive characteristics can be defined (Rangin, 1978) (Figure 1.19). The first of these is found in the central and western belts in the state and evolved over a permanently emergent band of Jurassic volcanic and volcanoclastic rocks. In this belt were developed lava extrusions, principally of andesite, which in the central and southern portions of the state contain intercalations of Lower Cretaceous marine sedimentary rocks (King, 1939; Roldán and Solano, 1978). The second realm, located in the eastern belt of the state, is composed of Lower Cretaceous marine sediments that afford evidence of a marine transgression coming from the Chihuahua basin during the Aptian–Albian (King, 1939; Rangin, 1978). These strata partly cover the volcanic and volcanoclastic Jurassic terranes.

In the Upper Cretaceous, both realms are affected by compressional deformation and by granitic plutonism accompanied by andesitic lava extrusions that become more intense toward the western part of the region of the Sierra Madre Occidental.

Evidence exists, in various outcrops, of volcanic activity that occurred in Sonora during the Early Cretaceous. In the Cerro Lista Blanca, south of Hermosillo, a unit of andesitic volcanic rocks of probable Early Cretaceous age crops out. Earlier, Dumble (1900) designated these beds the Lista Blanca Division and assigned an Upper Triassic position to them. Later King (1939) assigned to them the name Lista Blanca Formation and positioned them stratigraphically in the Lower Cretaceous. This same author noted numerous outcrops in the central and southern zones of the state where volcanic rocks of the Lower Cretaceous appear intercalated in marine sedimentary sequences also indicating that volcanic rocks of this epoch increase proportionally toward the west and southwest. In the northwest coasts of the state, Silver and Anderson (1978) recognized some volcanic rocks dated by radiometric methods as belonging to the Upper Jurassic and Lower Cretaceous.

Marine sedimentary sequences are seen to crop out in diverse localities and contain generally fossil faunas of the Aptian and Albian Stages. In the northwest of the state and southwest of Arizona, a sequence crops out that constitutes the Bisbee Group, formed in ascending stratigraphic order by the following: Glance Conglomerate; sandy shales and quartzose and feldspathic sandstones of the Morita Formation; limestones of the Mural Formation that vary from forereef to backreef; sandy shales and redbed sandstones of the Cintura Formation (Ransome, 1904; Rangin and Córdoba, 1976; Gamper and Longoria, 1980). In the Sahuaripa area, a sequence of more than 3000 m exists, composed of conglomerate, shale, sandstones, and limestone that represent the Palmar Formation in its lower part and the Potrero Formation in the upper part (King, 1939). Other marine sequences occur in the areas of Cucurpe, Sta. Ana, and Sierra Azul. These are formed chiefly of calcareous and sandy sediments of the Lower Cretaceous.

During the Late Cretaceous the territory of Sonora State underwent an uplift and general emersion resulting from a phase of compressional deformation that was active over a large part of western Mexico at the beginning of this epoch. The principal igneous activity consisted of granitic emplacements that migrated in time toward the east and the extrusion of lavas that varied from andesite to rhyolite, chiefly along the eastern belt of the state and toward the base of the Sierra Madre Occidental.

The exposures of Cretaceous batholithic bodies in Sonora constitute one of the outstanding characteristics of the region. These granite-granodiorite bodies

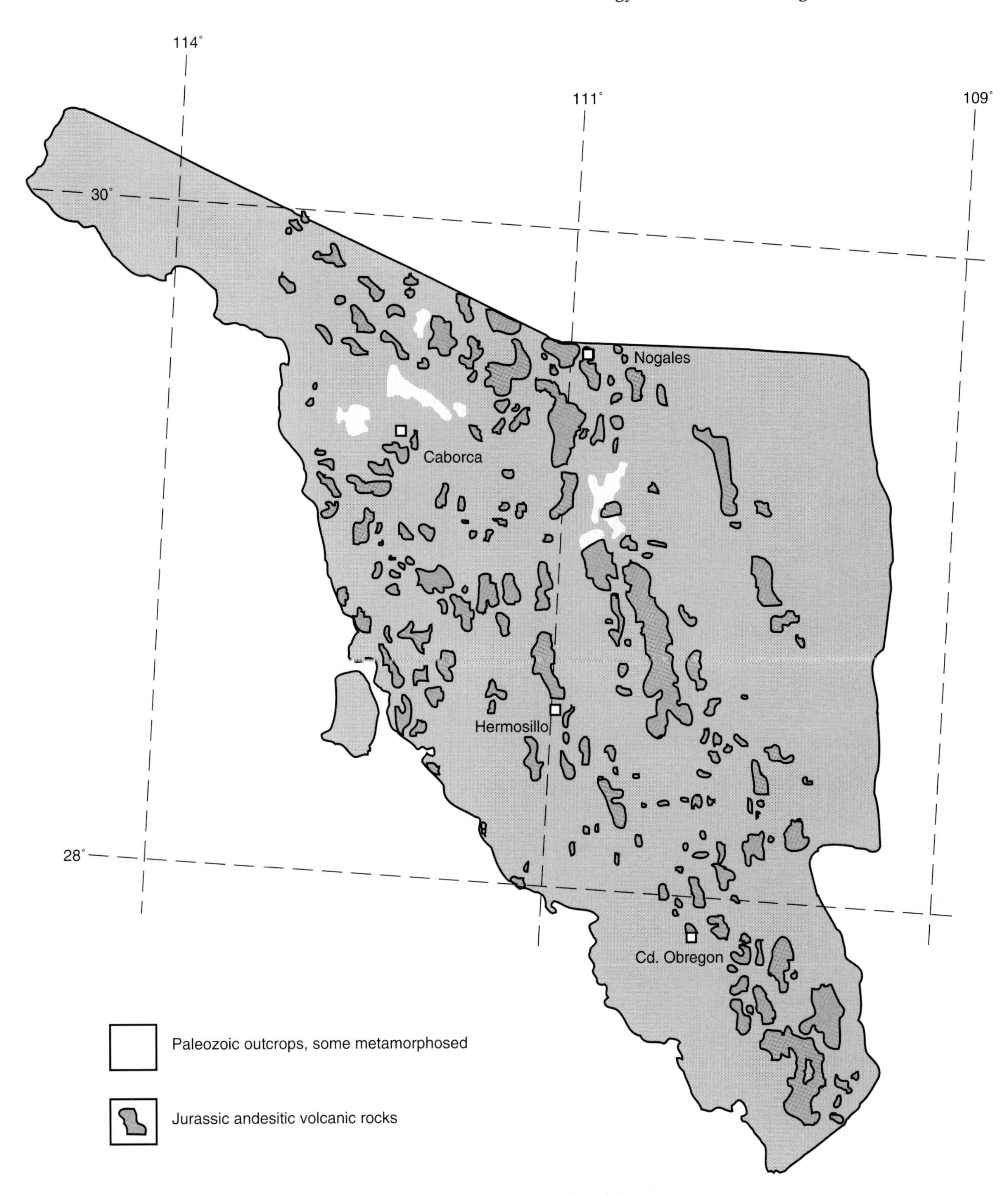

Figure 1.18. Distribution of outcrops of Mesozoic igneous rocks of Sonora.

have obscured in large part the deformational phenomena that occurred before their emplacement.

There is exposed in the Agua Prieta area a continental sedimentary sequence of Upper Cretaceous that covers with angular discordance the deformed units of the Bisbee Group. This sequence was designated by Taliefferro (1933) as the Cabullona Group and is made up of continental clastic sediments with intercalations of volcanic rocks, and contains dinosaur bones and an Upper Cretaceous flora (Rangin, 1978).

Over all the northeastern part of Sonora, numerous outcrops exist of Upper Cretaceous volcanic rocks

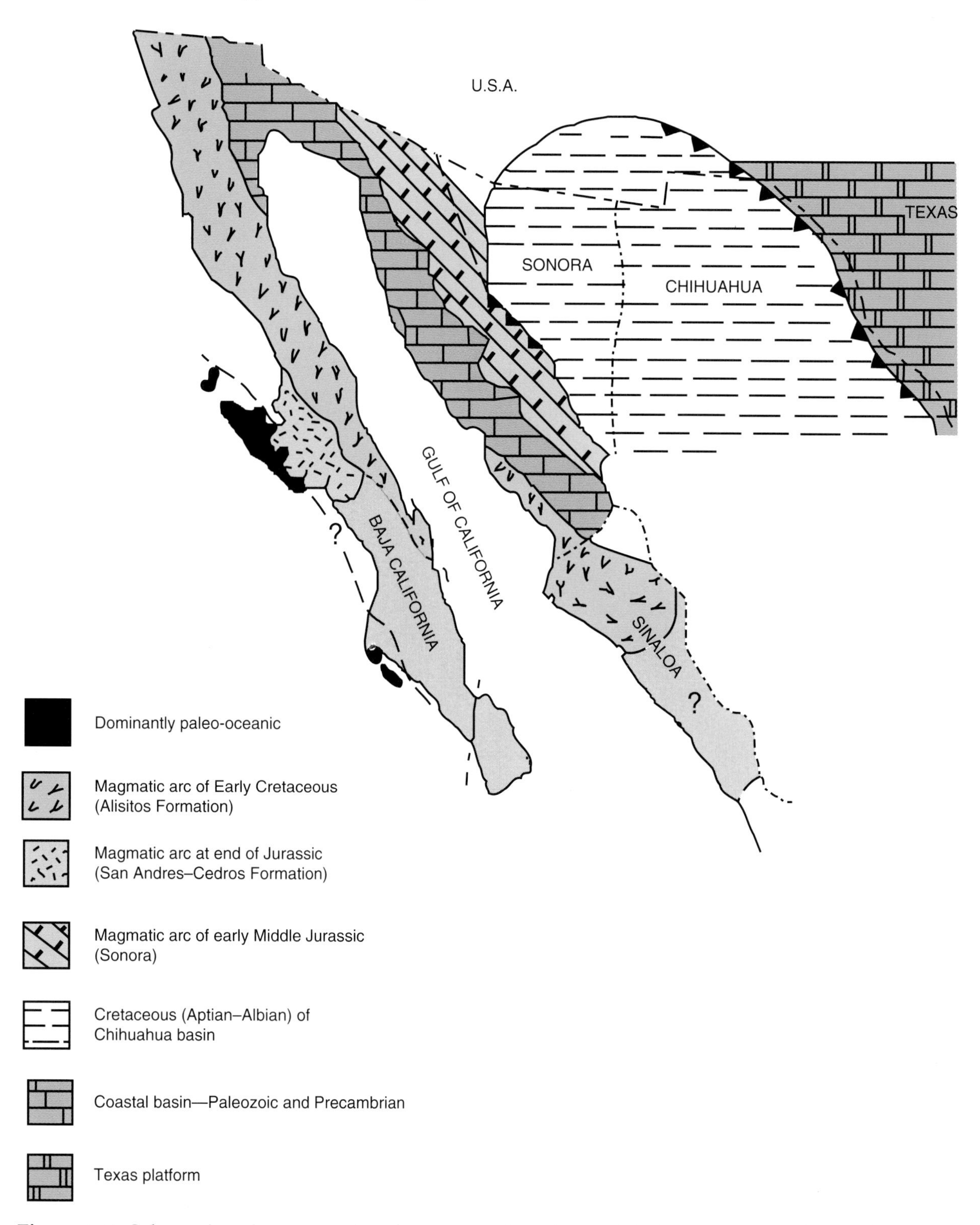

Figure 1.19. Schematic paleogeography of a portion of northwest Mexico during the Mesozoic. After C. Rangin (1978).

whose composition varies from rhyolite to andesite. The appearances of vulcanism in this Epoch seem to extend toward the base of the volcanic sequence of the Sierra Madre Occidental, where the presence of rocks 100 million years old has been reported (McDowell and Clabaugh, 1979).

According to Rangin (1978), at the beginning of the Tertiary an important assemblage of plutonic-volcanic rocks that is responsible for the mineralization of disseminated copper developed in northeastern Sonora. The volcanic rocks are generally related to intrusive bodies that affect and mineralize them (Sillitoe, 1973).

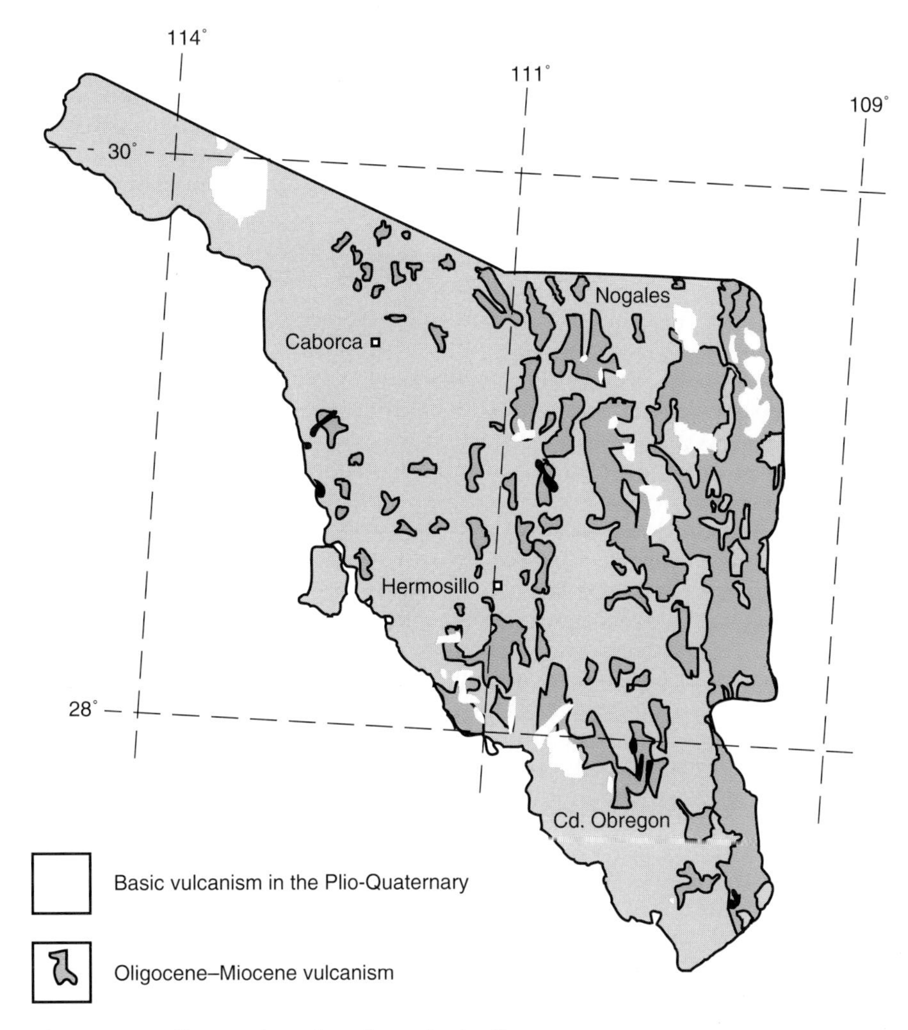

Figure 1.20. Cenozoic volcanic rocks in Sonora.

These volcanic rocks vary in composition from andesites and trachytes to dacites and rhyolites. In the central and southern part of the state, volcanic rocks of the lower Tertiary crop out whose composition is mainly intermediate. They cover with angular discordance the deformed Mesozoic sequences.

The principal volcanic event of the Tertiary of Sonora consists predominantly of ignimbrite extrusions of Oligocene–Miocene age that become a westward extension of the volcanic episodes responsible for the Sierra Madre Occidental (see Figure 1.20).

The exposures of this class of volcanic units generally form dissected tablelands that cover in large part older terranes and geological structures.

During the late Tertiary the whole region of Sonora was subjected to a series of normal faults that cut completely independently across all earlier structural units. This phenomenon resulted in a system of northwest-southeast faults and consequent formation of depressions that were filled by continental detrital sediments of the Baucarit Formation. This unit crops out in various localities of the state and is made up generally of lithic fragments of diverse composition that vary from subangular to rounded in only slightly consolidated strata (Dumble, 1900; King, 1939).

At the end of the Tertiary and beginning of the Quaternary, an important episode of alkaline basaltic vulcanism took place, together with tectonic distension of the normal faults, concurrent with episodes of opening of the Gulf of California (Clark et al., 1980; Rangin, 1978). This vulcanism has its clearest example in the Serrania del Pinacate located in the Altar Desert.

Pacific Coastal Plain

This region is characterized by the development of a plain constructed by the evolution of a system of deltas that have advanced gradually toward the west. These deltas were formed at the mouths of the rivers Mayo, Fuerte, Sinaloa, Culiacán, San Lorenzo, and Mocorito and have surrounded rocky prominences that formed old islands.

This zone is bordered on the west by a shoreline that developed sandy sediments, the sand being a product of the action of littoral currents, tides, and waves that have reworked the deltaic sediments and caused formation of bars, tombolos, and hooks. The eastern border of this zone is made up by the foothills of the Sierra Madre Occidental; here appears a mountain chain formed from rock units whose geochronologic range varies from Precambrian to lower Tertiary. These units are partially covered by the volcanic sequence of the Sierra Madre Occidental, a sequence that becomes dominant eastward.

The history of the pre-Tertiary terranes that are exposed on the eastern edge of Sinaloa share many affinities with the tectonic and paleogeographical styles that prevailed in Sonora and Baja California, which were united before the Pliocene.

It appears that the most ancient rocks that crop out in the State of Sinaloa are the metamorphics in the Sierra de San Francisco to the north of Los Mochis. A Precambrian age has been assigned to these rocks by previous authors (Rodríguez and Córdoba, 1978). De Cserna and Kent (1961) termed these rocks the Sonobari Complex. The unit consists of intercalations of muscovite and biotite gneiss with amphibolites. In addition, intrusions by bodies of gabbro and granodiorite are present and development of pegmatites and migmatites is observed. According to Rodríguez and Córdoba (1978), the gneisses were derived from sandy and argillaceous sedimentary rocks with possible intercalations of basic lava that have undergone at least two metamorphic events.

Along the eastern border of the Pacific Coastal Plain there exists a series of isolated outcrops of moderate extension that are of marine Paleozoic rocks. These sequences are composed principally of sandstones, shales, silts, and limestones; in some localities they have been affected by various grades of metamorphism. The stratigraphic relationship between this sequence and the Sonobari Metamorphic Complex has not been observed, and it is apparent that the contact with Mesozoic rocks is generally tectonic. Rodríguez and Córdoba (1978) report the discovery of the fusulinid Millerella sp., which indicates that the lower part of the sequence is probably of Late Mississippian to Early Pennsylvanian age.

These authors indicate that the Paleozoic sequences of Sinaloa were deposited in shallow water platform conditions. In general, one can consider that these sequences were deposited in a miogeosynclinal belt that could be a southward continuation of the Paleozoic Cordilleran geosyncline developed in the western United States.

The Mesozoic in Sinaloa exists as a very heterogeneous series of rock types that consists seemingly of a volcano-sedimentary assemblage which becomes a southeastward continuation of the volcano-volcanoclastic and sedimentary arc rocks of the Alisitos Formation of Baja California (Rangin, 1978). Along the eastern border of the coastal plain, extensive outcrops of volcanic rocks are observed. There are both lavas and pyroclastics whose composition varies from acidic to basic and that show the effects of both regional and contact metamorphism (Figure 1.21).

The Mesozoic sedimentary rocks are represented by sequences of limestones that in some localities are observed to be partially metamorphosed. Outcrops of these rocks are isolated; they are present above the intrusives in the form of roof pendants or of windows under the Tertiary cover. In some localities they are apparently intercalated within the Mesozoic metavolcanic sequence, but the contacts are not clearly observed.

The major part of the calcareous rocks that crop out in Sinaloa are seemingly of Cretaceous age, but Rodríguez and Córdoba (1978) consider that some of these rocks could be Jurassic, or perhaps older.

All the Mesozoic volcanic and sedimentary assemblages are affected by the emplacement of Mesozoic and Tertiary plutons. These intrusive bodies are the units that crop out most extensively in the State of Sinaloa. Their petrographic classification varies from granite to monzonite, with biotite and hornblend as principal mafic minerals. The extensive outcrops of this unit disappear under the volcanic cover of the Sierra Madre Occidental.

The periods of emplacement of the intrusive bodies appear to be similar to those occurring in Sonora. These emplacements migrate in age from Cretaceous in Baja California to early Tertiary at the border of Sinaloa and Chihuahua (Silver and Anderson, 1978).

During the Tertiary, important volcanic episodes occurred in the State of Sinaloa; chief among them are those that occurred in the middle Tertiary and that gave rise to the ignimbrite cover of the Sierra Madre Occidental. This ignimbrite sequence covers a large part of the Mesozoic rocks of the eastern border of the Pacific coastal plain and the intermediate and basic volcanics of the lower Tertiary.

Tectonic Summary

The outcrops of Precambrian metamorphic rocks in northern Sonora constitute one of the most characteristic features of this region. According to Anderson and Silver (1979), these metamorphic rocks form two magmatic and orogenic belts with a northeast-southwest orientation, truncated and juxtaposed by movement along a left lateral zone of slippage that was active during the Jurassic in a northwest-southeast direction. These two orogenic belts form part of the Precambrian terranes with orientation similar to that encountered in the southwest portion of the North American craton.

According to radiometric dates obtained by Anderson and Silver (1978), these sequences were deformed and metamorphosed 1650–1660 million years ago. However, periods of igneous intrusion between 1410 and 1440 million years ago have also been recognized, as well as one that is about 1100 million years old. This last episode of intrusion constitutes the first report of Grenvillian rocks in this region of the North American craton (Anderson et al., 1978).

These metamorphic terranes make up the basement, above which episodes of marine platform sedimentation occurred at the end of the Precambrian and during the Paleozoic. According to Fries (1962) this platform constituted a southern extension of the mio-

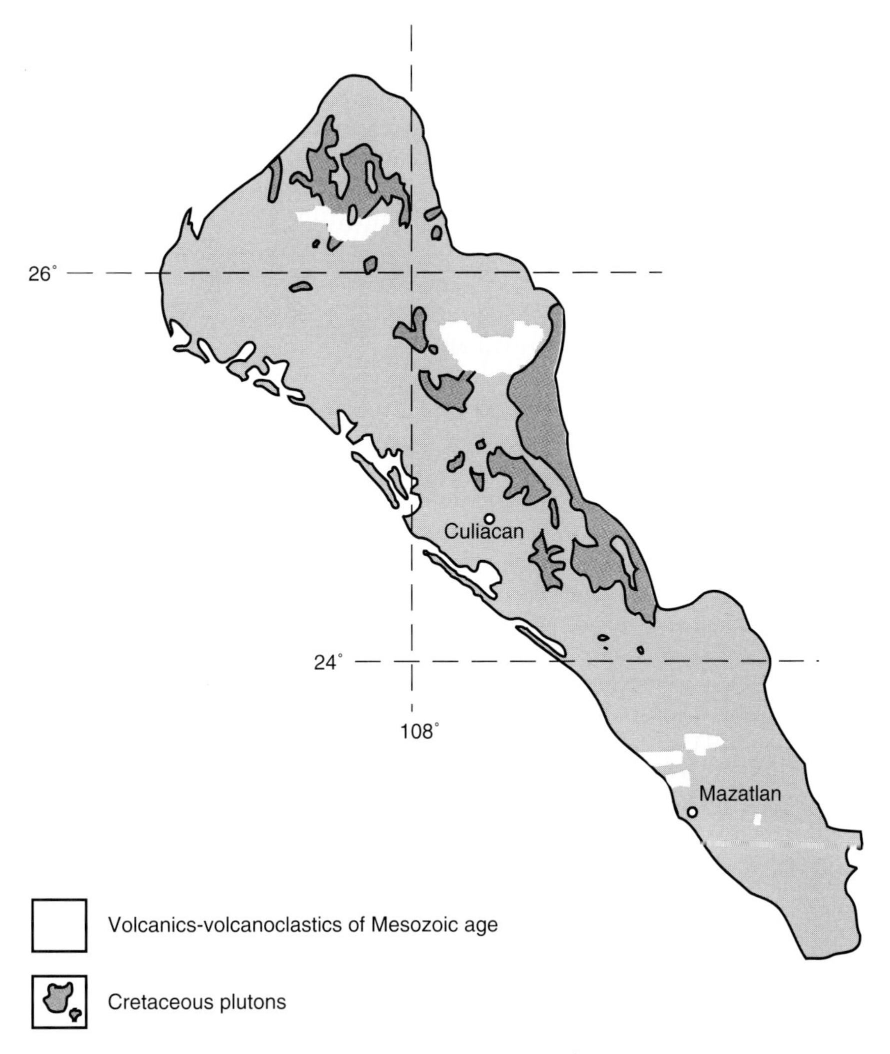

Figure 1.21. Outcrops of igneous rocks in Sinaloa.

geosyncline of the Cordilleran Geosyncline. He designated this area the Sonoran Trough. This trough underwent a slow subsidence during the whole Paleozoic, with some interruptions marked by hiatus in the sequences that crop out in Sonora and Sinaloa. Fries (1962) considers that at the end of the Permian, a period of gentle folding occurred, resulting in uplift and block-faulting that destroyed the earlier geosynclinal pattern.

In the Late Triassic–Early Jurassic, two paleogeographic elements provided the setting for processes of sedimentation in Sonora. On the one hand, the ancient bay of Antimonio existed in the present northwestern part of the state, in which a thick marine sedimentary sequence accumulated to the east, derived from positive areas. In contrast, there existed the swampy basin of San Marcial, located southeast of Hermosillo, in which beds of coal, gypsiferous limestones, sandstones, and shales accumulated (Alencaster, 1961).

Overlying the sedimentary sequence of Antimonio there is a package of volcanic and volcanoclastic rocks. These together with the Lower Jurassic volcanic intercalations crop out in the Cerro Rajón southeast of Caborca and indicate the initiation of volcanic activity in the Mesozoic. This volcanic activity has been attributed by many authors to the presence of a zone of convergence located to the west. The subsidence of a paleopacific plate beneath the continental crust of Mexico, and the partial fusion of a plate at the level of the asthenosphere, originated the construction of a magmatic arc that was active during the Mesozoic. The magmatic activity related to this arc is interrupted only by the development of a left-lateral strike-slip zone, the Mojave-Sonora Megashear of Silver and Anderson (1974).

The setting of convergence of the Paleopacific and North American plates is developed in two principal phases of deformation whose relationships can be clearly observed in the Cabullona area to the south of Agua Prieta and Naco. The first of these occurred at the beginning of the Late Cretaceous and is indicated by angular discordance between the sandy calcareous Lower Cretaceous sediments and the detrital continental sediments of the Upper Cretaceous that crop out in the Cabullona basin. The second phase corresponds to the compressive deformation from the end of the Cretaceous to the beginning of the Tertiary. This deformation originated folds with a northwest-southeast axial direction; they can be observed on the western flank of the Sierra Madre Occidental as well as in the overthrusting of the Lower Cretaceous and Paleozoic sequences above the Upper Cretaceous Cabullona Group of the Naco and Agua Prieta region (Rangin, 1977b). According to Rangin (1978) there seems to have developed between the Late Jurassic and the Early Cretaceous a phase of deformation, still not well known in Sonora, that correlates with the Nevadan orogeny developed in North America.

At the beginning of the Late Cretaceous, the continental history of Sonora and Sinaloa commenced. In this epoch the most important plutonic emplacements of the region occurred. These become younger eastward. Also in this epoch the first volcanic episodes occurred that form the base of the Sierra Madre Occidental, whose principal period of construction was ignimbrite activity occurring in the late Oligocene (McDowell and Clabaugh, 1979).

In the Miocene, activity of the convergence zone to the west seems to have ceased and the development of the Gulf of California was initiated. This is accompanied in the adjacent regions of Sonora and Sinaloa by distensive tectonics consisting of horsts and grabens that were active into the Quaternary and are responsible for the present-day distribution of the chief orographic elements of the Sonoran desert.

The setting of this type of tectonism gives rise to the deposition of great thicknesses of conglomeratic continental sediments, the Baucarit Formation.

Economic Resources

In the State of Sonora (Figure 1.22), the most important resources are the deposits of copper and molybdenum that are chiefly located in an eastern belt of the state. The origin of most of these resources has been attributed to the emplacement of granite and granodiorite porphyries that occurred at the end of the Late Cretaceous and the beginning of the Tertiary. The principal rocks encompassed by this mineralization are Cenozoic volcanic rocks of intermediate composition, intrusives of the same type, and in some cases sedimentary rocks of marine origin.

The chief reserves of copper porphyry in Sonora are encountered in the areas of Cananea and Nacozari (localities 1, 2, and 3); other more minor deposits are located farther south and west of these localities (loc. 4).

The origin of this order of resources has been attributed by Sillitoe (1975) to the partial fusion of the oceanic crust under the continent and the later ascension of magmatic material with solutions rich in copper and molybdenum that formed "stockwork" deposits in roof pendants of large plutons and breccia pipe deposits.

The deposits of lead and zinc in the State of Sonora are present in zones of metasomatic replacement and in hydrothermal veins. A major part of the first are of Laramide age while the second are generally associated with middle Cenozoic volcanic rocks (Echavarri et al., 1977). The principal localities with these types of deposits are: Cananea, San Felipe, El Tecolote, Sierra de Cabullona, Lampazos, and San Javier. The gold and silver deposits are located at shallow levels in hydrothermal veins that contain the above-mentioned reserves of lead and zinc. Principal localities with these types of deposits are El Tigre, Las Chispas, Lampazos, and San Javier.

Tungsten is an element with significant occurrence in the zones of contact metamorphism in the State of Sonora. Generally it occurs in the form of the mineral scheelite and on occasions is found associated with metasomatic deposits of copper, zinc, and collapse breccias associated with deposits of copper-bearing porphyry (Echavarri et al., 1977). The most important deposits of tungsten are found near Baviacora.

The most important nonmetallic deposits are fluorite and graphite. The first mineral is of hydrothermal origin and consists of veins that are exploited principally in Esqueda and Santa Rosa; the second is found associated with coal and is present as intercalations in the paludal sequence of the Barranca Group of the Upper Triassic.

In the State of Sinaloa (Figure 1.22), deposits of copper and molybdenum form resources of the copper-bearing porphyry type such as occur in Santo Tomás–Cuchicari and Tameapa. Deposits associated with stocks or veins of quartz with the presence of wolframite and tungsten are found in the mines of El Magistral, La Guadalupana, San José del Desierto, and El Guayabo. Deposits of molybdenum in stockwork form exist in the mines of Los Chicharrones and Las Higueras and breccia and hydrothermal vein deposits occur in El Magistral (loc. 5), Bahuita, Las Pastillas, La India, as well as the regions of Sinaloa de Leyva, Culiacán, San Ignacio, and Plomoso. There are important amounts of lead, zinc, and silver in deposits found in hydrothermal veins.

These latter veins are part of a belt that runs along the eastern half of the state and that includes, furthermore, deposits of the western borders of Chihuahua and Durango. Epithermal veins predominate in this belt. They contain gold, silver, lead, and zinc, which are most important in the State of Sinaloa. The enclosing rocks of this class of deposits are generally andesites at the base of the volcanic sequence of the Sierra Madre Occidental, and in places some plutonic rocks. The zones of Guadalupe and Calvo, Rosaratilo, Guadalupe de los Reyes, Pánuco, and Tayoltita contain these types of deposits. The latter site contains the richest gold mineral district in the nation.

SIERRA MADRE OCCIDENTAL

The Sierra Madre Occidental is formed from an extensive volcanic meseta affected by normal faults and grabens that detract from its homogeneous and pseudohorizontal appearance, especially along its flanks (Figure 1.23). The eastern border of the Sierra grades into the Basin and Range Province of

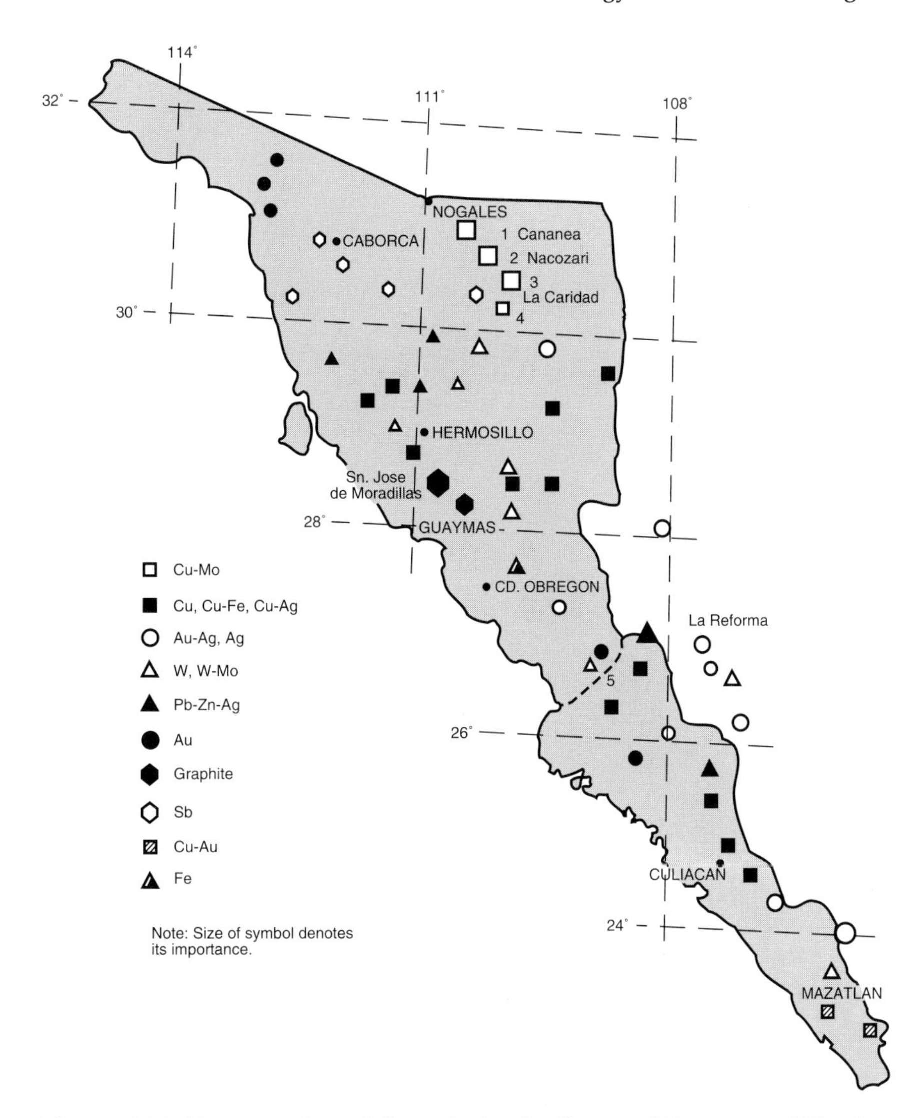

Figure 1.22. Known mineral deposits in the States of Sonora and Sinaloa (taken from the metalogenic map of the Republic of Mexico, G.P. Salas, 1975).

Chihuahua, while the western border consists of an abrupt termination with normal faults of major displacement and zones of deep barrancas (steep-walled canyons).

According to McDowell and Clabaugh (1979), the Sierra Madre Occidental is composed of two important igneous sequences, whose contact marks an intermediate period of volcanic calm. The older sequence is formed mainly from volcanic rocks of intermediate composition, and igneous bodies whose ages vary between 100 and 45 million years. The more recent is composed of rhyolitic and rhyodacitic ignimbrites in generally horizontal position or slightly inclined and whose ages vary between 34 and 27 million years.

The lower volcanic complex is dominantly in the form of lava flows and pyroclastic units and also contains intercalations of siliceous ignimbrites. This lower complex contrasts in large measure with the upper one, because of its slightly deformed character and intense faulting and disturbed aspect. The sequences that comprise it are, in general, the host rocks for the principal mineralization of a large part of this region of Mexico. The outcrops of this lower complex are, as expected, more restricted than those of the upper but have been recognized over all the slope toward the Pacific in Sonora and Sinaloa. The upper contact displays an irregular surface with strong relief and shows marked contrast in the degree of alteration of the sequences.

The upper complex constitutes the most continuous and extensive ignimbrite cover in the world and is observed to form an area elongated northwest-southeast about 250 km broad and more than 1200 km long. Toward the north, this blanket has its last outcrops about at the border with the United States and to the south it disappears beneath rocks of basic and intermediate composition that make up the Neovolcanic axis.

According to Demant and Robin (1975), the thickness of these ignimbrites in some localities approaches

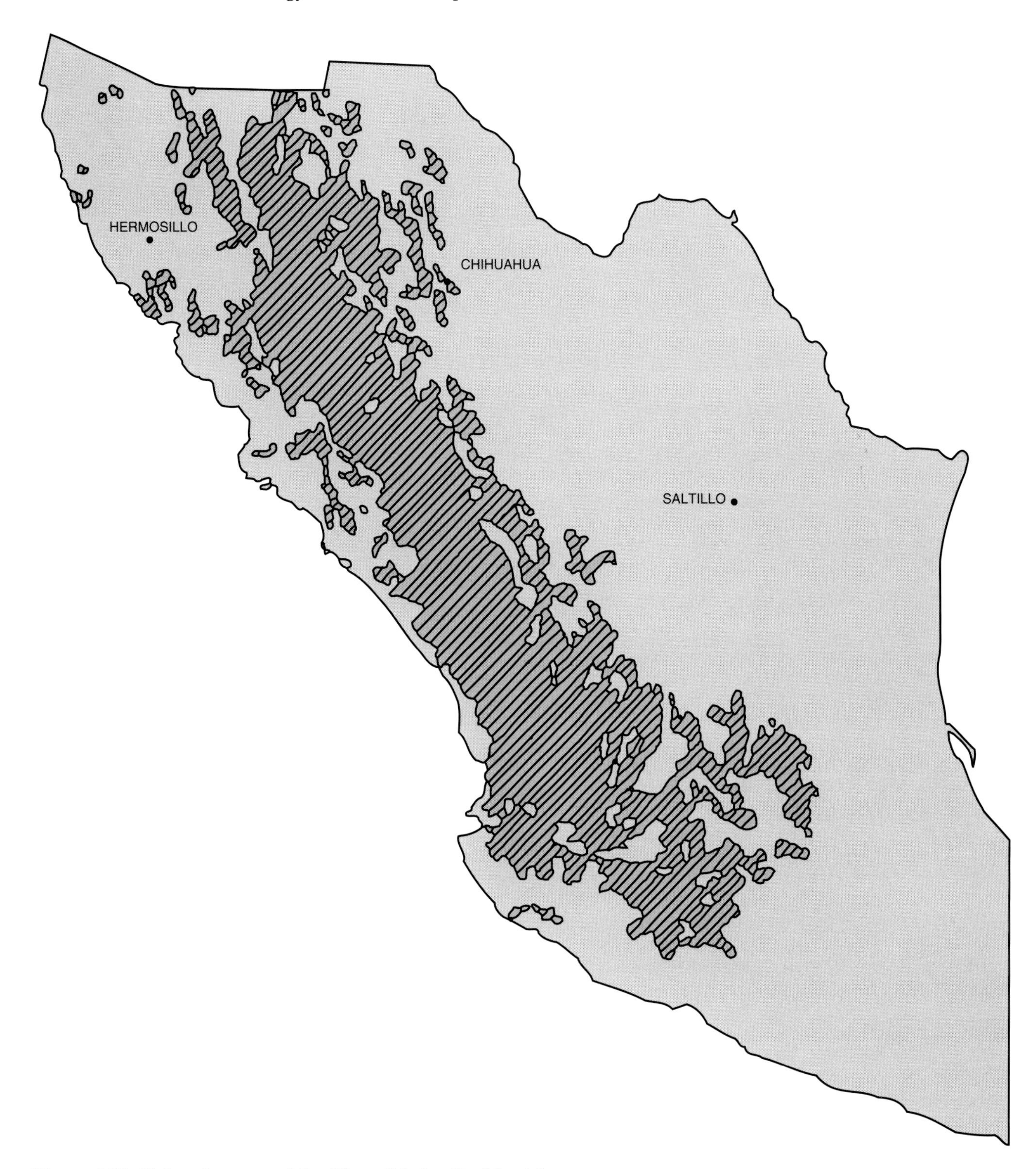

Figure 1.23. Volcanic cover of the Sierra Madre Occidental.

more than 1000 m. McDowell and Clabaugh consider that the number of calderas originated during the emission of this great volume of rock ought to have been between 200 and 400. Many of these have a diameter greater than 40 km, although the semicircular configuration today is obscured by the presence of normal faults and recent alluvial deposits.

The lower volcanic complex constitutes a typical calcalkaline magmatic arc related to a convergent continental margin where the Farallon plate is buried under the continental crust of Mexico. This phenomenon of convergence lasted until 29 million years ago, when the system of the expanding eastern Pacific impinged against the western margin of Mexico

(Atwater, 1970). Nevertheless, the interruption of magmatism in the interval of 45 to 34 million years demonstrates a break in the continuity of these processes. McDowell and Clabaugh (1979) consider that this period of calm has two possible causes: one of these is a lowered amount of convergence or a change of inclination of the plate being subducted; the other is the subduction of an active oceanic ridge. These same authors do not set forth a satisfactory tectonic explanation for the sudden volcanic activity indicated by the upper complex and the bimodal character of this volcanic sequence when compared with a silica and anorthosite standard.

Demant and Robin (1979) explain the origin of the ignimbrite blanket as the typical vulcanism of a rift zone behind an andesitic arc caused by the reaction of the crust to the subduction movements and indicate the coexistence of compressive and distensive vulcanism.

The principal mineralization within the Sierra Madre Occidental was partly discussed in the sections above, but the discussion is complemented by consideration of the whole Chihuahua area. It is thus convenient to indicate some generalities related to this subject.

The larger part of the mineral masses that are located in the Sierra Madre Occidental are strictly related to the lower volcanic complex. The copper-bearing porphyries of Cananea and Nacozari are related to the episodes of intrusive emplacement at the Cretaceous–Tertiary boundary and the hydrothermal deposits belong to a period that fluctuates between 49 and 28 million years (Clark et al., 1980). With respect to the hydrothermal veins, belts are encountered situated on both flanks of the Sierra Madre Occidental. The first of them, located to the west, comprises the deposits of gold and silver of Sinaloa and Sonora, such as those of Tayoltita and San José de Grácia. The second belt, on the east side of the Sierra, includes areas with lead, zinc, and silver, such as Santa Barbara and San Francisco del Oro.

BIBLIOGRAPHY AND REFERENCES

Abbreviation UNAM is Universidad Nacional Autónoma de México

Alencaster, Gloria, 1961, Estratigrafía del Triásico Superior de la parte central del Estado de Sonora. Paleontología Mexicana 11, parte 1, Instituto de Geología, UNAM, 18 p.

Anderson, D.L., 1971, La Falla de San Andrés, in Tuzo Wilson, 1974, Deriva Continental y Tectónica de Placas. Selecciones de Scientific American, W.H. Freeman and Company, San Francisco y Londres, p. 163–179.

Anderson, T.H., and L.T. Silver, 1978, Jurassic magmatism in Sonora, Mexico: Geological Society of America Abstract with Programs, v. 10, p. 359.

Anderson, T.H., and L.T. Silver, 1979, The role of the Mojave Sonora Megashear in the tectonic evolution of northern Sonora, Guidebook Field Trip no. 27, Geology of Northern Sonora: Geological Society of America. Annual Meeting in San Diego, p. 59–66.

Anderson, T.H., J.H. Eels, L.T. Silver, 1978, Rocas Precámbricas y Paleozóicas de la región de Caborca, Sonora, México. Libreto Guia del Primer Simposio sobre la Geología y Potencial Minero del Estado de Sonora. Hermosillo, Sonora, Instituto de Geología, UNAM, p. 5–34.

Atwater, T., 1970, Implications of plate tectonics for the Cenozoic tectonic evolution of western North America: Geological Society of America Bulletin v. 81, p. 3513–3536.

Beal, C.H., 1948, Reconnaissance of the geology and oil possibilities of Baja California: Geological Society of America Memoir 3l, 138 p. (Original not consulted, cited in R.G. Gastil et al., 1975, Reconnaissance geology of the State of Baja California, Geological Society of America Memoir 140, 170 p.)

Clark, K.F., P.E. Damon, S.R. Schutte, and M. Shaffiquillah, 1980, Magmatismo en el norte de México en relación con los yacimientos metalíferos. Revista, Geomimet, no. 106, p. 49–71.

Cooper, G.A., and A.R. Arellano, 1946, Stratigraphy near Caborca, northwest Sonora, Mexico. AAPG Bulletin, v. 30, p. 606–611.

Cooper G.A., et al., 1952, Cambrian stratigraphy and paleontology near Caborca, northwest Sonora, Mexico: Smithsonian Miscellaneous Collections, v. 119, 184 p.

Corona, F., 1979, Preliminary reconnaissance geology of Sierra La Gloria and Cerro Basura, northwestern Sonora, Mexico. Guidebook Field Trip 27, Geology of Northern Sonora: Geological Society of America Annual Meeting in San Diego, p. 41–58.

Darton, N.H., 1921, Geologic reconnaissance in Baja California: Journal of Geology, v. 29, p. 720–748. (Original not consulted, cited in F. Lozano, 1976, Evaluación petrolifera de la peninsula de Baja California, Mexico. Boletin Asociación Mexicana Geólogos Petroleros, v. 27, no. 4–6, p. 106–303.)

De Cserna, Z., and B.H. Kent, 1961, Mapa geólogico de reconocimiento y secciones estructurales de la región de San Blas y el Fuerte, Estados de Sinaloa y Sonora. Instituto de Geologia, UNAM, cartas geologicas y minerales, no. 4.

Demant, A., and C. Robin, 1975, Las fases del volcanismo en México; una síntesis en relación con la evolución geodinámica desde el Cretácico: Revista Instituto de Geologia, UNAM, v. 75, p. 70–83.

Dickinson, W.R., 1979, Plate tectonics and the continental margin of California, *in* W.G. Ernest, ed., The Geotectonic Development of California: Rubey volume 1, Prentice Hall, p. 1–28.

Dumble, E.T., 1900, Notes on the geology of Sonora: Geological Society of America Bulletin, v. 11, p. 122–152.

Echavarri, A., A.O. Saitz, and G.A. Salas, 1977, Mapa Metalogenético de Estado de Sonora: Revista Geomimet 2a Epocha July–August, no. 88, Consejo de Recursos Minerales.

Finch, J.W., and P.L. Abbott, 1977, Petrology of a Triassic marine section, Vizcaíno Peninsula, Baja California Sur, México: Sedimentary Geology, v. 19, p. 253–273.

Finch, J.W., E.A. Pessagno, and P.L. Abbott, 1979, San Hipolito Formation: Triassic marine rocks of the Vizcaíno Peninsula. Field Guides and Papers of Baja California, Geological Society of America Annual Meeting in San Diego, p. 117–120.

Fries, C., 1962, Reseña geológica del Estado de Sonora, con énfasis en el Paleozóica: Boletín Asociación Mexicana Geólogos Petroleros, v. 14, p. 257–273.

Gamper, M., and F.J. Longoria, 1980, Bioestratigrafía y facies sedimentárias del Cretácico Inferior de Sonora: Resúmenes de la V Convención Geológica Nacional, México, D.F., p. 14–15.

Gastil, R.G., and D. Krummenacher, 1978, The migration of the axis of Pacific margin magmatism across Baja California, Sonora and Chihuahua. Resúmenes del Primer Simposio sobre la Geología y Potencial Minero del Estado de Sonora, Hermosillo, Sonora: Instituto de Geologia, UNAM, p. 63–64.

Gastil, R.G., R.P. Phillips, and E.C. Allison, 1975, Reconnaissance geology of the State of Baja California: Geological Society of America Memoir 14, 170 p.

Gastil, R., G. Morgan, and D. Krummenacher, 1981, The tectonic history of peninsular California, *in* W.G. Ernest, ed., The Geotectonic Development of California: Rubey volume 1, Prentice Hall, p. 285–305.

González, C., 1979, Geology of the Sierra del Alamo. Guidebook Field Trip no. 27, Geology of Northern Sonora: Geological Society of America Annual Meeting in San Diego, p. 23–31.

Heim, A., 1922, Notes on the Tertiary of southern Lower California: Geological Magazine. (Original not consulted, cited in F. Mina, 1956, Bosquejo geológico de la parte sur de la peninsula de Baja California, Excursion A-7 of the XX Congreso Geológico Internacional, México, p. 11–42.)

Imlay, R.W., 1939, Paleogeographic studies in northeastern Sonora: Geological Society of America Bulletin, v. 50, p. 1723–1744.

Keller, W.T., and F.E. Wellings, 1922, Sonora: Cía Petrolera el Aguila, Geological Report no. 180, 38 p. (unpublished). (Original not consulted, cited in T.H. Anderson, J.H. Eells, and L.T. Silver, 1978, Rocas Precámbricas y Paleozóicas de la región de Caborca, Sonora, México. Guia del Primer Simposio sobre la Geología y Potencial Minero del Estado de Sonora, Hermosillo, Instituto de Geología, UNAM, p. 5–34.)

King, R.E., 1939, Geological reconnaissance in northern Sierra Madre Occidental of Mexico. Geological Society of America Bulletin, v. 50, p. 1625–1722.

Longoria, F.J., and V.A. Peréz, 1978, Bosquejo geológico de los cerros Chino y Rajón, cuadrángulo Pitiquito-La Primavera (NO de Sonora): Bulletin Department of Geology, UNI-SON, v. 1, no. 2, p. 119–144.

Longoria, F.J., M.A. González, J.J. Mendoza, and V.A. Pérez, 1978, Consideraciones estructurales en el cuadrángulo Pitiquito-La Primavera, (Noroeste de Sonora): Bulletin Department of Geology, UNI-SON, v. 1, no. 1, p. 61–67.

López-Ramos, E., 1979, Geología de México, 2a edición, México D.F., Edición escolar, 3 volúmenes.

Lozano, F., 1976, Evaluación petrolífera de la peninsula de Baja California, México. Boletín Asociación Mexicana Geólogos Petroleros, v. 27, n. 4–6, p. 106–303.

Márquez-Castañeda, B., 1984. Estudio geológico del área de Santa Barbara, Chihuahua. Unpublished Report of the Faculty of Engineering, UNAM.

McDowell, F.W., and S. Clabaugh, 1979, Ignimbrites of the Sierra Madre Occidental and their relation to the tectonic history of western Mexico, *in* C.E. Chapin and W.E. Elston, eds., Ashflow Tuffs: Geological Society of America Special Paper 180.

McEldowney, R.C., 1970, An occurrence of Paleozoic fossils in Baja California, Mexico: Geological Society of America Abstracts with Programs, v. 2, p. 117. (Original not consulted, cited in R.G. Gastil et al., 1975, Reconnaissance geology of the State of Baja California: Geological Society of America Memoir 140, 170 p.)

McKenzie, D.P., and W.J. Morgan, 1969, Evolution of triple junctions: Nature, v. 224, p. 125–133.

Mina, F., 1956, Bosquejo geológico de la parte sur de la península de Baja California: Excursion A-1 of the XX Congreso Geológico Internacional, México. p. 11–42.

Mina, F., 1957, Bosquejo geológico de la parte sur de la península de Baja California, Boletín Asociación Mexicana Geólogos Petroleros, v. 9, p. 139–269.

Mulchay, R.B., and J.R. Velasco, 1954, Sedimentary rocks at Cananea, Sonora, Mexico with the sections at Bisbe and Swisshelm Mountains, Arizona: AIME Transactions, v. 199, p. 628–632. (Original not consulted, cited in Cía Minera Cananea, S.A., 1978, Geología del Distrito Minero de Cananea, Sonora, Libreto Guia del Primer Simposio sobre la Geología y Potencial Minero del Estado de Sonora, Instituto de Geología, UNAM, p. 57–70.)

Ortega-Gutiérrez, F., 1982, Evolución magmática y metamórfica del complejo cristalino de La Paz, B.C.S.: Resúmenes de la VI Convención Geológica Nacional de la Sociedad Geológica Mexicana, p. 90.

Ortlieb, L., 1978, Reconocimiento de las terrazas marinas Cuaternárias de la parte central de Baja California: Revista, Instituto de Geología, UNAM, v. 2, no. 2, p. 200–211.

Pantoja-Alor, J., and J. Carrillo-Bravo, 1966, Bosquejo geológico de la región de Santiago, San Jose del Cabo, Baja California: Boletin Asociación Mexicana Geólogos Petroleros, v. 17, nos. 1–2, p. 1–11.

Patterson, D.L., 1979, The Valle Formation: physical stratigraphy and depositional model, southern Vizcaíno Peninsula, Baja California Sur: Field Guides and Papers of Baja California: Geological Society of America Annual Meeting in San Diego, p. 73–76.

Rangin, C., 1977a, Sobre la presencia del Jurásico Superior con amonitas en Sonora septentrional, México: Revista, Instituto de Geología, UNAM, v. 1, p. 1–4.

Rangin, C., 1977b, Tectónicas sobrepuestas en Sonora septentrional: Revista, Instituto de Geología, UNAM, v. 1, p. 44–47.

Rangin, C., 1978, Consideraciones sobre la evolutión geológicas de la parte septentrional del Estado de Sonora. Libreto Guia del Primer Simposio sobre la Geología y Potential Minero del Estado de Sonora, Hermosillo, Sonora, Instituto de Geología, UNAM, p. 35–56.

Rangin, C., 1979, Evidence for superimposed subduction and collision processes during Jurassic–Cretaceous time along Baja California continental borderland. Field Guides and Papers of Baja California: Geological Society of America Annual Meeting in San Diego, p. 37–52.

Rangin, F., 1978, Consideraciones sobre el Paleozóico sonorense: Resúmenes del Primer Simposio sobre de Geología y Potential Minero del Estado de Sonora, Hermosillo, Sonora: Instituto de Geología, UNAM, p. 35–56.

Rangin, F., and D.A. Córdoba, 1976, Extensión de la Cuenca Cretácica Chihuahuense en Sonora septentrional y sus deformaciones: Memória del Tercer Congreso Latinoamericano de Geología, México, 14 p.

Ransome, F.L., 1904, Description of the Bisbee quadrangle, Arizona: U.S. Geological Survey, v. 112, 17 p.

Rodríguez, R., and D.A. Córdoba, 1978, eds., Atlas geológico y evaluación geológico-minero del Estado de Sinaloa: Instituto de Geología, UNAM y Secretaría del Desarrollo Económico del Estado de Sinaloa, 702 p.

Roldán, J., and Solano B., 1978, Contribución a la estratigrafía de las rocas volcánicas del Estado de Sonora: Bulletin of the Department of Geology, UNI-SON, v. 1, p. 19–26.

Salas, G.P., 1980, Carta y provincias metalogenéticas de la Republica Mexicana, Consejo de Recursos Minerales 2nd Edición, 199 p. (Mapa con texto.)

Santillán, M., and T. Barrera, 1930, Las posibilidades petrolíferas en la costa occidental de la Baja California, entre los paralelos 30 y 32 de latitud norte: Anales de Instituto de Geología, v. 5, p. 1–37.

Sillitoe, R.H., 1973, The tops and bottoms of porphyry copper deposits: Economic Geology, v. 68, p. 709–715.

Sillitoe, R.H., 1975, A reconnaissance of the Mexican porphyry copper belt.

Silver, L.T., and T.H. Anderson, 1974, Possible left-lateral early to middle Mesozoic disruption of the southwestern North American craton margin: Geological Society of America Abstracts with Programs, v. 6, p. 955.

Silver, L.T., and T.H. Anderson, 1978, Mesozoic magmatism and tectonism in northern Sonora and their implications for mineral resources, Resúmenes del Primer Simposio sobre la Geología y Potencial Minero del Estado de Sonora, Hermosillo, Sonora: Instituto de Geología, UNAM, p. 117–118.

Stoyanow, A., 1942, Paleozoic paleogeography of Arizona: Geological Society of America Bulletin, v. 53, p. 1255–1282.

Taliefferro, N., 1933, An occurrence of Upper Cretaceous sediments in northern Sonora, Mexico: Journal Geology, v. 41, p. 12–37.

Valentine, W.G., 1936, Geology of the Cananea Mountains, Sonora, Mexico: Geological Society of America Bulletin, v. 47, p. 53–86.

2. Geology of the Northern and Northeastern Regions of Mexico

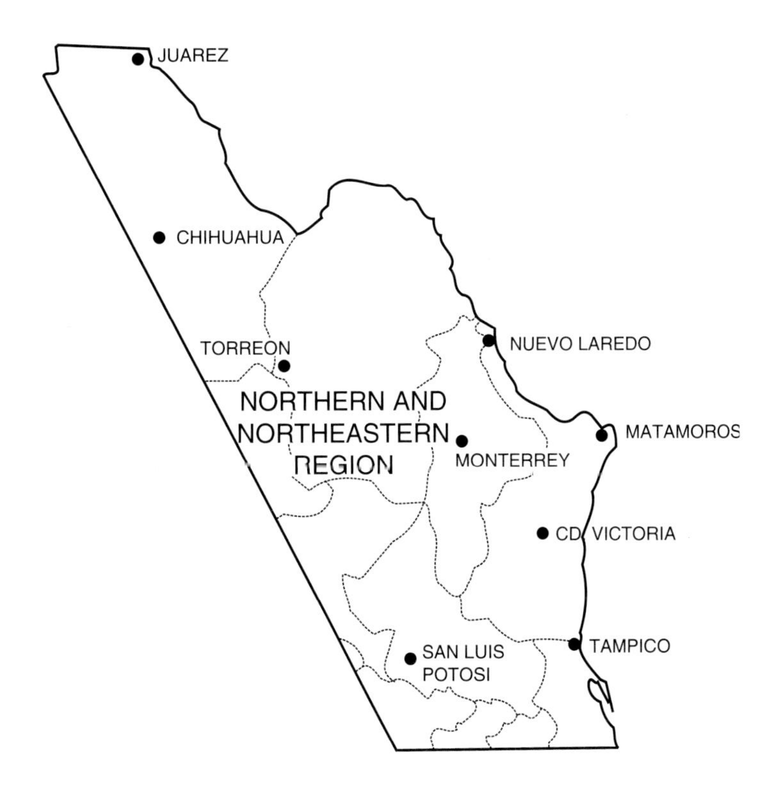

GENERAL CONSIDERATIONS

For description of the northern and northeastern regions of Mexico, one can use the following natural limits: to the west, the Sierra Madre Occidental; to the east, the Gulf of Mexico coast; and to the south, the northern edge of the Neovolcanic axis.

The region comprises, according to the physiographic division of the DGG (see Figure 1.1), the provinces of Sierras and Plains of the North (Basin and Range), the Sierra Madre Oriental, the Great Plain of North America, the Central Mesa, and the Coastal Plain of the northern Gulf of Mexico. Nevertheless, the division that is used is based fundamentally on the paleogeographical elements of the Mesozoic in this part of Mexico. In various forms, these elements have a general correspondence with the physiographic provinces mentioned above, chiefly in the origin of orographic forms that are particular expressions of the types of geological phenomena that generate them.

The climate of all the region varies in general from hot to temperate, and regular summer rains occur. Additionally, precipitation records reveal that the climates vary from dry to semi-arid in the west of this zone and from humid to subhumid in the Sierra Madre Oriental and the Coastal Plain of the northern Gulf of Mexico.

CHIHUAHUA AREA

General Geology

The area of the State of Chihuahua is characterized, particularly in its eastern part, by the presence of folded mountains formed from marine Mesozoic strata. These mountains make up prominent topographic peaks that occur separated by great plains that were elevated as they were filled. These are tectonic troughs filled with continental sediments and some lava debris blocks. They originated as local filled basins termed *bolsones*. The folded sedimentary sequences gradually disappear toward the western margin of the state, to an edge under the ignimbrite cover of the Sierra Madre Occidental.

The folded sedimentary rocks, which crop out in the major part of the area, developed above a Precambrian and Paleozoic basement that is exposed in some localities and that also has been reported in wells drilled by Pemex. In the area of the Sierra del Cuervo, Mauger and co-workers (1983) obtained a K-Ar age corresponding to the Grenville from a metamorphic block included in a Permian sequence. Quintero and Guerrero (1985), in addition, reported the exposure of a similar metamorphic unit to the south of Mina Plomosas that could be an outcrop of the Chihuahua Precambrian basement.

The Paleozoic rocks that crop out widely in some areas of Texas have in northeastern Mexico very restricted exposures with the result that it is difficult to reconstruct the paleogeographic elements of that area (Figure 2.1). Gonzáles (1976) considers that the outcrops of limestones and dolomites of the lower Paleozoic of Chihuahua reflect a platform environment similar to the facies developed across the North American craton and considers it logical that this element continues toward Mexico. In contrast, he indicates that in the Pennsylvanian–Permian interval, the sedimentological pattern presents more irregular conditions caused by the action of block faulting that developed intracratonic platforms and basins on which were deposited respectively carbonates and terrigenous sediments. The Diablo Platform dates from this time; its southwest border forms a marked lineament that coincides approximately with the course of the Rio Grande along a belt from Ciudad Juárez to Ojinaga. This tectonic feature has maintained its influence on sedimentological events and deformation during the Mesozoic and even the Cenozoic. DeFord (1969) makes note of the sudden disappearance, at the Mexican frontier, of the Ouachita belt composed of deformed Paleozoic terrigenous sediments. This leads one to believe that this

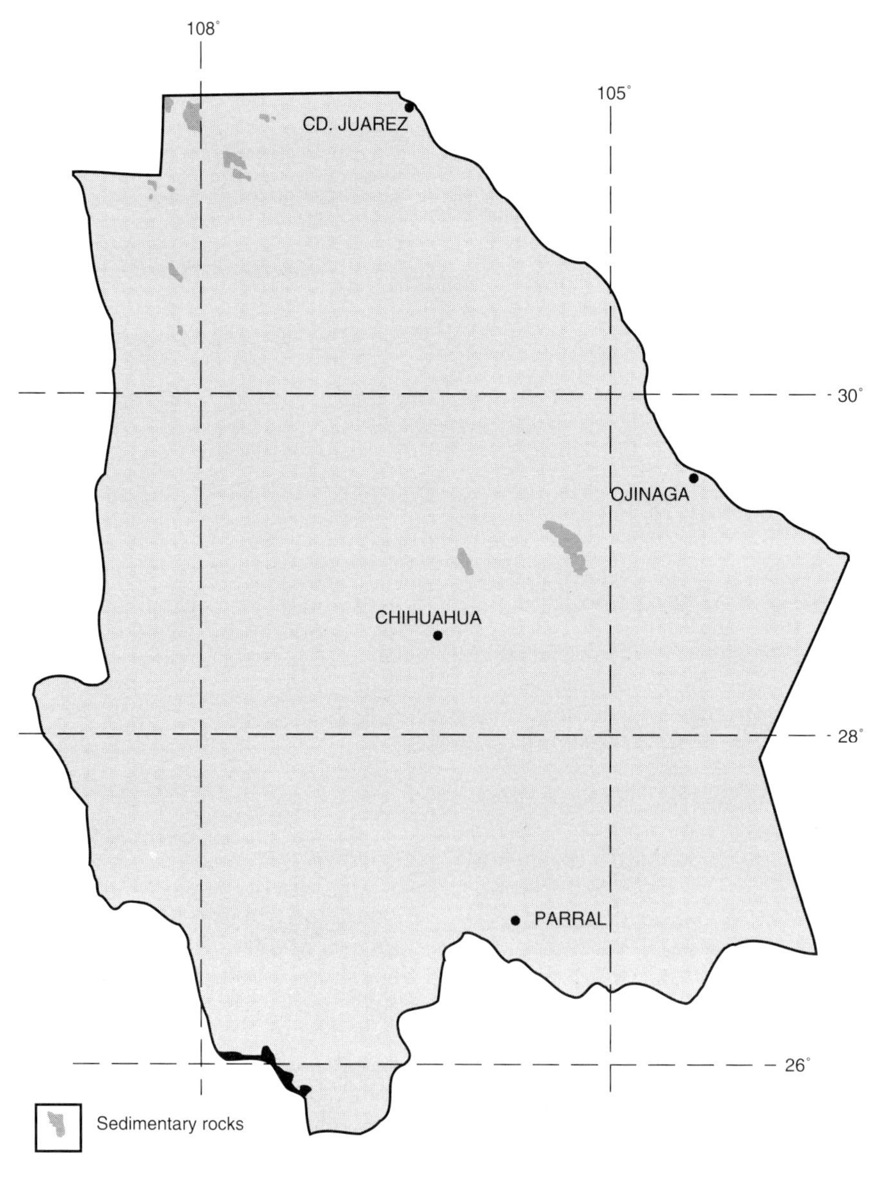

Figure 2.1. Paleozoic sedimentary rocks of Chihuahua.

belt continues under the Mesozoic sequence of Chihuahua to the east of the calcareous outcrops of the Paleozoic at Placer de Guadalupe–Mina Plomosas. Nevertheless, one cannot rule out the possibility that this belt passes to the west of the above-cited locality, since in the area of Aldama the presence of a considerable sequence of dark gray siltstones similar to those outcropping on the northern flank of the Ouachita belt in Texas has been reported.

The difficulty in defining the Paleozoic tectonic elements in Chihuahua is caused by the very scarce outcrops and the fact that this region encompasses the confluence of the North American craton, the Ouachita belt, and the Cordilleran miogeosyncline, all of whose interrelationships are still somewhat confusing.

The principal Paleozoic outcrops of Chihuahua include the sequences exposed in the area around Mina Plomosas northeast of Chihuahua (mostly limestones); Aldama siltstones, which are present to the north of Chihuahua City; and various minor localities that occur in the northwest corner of the state where limestone and dolomite platform facies appear. In the oil wells Moyotes no. 1 and Chinos no. 1, Paleozoic sequences were drilled. In the latter well Ordovician, Cambrian, and metamorphic Precambrian rocks were encountered (Navarro and Tovar, 1973).

The orogenic deformation of the Marathon-Ouachita geosyncline during the Mississippian Pennsylvanian interval and the normal faulting of the southern portion of the North American craton were followed by a long period of emersion (Permian to Middle Jurassic) during which redbeds were deposited within a tectonic setting of intense normal faulting. This episode of continental deposition is widely known over the neighboring regions of Torreon and northern Zacatecas.

At the beginning of the Kimmeridgian, the eastern portion of Chihuahua formed a marine basin as a consequence of a transgression initiated during this time (DeFord, 1969). This basin is limited to the northeast by the Diablo Platform, to the southwest by the Aldama Peninsula, and to the east by the Coahuila Peninsula or Island (see Figure 2.4). Gonzáles (1976) cites an unpublished work by R. Garza of Pemex in which it is suggested that the Aldama Peninsula and the Coahuila Island might have constituted a single positive element beside which the Chihuahua basin might have had communication across the Gulf of Sabinas in Coahuila. The first stages of the Upper Jurassic marine transgression covering the Chihuahua basin gave rise to deposition of an evaporite sequence that is now manifested by diapiric structures of salt and gypsum located to the south of Ojinaga and Ciudad Juárez, as well as encountered in oil wells drilled by Pemex in the Sierra de Cuchillo Parado (DeFord, 1969). The chief outcrops of the Upper Jurassic contain argillaceous and calcareous sequences and are located principally between Ciudad Juárez and Chihuahua, mainly to the north of Sierra de Samalayuca, in the Sierra de la Alcaparra, in the Sierra El Kilo, and in the Sierra La Mojina.

At the beginning of the Cretaceous, during the Neocomian, marine sedimentation continued in the Chihuahua basin, chiefly with deposits of limestone and gypsum of the Alcaparra Formation as well as with the clays and sandstones of the Las Vigas Formation. During this epoch the Coahuila Island remained still emergent and the Aldama Peninsula was covered by a marine transgression. At the end of the Neocomian and beginning of the Aptian, the seas began a very important transgression over the Coahuila Island and the internal terranes of Sonora and Sinaloa (Rangin and Córdoba, 1976). In the Chihuahua basin a very thick, fundamentally calcareous sequence then developed that includes the Cuchillo Formation and the Chihuahua Group (Benigno and Lágrima formations, the Finlay Limestone, and Benavides Formation) (Córdoba, 1970). This transgression across the positive elements reaches its maximum development in the Albian–Cenomanian interval, during which important reef facies were developed over the Aldama Platform (Franco, 1978). During the Late Cretaceous, terrigenous sedimentation in the Chihuahua region marks the uplift and volcanic activity of the western part of Mexico. These terrigenous deposits constitute the Ojinaga Group, recognized in the area of the city of this name; they reflect a deltaic environment that in the Campanian marks the advance of the coast line to the east (González, 1976). The absence of Upper Cretaceous sediments in the area of the Aldama Peninsula suggests that this region remained emergent during most of Late Cretaceous time.

The end of the Mesozoic Era is marked by folding of the Mesozoic cover (Figure 2.2) as a result of a "decollement" or sliding at the level of the basal evaporite sequence. On the other hand, the structural axes present generally a northwest-southeast orientation. The recumbence of the inverse faults, which have opposite vergence on both sides of the basin, has been interpreted to be due to the up-arching of a central belt at the level of the basement, which originated the slippage toward the Diablo and Aldama platforms, respectively (Gries and Haenggi, 1970).

In the Cenozoic Era the Chihuahua region evolved as an emergent zone that was partly covered (almost completely so in the western portion) by emissions of ignimbrites of Oligocene–Miocene age. McDowell and Clabaugh (1979) indicate that the volcanic rocks to the east side of Chihuahua (Figure 2.12) have a different chemical composition from the major volcanic areas and constitute an intermediate province between the calcalkaline series of the Sierra Madre Occidental (west of Chihuahua) and the alkaline series of Trans-Pecos Texas. The distentional tectonics of the late Tertiary caused the formation of grabens in which considerable thicknesses of continental sediments were deposited.

Economic Resources

According to the map of Metallogenic Provinces of the Mexican Republic (Salas, 1975), the State of Chihuahua belongs to the Metallogenic Province of

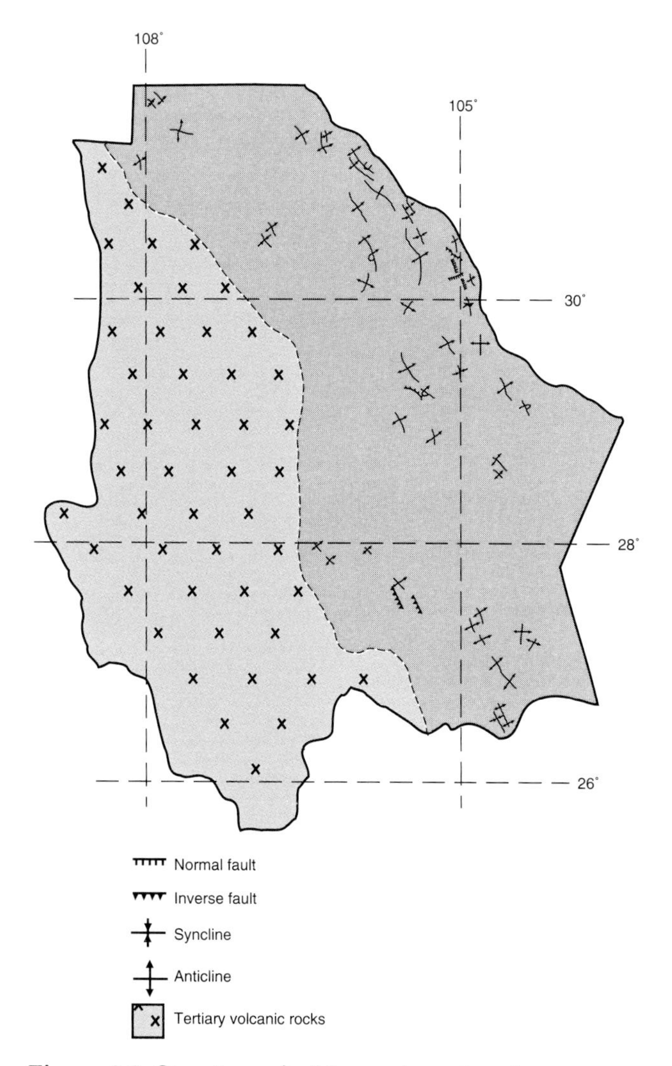

Figure 2.2. Structures in Mesozoic rocks of Chihuahua.

the Sierra Madre Oriental. The principal deposits of this region are the hydrothermal deposits of silver, lead, zinc, and gold that are localized principally in the central belt of the state and that follow in large part the axis of the Mexican Republic (see Figure 2.3). The mineral districts of Santa Eulalia, Naica, Hidalgo del Parral, Santa Barbara, and San Francisco del Oro belong to this central belt of Chihuahua state. According to Clark et al. (1980), during the end of the Mesozoic and a large part of the Cenozoic, there occurred an eastward migration in time and space and a later return of the magmatic arc related to the convergent margin that was developing along the western border of the country. The episodes of hydrothermal mineralization mentioned above occurred 28–40 million years ago (Clark et al., 1980). These episodes were related to the magmatic activity that is a product of the partial fusion induced by the Farallon plate under the continental crust of Mexico during which time a return to westward migration of the magmatic arc was occurring.

The hydrothermal deposits of manganese, contained generally in ignimbrites, also form a belt of deposits within which the localities of Talamantes, Terrantes, and Casas Grandes occur.

The iron deposits of eastern Chihuahua belong to a belt associated with the return to the west of magmatism related to arc migration (Clark et al., 1980). The chief deposits are encountered in the mines of La Perla and Hercules in the vicinity of Coahuila.

The uranium-bearing volcanogenic deposits of Chihuahua are linked to the migration of magmatism toward the east and are related to rhyolitic and trachytic lavas about 40 million years in age. The principal location of this type of deposit is the Sierra de Peña Blanca, located in the north of Chihuahua.

With respect to petroleum reservoirs, the calcareous Paleozoic rocks that possess platform facies, chiefly those that crop out in the northwest part of the state, occur at depth with petroleum possibilities. They show considerable porosity and resemble productive rocks of western and central Texas (González, 1976). Information from petroleum wells has revealed possibilities in the Jurassic of Chihuahua since they have penetrated calcareous-argillaceous sequences with high organic content, which could serve as source rocks as well as some porous sequences of platform facies that could be reservoirs (González, 1976).

COAHUILA AND NUEVO LEÓN

General Geology

This region is characterized by the predominant presence of folded Mesozoic sedimentary rocks that rest upon a Paleozoic and Precambrian basement. The most significant physiographic feature is the flexure that the Sierra Madre Oriental undergoes at the latitude of Monterrey; from here the fold belt acquires a general east-west orientation. To the north of this flexure the orographic elements become more widely separated and the geologic structures less narrow. Furthermore, the topographic relief diminishes gradually to the east grading into the coastal plain of the Gulf of Mexico.

The Paleozoic basement, above which the Mesozoic sequence of this region developed, has been interpreted as a continuation of the Ouachita belt of the southeastern United States. Denison et al. (1970) have indicated that the Granjeno schists of Peregrina Canyon maintain a great similarity to the eastern internal zones of the Ouachita belt. This affirmation seems to be corroborated by the metamorphic basement reported in the petroleum wells of the States of Nuevo León and Tamaulipas. In contrast, the detrital Permian sediments reported from the Delicias-Acatita area are similar to those of the frontal belt east of the Ouachita geosyncline.

Periods of emersion and normal faulting occurring during the Triassic and part of the Jurassic controlled the paleogeography of the upper part of the Mesozoic and gave way to continental redbed deposition that has been reported chiefly within and to the south of the Monterrey-Torreon transverse sector.

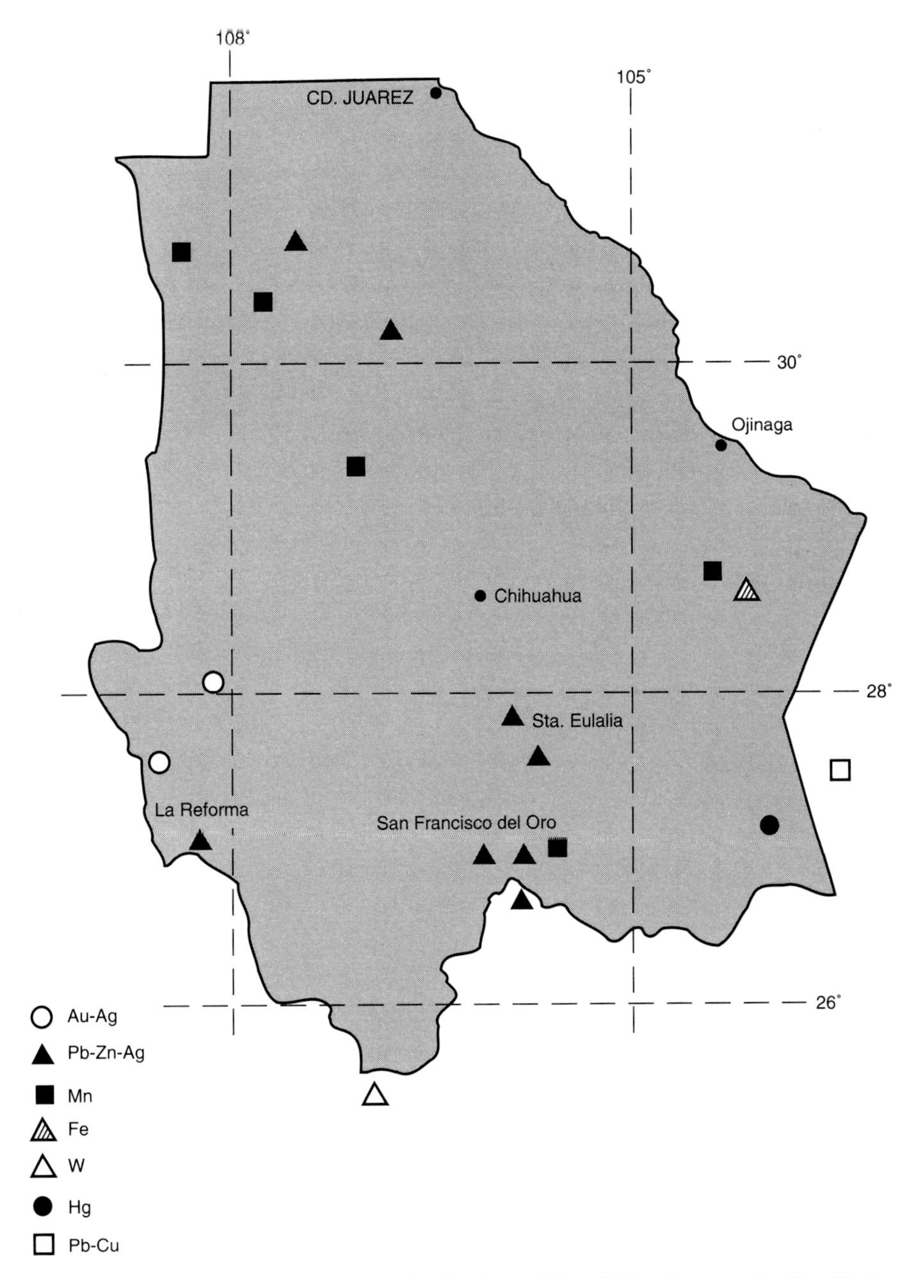

Figure 2.3. Distribution of principal mineral localities known in the State of Chihuahua (taken from the Metallogenic Map of the Republic of Mexico by G.P. Salas, 1975).

In the Late Jurassic, a transgression occurred over northeastern Mexico; it concurred with the formation of the Sabinas Gulf, the Coahuila Island, and the Peninsula and Archipelago of Tamaulipas (see Figure 2.4). This phenomenon has been related by various authors to the opening of the western extremity of the Tethyan Sea during the initial disintegration of the supercontinent Pangea. During this process the Paleogulf of Sabinas was defined in the Oxfordian. It possesses the characteristics of an intracratonic basin developed in the more tectonically stable southern North American craton. Terrigenous and calcareous evaporite deposits were developed in the first stages of transgression in the Sabinas Gulf. They formed under conditions of strong evaporation (González, 1976), mainly in the Oxfordian (see Figure 2.5). The Minas Viejas, Novillo, Olvido, Zuloaga, and La Gloria formations belong to this epoch. The last two formations represent respectively the extra-littoral and near-coastal facies of the upper Oxfordian (Rogers et al., 1961). The advance of the marine transgression during the Kimmeridgian and Tithonian created deposits of the open sea, the La Caja and Pimienta formations, which are composed of calcareous-argillaceous sequences with carbonaceous horizons, just as in the La Casita Group (see Figure 2.6).

During the beginning of the Early Cretaceous, the marine transgression of the Late Jurassic continued

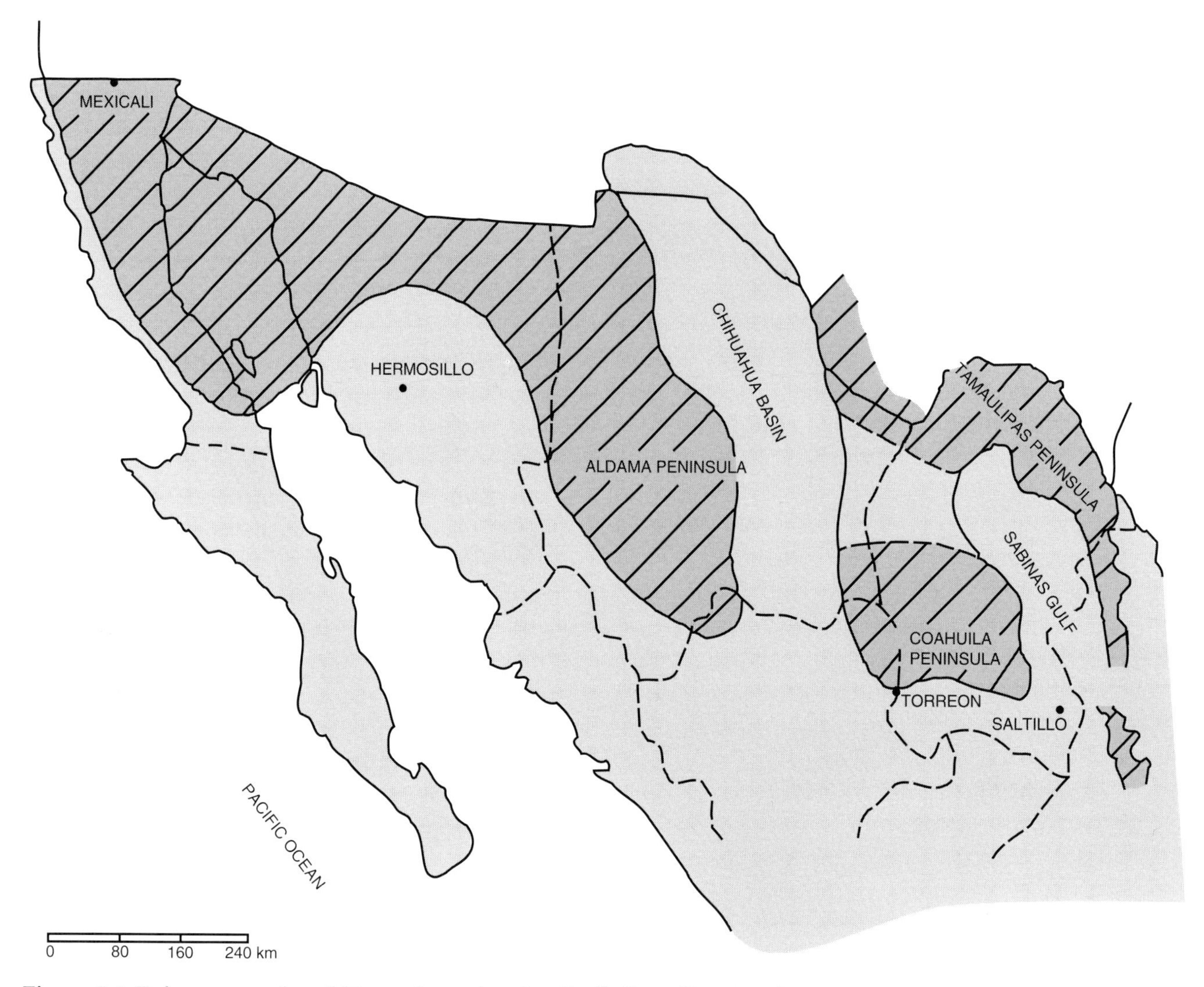

Figure 2.4. Paleogeography of Upper Jurassic, after Raúl González-Garcia, 1976.

and gave way throughout all the Neocomian to deposition of a heterogeneous sequence composed of various formations. The San Marcos arkose constitutes a littoral and continental facies encompassing a major part of the Neocomian deposited simultaneously with other diverse formations. The configuration of the San Marcos arkose permits the observation of intercalations of the unit between formations deposited contemporaneously in marine platform environments. This is indicated by various lentils formed off the coastline of the Coahuila Island in different stratigraphic levels. The Menchaca Formation is formed by a sequence of limestones and some intercalations of marl and shale. This formation constitutes the base of the platform sequence of the Neocomian that is represented higher up by the shales and sandstones of the Barril Viejo Formation; the limestones and shales of the Padilla Formation; the calcareous-argillaceous sequence of the La Mula; and the limestones, dolomites, and evaporites of the La Virgen Formation. In the southeast sector of the Sabinas Gulf, the argillaceous limestones of the Taraises Formation were deposited during the Berriasian–Hauterivian interval.

From the Hauterivian to the Aptian, in all of northeast Mexico, calcareous deposits formed; these occur in various facies. In a large part of the Sabinas Gulf, limestones of the Cupido Formation were deposited in a platform environment. Furthermore, there developed a reefal lineament that trends from Laredo to Monterrey and from there to the west toward Torreon. This is considered an integral part of the Cupido. Finally, outside of the reefal margin that bounds this formation, facies of the open sea developed, constituting the upper Tamaulipas Formation (see Figure 2.7).

At the Aptian–Albian boundary, there suddenly appeared a general influx of fine terrigenous clastics into the Gulf of Sabinas; this forms the La Peña Formation. This influx could be a response to epeirogenic uplifts of the surrounding positive elements or to a eustatic fall in sea level (Smith, 1970; Charleston, 1973).

During the Albian–Cenomanian interval, an important marine transgression, completely covering elements that until then had remained positive, began the development of thick sequences of carbonates

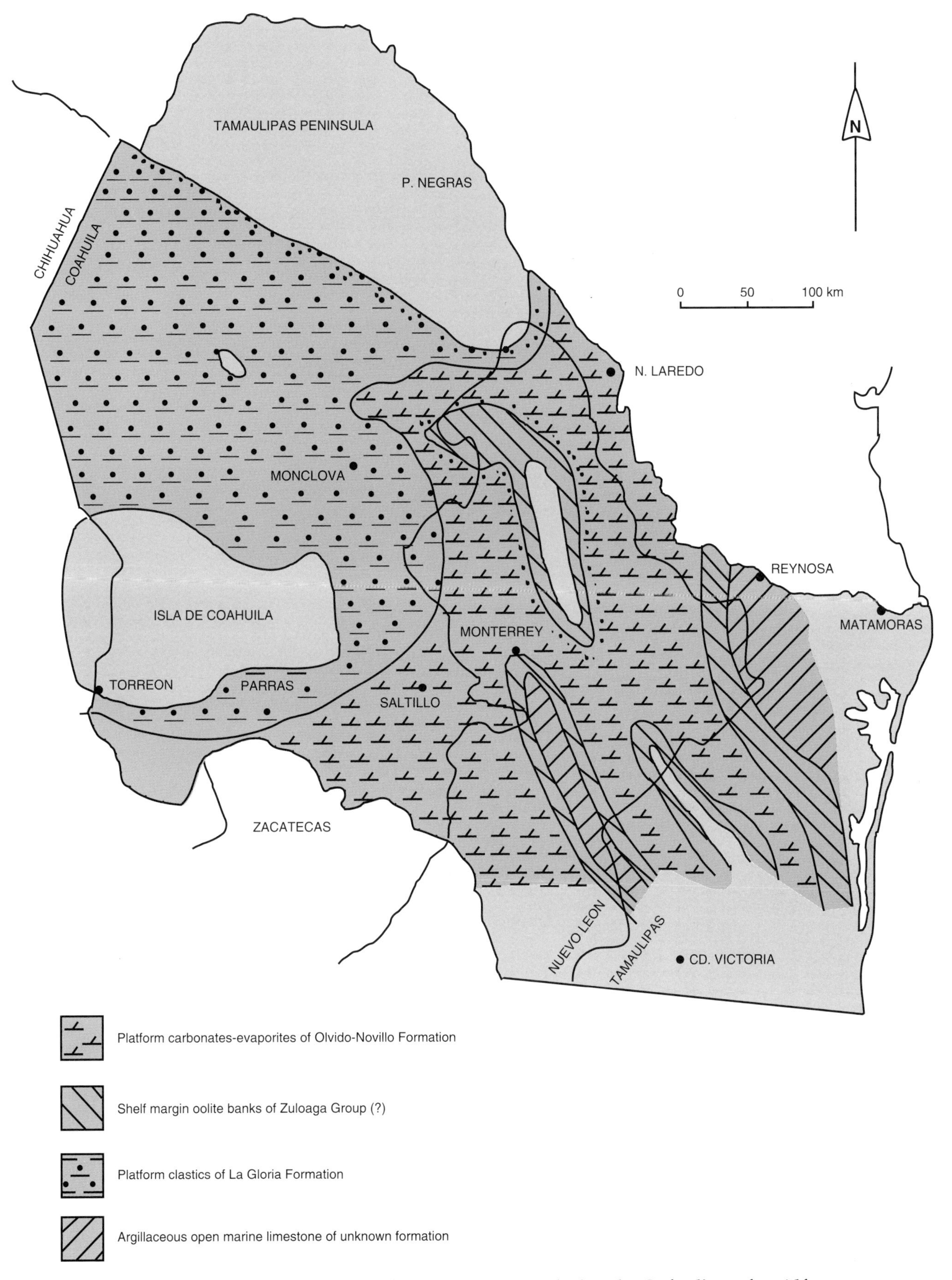

Figure 2.5. Paleogeographical configuration of northeast Mexico during the Oxfordian, after Alfonso-Zwanziger, 1978.

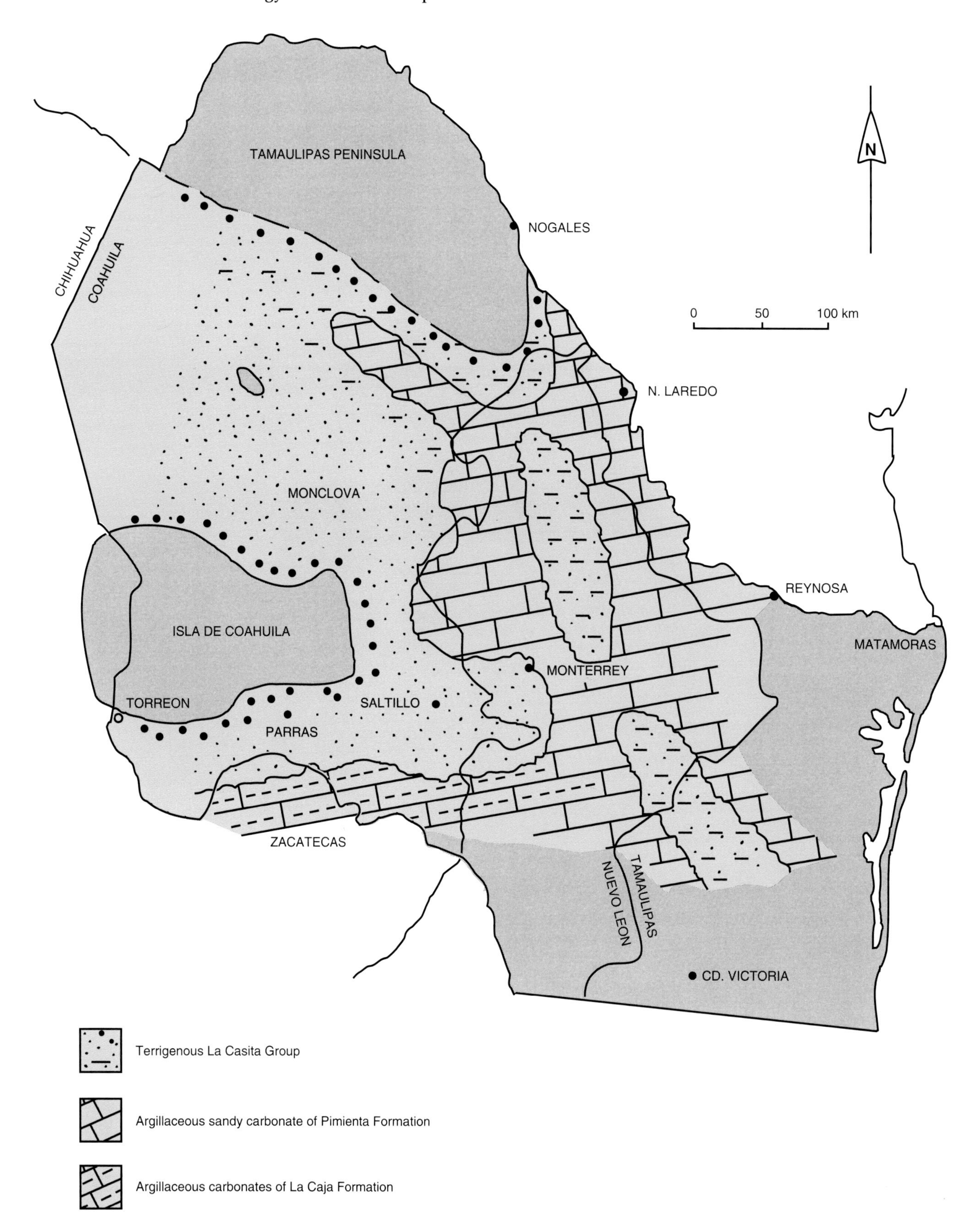

Figure 2.6. Paleogeographical configuration of northeast Mexico during the Kimmeridgian and Tithonian, after Alfonso-Zwanziger, 1978.

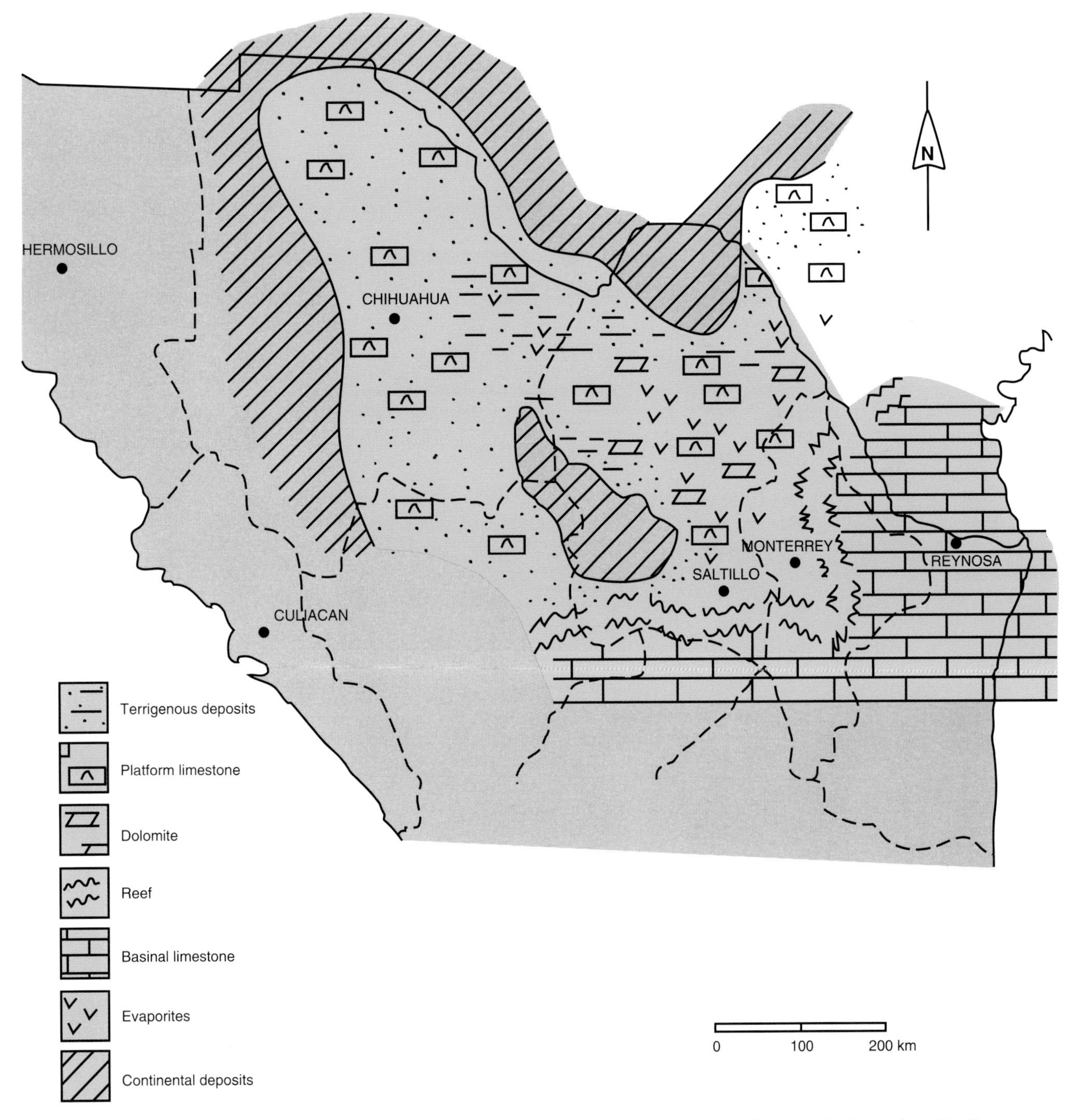

Figure 2.7. Paleogeography of Neocomian–lower Aptian for northern and northeast Mexico, after Raúl González-Garcia, 1976.

over all of northeast Mexico. Over the Burro platform (Tamaulipas Peninsula) and the Coahuila Island, sequences of shallow marine and evaporite facies were deposited owing to the presence of reefs that bordered the tectonic elements. The Aurora, Acatita, and upper Tamaulipas formations belong to this interval (see Figure 2.8).

During the Late Cretaceous, over all the region in general, terrigenous sediments coming from western Mexico were deposited. These underwent orogenic deformation at the beginning of this epoch and later a general uplift. With the gradual retreat of the seas toward the east, successive coastlines and deltas developed with consequent detrital clastic deposition (see Figure 2.9). In the La Popa and Parras basins slow subsidence induced the accumulation of great thicknesses of shales and sandstones. The formations of Del Rio, Buda, Indidura, Eagle Ford, Caracol, Austin, Parras, Upson, San Miguel, Olmos, Escondido, and Difunta are units belonging to the Upper Cretaceous. Sediments that constitute the last formation have been considered by Tardy et al. (1974) as flysch deposits that precede the orogenic deformations.

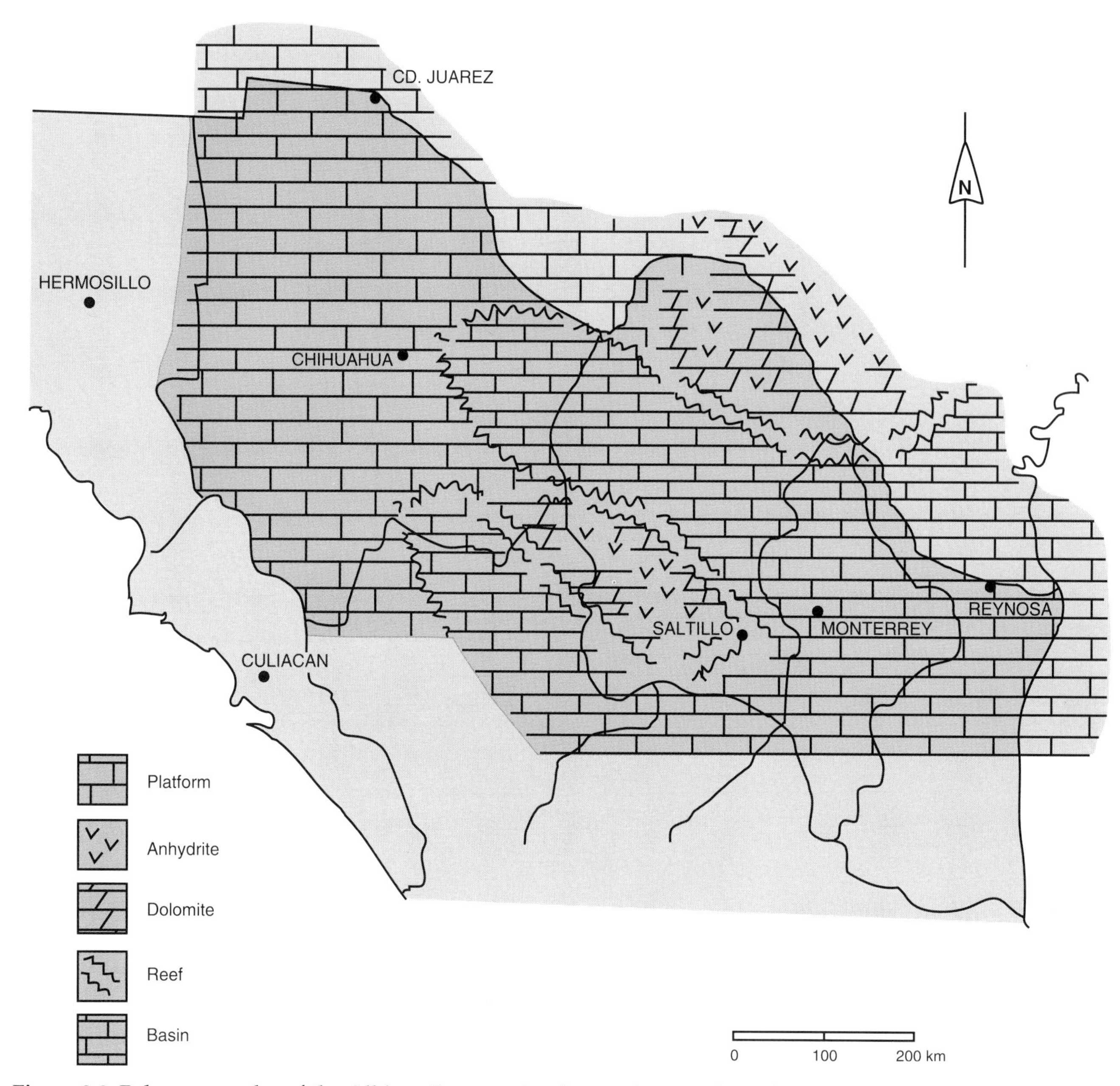

Figure 2.8. Paleogeography of the Albian–Cenomanian for northern and northeastern Mexico, after Raúl González-Garcia, 1976.

Deformations of the Laramide Orogeny occurred mainly in the early Cenozoic. The anticlinal and synclinal structures so characteristic of the Coahuila landscape belong to these epochs. The style of deformation of the Sabinas Gulf area is less intense than that observed in the front of the Parras basin where recumbent folds and overthrusts have significant development. The folds are somewhat narrow and only are recumbent and overthrust toward the positive elements at the edges of the Sabinas Gulf. Over the old positive elements the structures are even more gentle and can be observed to have the form of large periclines. From the time of these orogenic deformations the continental evolution of the region began with important continental deposits induced by Late Cretaceous normal faulting.

During the Cenozoic, isolated pulses of igneous activity occurred in this part of the country, chiefly in the Oligocene, when intrusions of nepheline syenite were emplaced (Bloomfield and Cepeda, 1973). Clark and co-workers (1980) consider these igneous bodies as part of an alkalic igneous belt that is extended from New Mexico into Mexico. These authors consider that this alkaline magmatism was caused by the subduction that was occurring in western Mexico and that constitutes the most distant manifestation of the ancient ocean trench that was forming more than 1000 km away. There also exist Oligocene volcanic occurrences similar to the siliceous rocks east of Chihuahua, in addition to small basaltic emissions in the Pliocene and Quaternary.

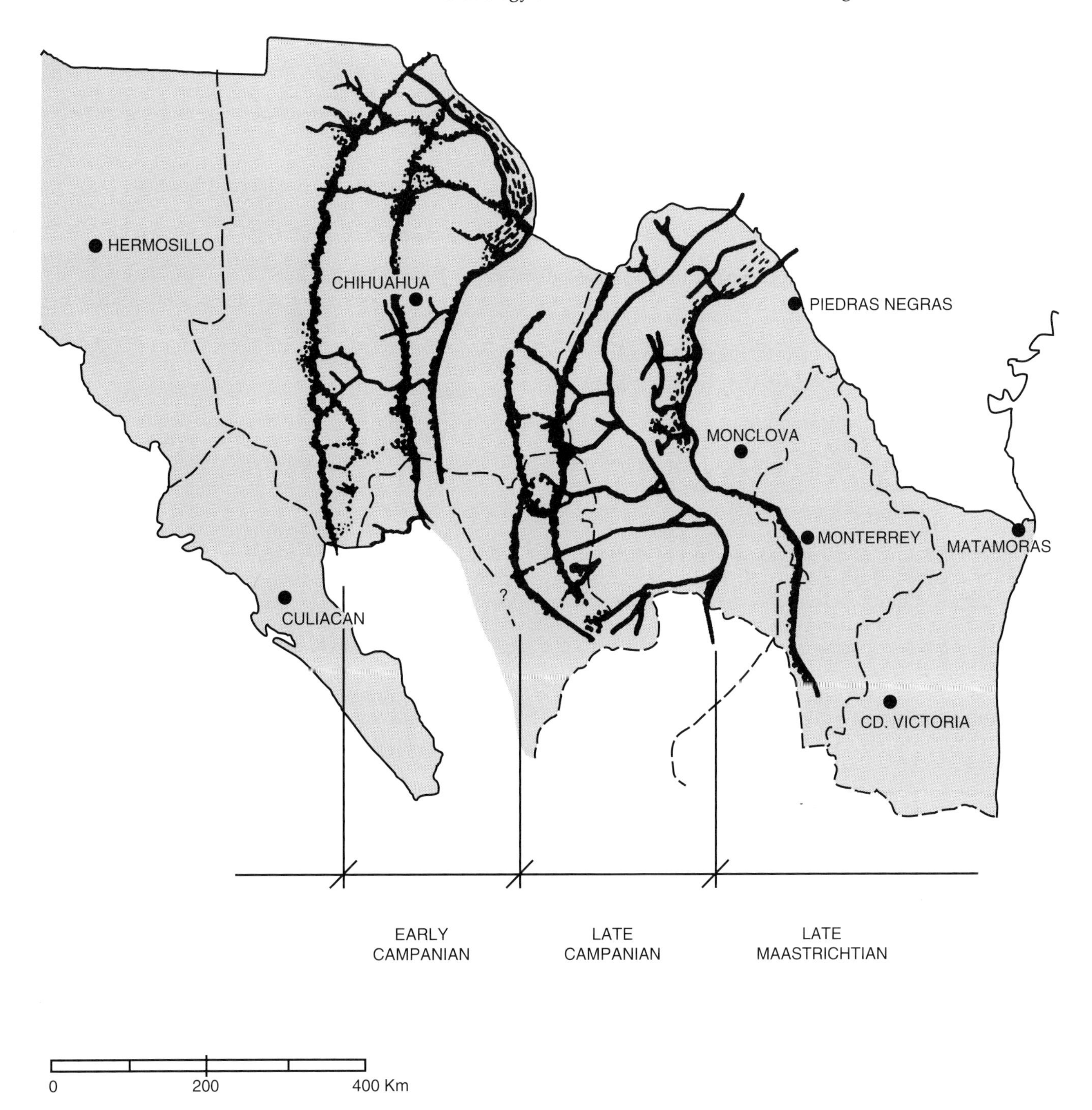

Figure 2.9. Paleogeography of the Upper Cretaceous for northern and northeastern Mexico, after Raul González, 1976.

Economic Resources

The area of the Sabinas paleogulf and the platforms of Coahuila and Tamaulipas have been the object of very extensive petroleum exploration because they contain favorable characteristics for the development of this resource, and, furthermore, abundant commercial production has been obtained in adjacent regions. The stratigraphic levels with major possibilities are in the Upper Jurassic and Lower Cretaceous, since in these rocks both source beds and strata with reservoir characteristics appear. Petróleos Mexicanos has drilled numerous exploration wells and has encountered important shows of hydrocarbons in the Sabinas paleogulf area.

The coal-bearing zone of Sabinas is formed by sedimentary deposits located within the deltaic sequence of the Upper Cretaceous. Specifically, these beds belong to the Olmos Formation of Maastrichtian age that was deposited in a dominantly swampy environment. This zone is the principal producer of coal in the nation and its major reserve (see Figure 2.13).

There also exist numerous deposits of fluorite and barite distributed in belts generally oriented northwest-southeast. These belts clearly parallel mineralization belts developed in northern Mexico as a consequence of the magmatism associated with subduction in the west. These fissures are hydrothermal veins that occur generally within the host rock of Lower Cretaceous limestone. This area is the principal producer of fluorite in Mexico. The State of Coahuila also contains resources of phosphorite of sedimentary origin that forms horizons in strata of the La Caja Formation of the Upper Jurassic; this constitutes one of the most important sources of this material in the nation.

SIERRA MADRE ORIENTAL, GULF COASTAL PLAIN, AND MESA CENTRAL

General Geology

The Eastern Sierra Madre and the adjacent areas are composed chiefly of Mesozoic sedimentary rocks that were deposited and developed over a Paleozoic and Precambrian basement. The Sierra consists of an orogenic mountain belt that follows, in its southern segment, a general northwest-southeast trajectory, and at the latitude of Monterrey, turns to follow an east-west trajectory toward Torreon. The Sierra Madre is composed of narrow folds with an orientation that follows the general strike of the mountain belt. In the direction of the Mesa Central the valleys are wider, the Sierra anticlines less narrow, and toward the west they become gradually covered by the volcanic rocks of the Sierra Madre Occidental.

The Precambrian and Paleozoic basement can be observed in isolated outcrops that occur in erosional windows in the folded Mesozoic sequence (Table 2.1).

In the area of Ciudad Victoria, numerous authors have described an important sequence of upper Paleozoic rocks that rests on metamorphic rocks of early Paleozoic and Precambrian age (Carrillo-Bravo, 1961; Fries and Rincón, 1965). The lower metamorphic sequence is composed of the Novillo gneiss of Precambrian age, La Presa quartzite of the Cambrian, and by the Granjeno schist, later than both of the above units. According to radiometric dating of Denison et al. (1971), the Granjeno schist originated by metamorphic processes that occurred in the Pennsylvanian–Permian interval. These authors suggest that the schist was positioned tectonically in juxtaposition with the sedimentary sequence of the upper Paleozoic that is contemporaneous with it. De Cserna et al. (1977) consider this unit to be allochthonous, positioned in tectonic contact on the Novillo gneiss in an episode later than the formation of the La Presa quartzite and before the deposition of the upper Paleozoic sequence. Furthermore, these authors have dated the metamorphism of the Granjeno schist and have assigned it to the Ordovician (446 million years). Ramírez-Ramírez (1978) has suggested that the tectonic emplacement of the Granjeno schist occurred in the late Paleozoic, at the time of intense folding of the upper Paleozoic sedimentary sequence. According to the model of this author, the Granjeno schist comes from an eastern belt belonging to an internal zone of the Ouachita trend that was metamorphosed during the Carboniferous. The sedimentary sequence deposited in the Silurian–Permian interval has been considered by most authors as a tectonic autochthon developed above continental basement represented by the Novillo Gneiss and belonging to the Ouachita belt of southern North America.

Other Paleozoic outcrops of the Sierra Madre Oriental are located in the area of Huayacocotla at the latitude of the 21st parallel. At this locality Paleozoic rocks are seen exposed in the nucleus of a major anticlinorium whose flanks are composed of a thick sequence of Mesozoic sediments. Here the Paleozoic constitutes a metamorphic sequence of gneiss, schist, and metaconglomerate probably belonging to the early part of the era and a flysch sequence more than 2000 m thick that is Permian in age. In addition, Mississippian shales, sandstones, and conglomerates have been reported in the area of Calnali, Hidalgo (Carrillo-Bravo, 1965).

The Triassic is represented in the Sierra Madre Oriental and in neighboring areas by redbeds belonging to the Huizachal Formation. These continental sediments attest to a long period of emersion in this part of the country. This originated later than the orogenic deformation at the end of the Paleozoic (Table 2.1).

The Mesa Central contains numerous outcrops of metamorphic sequences that could belong to the Triassic or the end of the Paleozoic. In the area of Zacatecas City, above a metamorphic sequence, there rest marine partially metamorphosed sedimentary rocks that contain fossils of Carnian (Late Triassic) age (Burckhardt, 1930). These, together with those at Peñon Blanco and Charcas, S.L.P., constitute the only known outcrops of marine Triassic in this part of Mexico. Other outcrops of schistose rocks of probable late Paleozoic or Early Triassic age are located in the areas of Caopas, Zacatecas, and Guanajuato.

During the beginning of the Jurassic, continental deposition continued in this part of Mexico with redbed sedimentation, except in the region of the Huayacocotla anticlinorium, where an advance of the seas that induced marine sedimentation of an argillaceous and sandy sequence is recorded. Carrillo-Bravo (1971) termed this area the "Liassic Basin of Huayacocotla." These sedimentary strata, named by the same author the Huayacocotla Formation, were deformed at the end of the Early Jurassic leading to the predominance of continental deposition in the whole region during the Middle Jurassic.

In the Late Jurassic, there is recorded generally in all of northern and northeastern Mexico a marine transgression that Tardy (1980) related to the western opening of the Tethyan Sea during the disintegration of the supercontinent Pangea (Table 2.1). Pilger (1978) indicated that the opening of the Gulf of Mexico was earlier than the opening of the Atlantic, from which it might be supposed that the marine transgression of the first half of the Mesozoic ought to have come from the Pacific rather than from the east. The affinity of the faunas of eastern Mexico with those of the Pacific

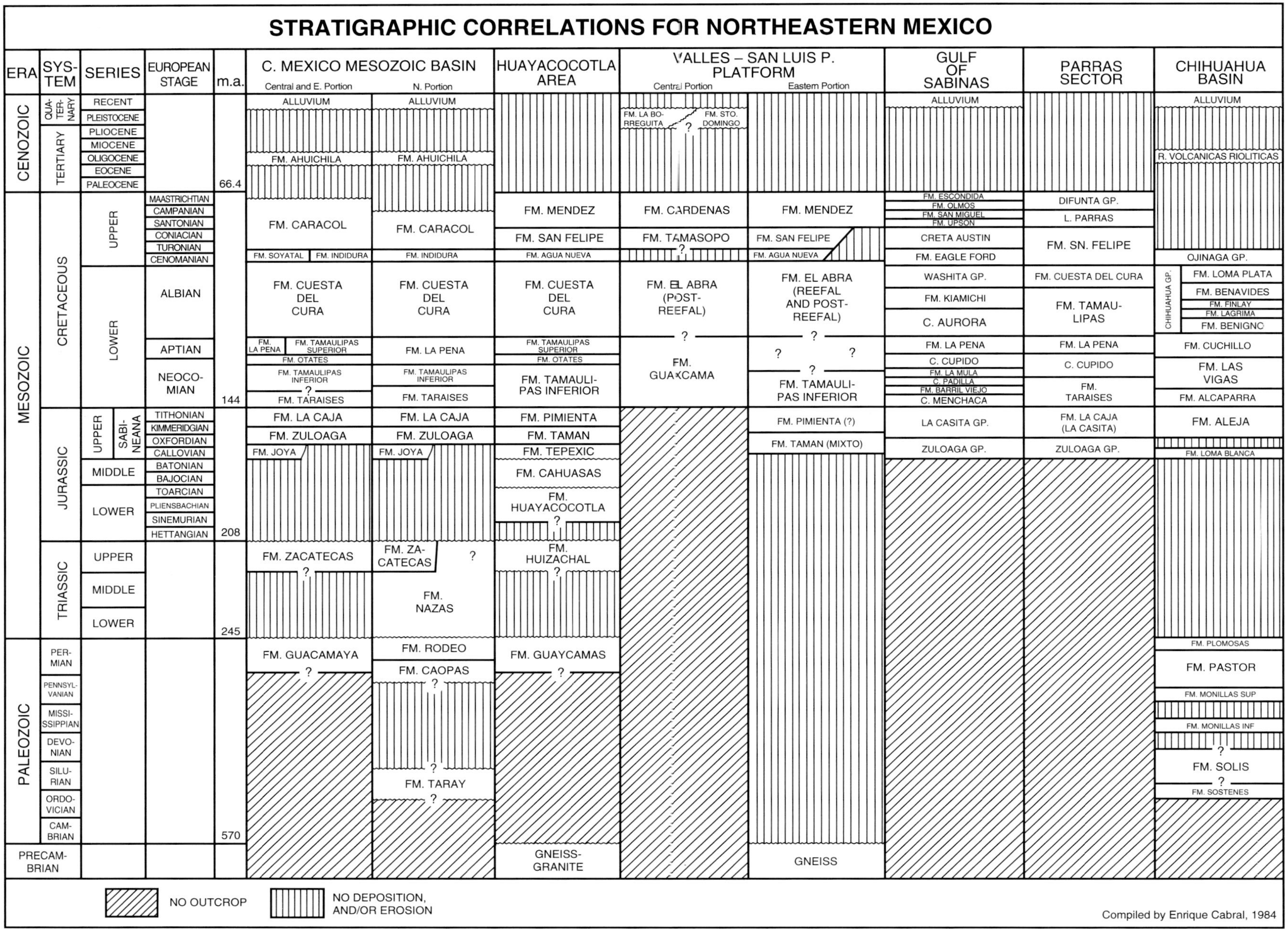

STRATIGRAPHIC CORRELATIONS FOR NORTHEASTERN MEXICO

ERA	SYSTEM	SERIES	EUROPEAN STAGE	m.a.	C. MEXICO MESOZOIC BASIN – Central and E. Portion	C. MEXICO MESOZOIC BASIN – N. Portion	HUAYACOCOTLA AREA	VALLES – SAN LUIS P. PLATFORM – Central Portion	VALLES – SAN LUIS P. PLATFORM – Eastern Portion	GULF OF SABINAS	PARRAS SECTOR	CHIHUAHUA BASIN
CENOZOIC	QUATERNARY	RECENT			ALLUVIUM	ALLUVIUM		FM. LA BOREGUITA ? FM. STO. DOMINGO		ALLUVIUM		ALLUVIUM
		PLEISTOCENE										
	TERTIARY	PLIOCENE										
		MIOCENE										
		OLIGOCENE			FM. AHUICHILA	FM. AHUICHILA						R. VOLCANICAS RIOLITICAS
		EOCENE										
		PALEOCENE		66.4								
MESOZOIC	CRETACEOUS	UPPER	MAASTRICHTIAN		FM. CARACOL	FM. CARACOL	FM. MENDEZ	FM. CARDENAS	FM. MENDEZ	FM. ESCONDIDA	DIFUNTA GP.	
			CAMPANIAN							FM. OLMOS		
			SANTONIAN							FM. SAN MIGUEL	L. PARRAS	
			CONIACIAN				FM. SAN FELIPE	FM. TAMASOPO	FM. SAN FELIPE	FM. UPSON	FM. SN. FELIPE	
			TURONIAN							CRETA AUSTIN		
			CENOMANIAN		FM. SOYATAL / FM. INDIDURA	FM. INDIDURA	FM. AGUA NUEVA	?	FM. AGUA NUEVA	FM. EAGLE FORD		OJINAGA GP.
		LOWER	ALBIAN		FM. CUESTA DEL CURA	FM. CUESTA DEL CURA	FM. CUESTA DEL CURA	FM. EL ABRA (POST-REEFAL)	FM. EL ABRA (REEFAL AND POST-REEFAL)	WASHITA GP. / FM. KIAMICHI / C. AURORA	FM. CUESTA DEL CURA / FM. TAMAULIPAS	CHIHUAHUA GP.: FM. LOMA PLATA / FM. BENAVIDES / FM. FINLAY / FM. LAGRIMA / FM. BENIGNO
			APTIAN		FM. LA PENA / FM. TAMAULIPAS SUPERIOR / FM. OTATES	FM. LA PENA	FM. TAMAULIPAS SUPERIOR / FM. OTATES	?	? ?	FM. LA PENA	FM. LA PENA	FM. CUCHILLO
			NEOCOMIAN	144	FM. TAMAULIPAS INFERIOR ? FM. TARAISES	FM. TAMAULIPAS INFERIOR / FM. TARAISES	FM. TAMAULIPAS INFERIOR	FM. GUAXCAMA	? FM. TAMAULIPAS INFERIOR	C. CUPIDO / FM. LA MULA / C. PADILLA / FM. BARRIL VIEJO / C. MENCHACA	C. CUPIDO / FM. TARAISES	FM. LAS VIGAS / FM. ALCAPARRA
	JURASSIC	UPPER (SABINEANA)	TITHONIAN		FM. LA CAJA	FM. LA CAJA	FM. PIMIENTA		FM. PIMIENTA (?)	LA CASITA GP.	FM. LA CAJA (LA CASITA)	FM. ALEJA
			KIMMERIDGIAN		FM. ZULOAGA	FM. ZULOAGA	FM. TAMAN		FM. TAMAN (MIXTO)			
			OXFORDIAN							ZULOAGA GP.	ZULOAGA GP.	FM. LOMA BLANCA
			CALLOVIAN		FM. JOYA	FM. JOYA	FM. TEPEXIC					
		MIDDLE	BATONIAN				FM. CAHUASAS					
			BAJOCIAN									
		LOWER	TOARCIAN				FM. HUAYACOCOTLA					
			PLIENSBACHIAN									
			SINEMURIAN				?					
			HETTANGIAN	208								
	TRIASSIC	UPPER			FM. ZACATECAS ?	FM. ZACATECAS ?	FM. HUIZACHAL ?					
		MIDDLE				FM. NAZAS						
		LOWER		245								
PALEOZOIC	PERMIAN				FM. GUACAMAYA ?	FM. RODEO / FM. CAOPAS ?	FM. GUAYCAMAS ?					FM. PLOMOSAS / FM. PASTOR
	PENNSYLVANIAN											FM. MONILLAS SUP
	MISSISSIPPIAN											FM. MONILLAS INF
	DEVONIAN					?						? FM. SOLIS ?
	SILURIAN					FM. TARAY ?						
	ORDOVICIAN											FM. SOSTENES
	CAMBRIAN			570								
PRECAMBRIAN							GNEISS-GRANITE		GNEISS			

NO OUTCROP — NO DEPOSITION, AND/OR EROSION

Compiled by Enrique Cabral, 1984

Table 2.1. Stratigraphic correlations for northeast Mexico.

(Longoria, personal communication) is a fact that supports this supposition. In a time earlier than the Jurassic transgression, during Triassic continental deposition, a large part of what is now Mexico belonged to the western sector of the above-mentioned continent Pangea.

With the invasion of the Late Jurassic seas over most of northern and northeastern Mexico, the major paleogeographic elements began to be defined—elements active during the whole Mesozoic and that controlled sedimentation at this time and later tectonic deformation. Among the principal elements that were active during the Mesozoic in the present area of the Sierra Madre Oriental and adjacent regions are the Mesozoic basin of Mexico or Mexican geosyncline, the Valles–San Luis Potosí platform, the Peninsula or Island of Coahuila, the Peninsula or Archipelago of Tamaulipas, and the ancient Gulf of Mexico.

The Mesozoic basin, developed in the zone of the Mesa Central and the Sierra Madre Oriental, has been considered by numerous authors as a geosyncline, in the sense of a linear belt of subsidence where a considerable thickness of sediment accumulated and that later was destroyed by orogeny. Burkhardt (1930) considered that for the Late Jurassic in this region a major entrance of the sea that bordered the positive lands took place, except in the southeast toward the State of Veracruz.

Imlay (1938) mentions the existence in this region of a Mexican geosyncline, separated from that developed in western Mexico, that was termed the Pacific geosyncline. This author indicates, furthermore, that in the Jurassic and Cretaceous periods communication developed with portals between both geosynclines; these are evidenced by the migration of characteristic Mediterranean biota toward the Pacific province of the north and vice versa.

More recently, Tardy (1980) considered that the east-central portion of Mexico evolved as a geosyncline in which there were individualized two basins (Ancestral Gulf of Mexico and the Mesozoic basin of Mexico). These possessed a NNW–SSE orientation, were sites of calcareous pelagic sedimentation, and were separated by a crestal area (the Valles–San Luis Potosí platform) over which neritic sedimentation developed. Schmidt-Effing (1980) believed in the presence of an aulacogen in the area of Huayacocotla during the Early Jurassic; the possibility was suggested that this part of the country might have evolved as an aulacogen system during the first half of the Mesozoic, i.e., as a series of marine-invaded tectonic depressions associated with the initial expansion of the Atlantic. The constant activity of these troughs caused drastic changes in the bathymetry that provoked, in certain areas, deposition of pelagic strata established on continental crust without reaching an ultimate process of oceanization. In a large part of the Mesozoic basin of Mexico (Mesa Central and the western belt of the Sierra Madre Oriental), the Oxfordian marine transgression is marked by an initial deposit of the gypsum Minas Viejas, indicating a shallow water basin under climatic conditions of strong evaporation. These gypsum deposits play a very important role in the orogenic deformation at the end of the Mesozoic, according to the models of various authors. In the Ancestral Gulf of Mexico, the Oxfordian transgression also is initiated by evaporite deposits. Only at this time did shallow water deposits occur in both basins, since during the rest of the Mesozoic sedimentary conditions were pelagic in the Gulf basin in contrast to the inland basin where neritic deposits developed on the Valles–San Luis Potosí, Tamaulipas, and Coahuila platforms.

The Upper Jurassic deposits in the Mesozoic basin of Mexico consist in stratigraphic order of the Minas Viejas Gypsum, the Zuloaga Limestones, and the siltstones, limestones, and shales of the La Caja Formation. The clastic strata equivalent to near coastal facies of the last two formations are respectively the La Gloria and La Casita formations. The marine transgression initiated in the Oxfordian did not completely cover the Valles–San Luis Potosí platform, and some areas remained emergent during all of the Late Jurassic (Carrillo-Bravo, 1971). In addition, the platform or archipelago of Tamaulipas was partly emergent and the Coahuila platform totally so.

During the first part of the Early Cretaceous (Neocomian–Aptian), open sea deposition occurred in the Mesozoic basin of Mexico (Taraises and lower Tamaulipas formations) while over the Valles–San Luis Potosí platform a sequence that was chiefly evaporitic (Gauxcamá Formation) occurred. With the Albian there came a general marine transgression that covered the last positive elements and that expanded over the western portions of Mexico (Rangin and Córdoba, 1976). On the perimeter of the Valles–San Luis Potosi platform there developed at this time a reefal belt flanked by back-reef and forereef deposits (see Figure 2.10). This whole facies assemblage was termed the El Abra Formation (Carrillo-Bravo, 1971). Similar reefal developments have been interpreted to occur around the Coahuila and Burro platforms contemporaneous with the Aurora and Cuesta del Cura formations, and also in the area of Tuxpan where a reef in the form of an atoll forms the reservoir for hydrocarbons in the so-called Golden Lane (Figure 2.11).

At the beginning of the Late Cretaceous, the regimen of sedimentation in eastern Mexico changed drastically with a influx of detrital sediments coming from the west where an uplift associated with volcanic and plutonic activity was taking place. During this epoch the seas retreated gradually toward the east, with associated prograding deltas. In the areas of the Mesozoic basin, deposits of the Turonian Indidura Formation (limestones and shales) were laid down, followed by the Caracol Formation in the Coniacian–Maastrichtian (shales and sandstones), the Parras Shale from the Santonian, and the Difunta in the Campanian to Maastrichtian (shales and sandstones). Over the Valles–San Luis Potosí platform during most of the Late Cretaceous, a calcareous complex of platform type developed. This is composed of pure limestones and argillaceous limestones of the Tamasopa Formation belonging to the interval from

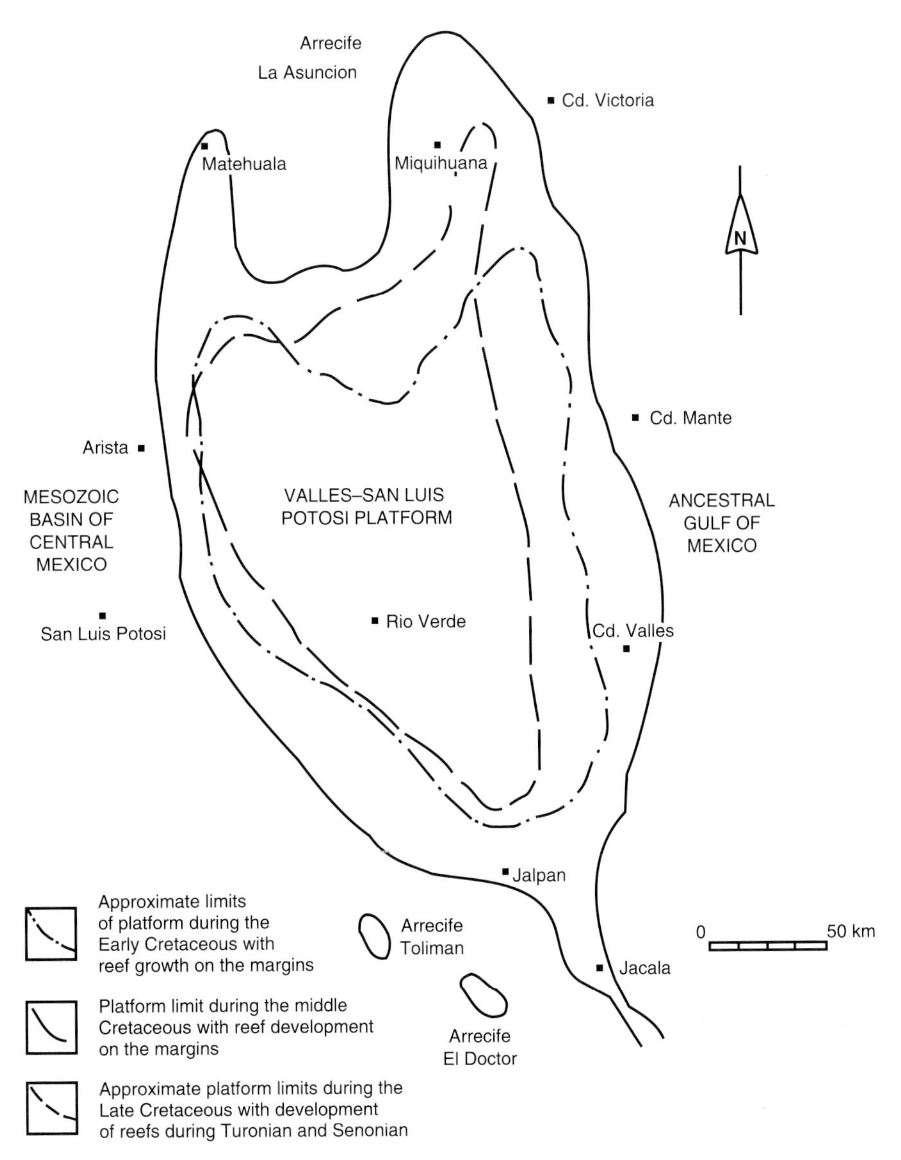

Figure 2.10. Limits of the Valles–San Luis Potosí platform during the Cretaceous.

upper Turonian to upper Senonian. This underlies the shales, sandstones, and limestones of the Cárdenas Formation of Campanian–Maastrichtian age. These formations are the platform equivalents of the Agua Nueva, San Felipe, and Mendez formations of the Ancestral Gulf of Mexico.

The first manifestations of the orogenic deformation at the beginning of the Cenozoic are flysch deposits associated with prograding deltas of the Late Cretaceous and with the foredeeps formed in the areas of Parras (Campanian–Maastrichtian) and the Chicontepec (Paleocene) where thick terrigenous sequences were deposited in deep water. These deformations began the construction of the Sierra Madre Oriental and initiated the continental history of a large part of this sector of the country.

In the period of maximum orogenic deformation in the Mesa Central area, deposits of molasse-type conglomerates belonging to the Ahuichila Formation and the Red Conglomerates of Guanajuato began to form. Generally these are polymictic conglomerates derived from erosion of the folded Mesozoic formations. Great thicknesses of continental alluvium accumulated in the synclinal depressions and tectonic troughs that imprinted a characteristic geomorphology on the landscape of the Mesa Central. The extreme western part of this zone of Mesozoic folds appears covered by mesetas of ignimbrites dissected and split into tributaries from the Sierra Madre Occidental. These sediments originated chiefly in the Oligocene.

In the Ancestral Gulf of Mexico, two principal Tertiary sedimentary basins are separated from each other by the Laramide folds of the Sierras de Tamaulipas and San Carlos. The Burgos basin, located to the north, contains marine sequences that are chiefly detrital and more than 1500 m thick, with the development of numerous growth faults contemporaneous with sedimentation, recognizable in the wells

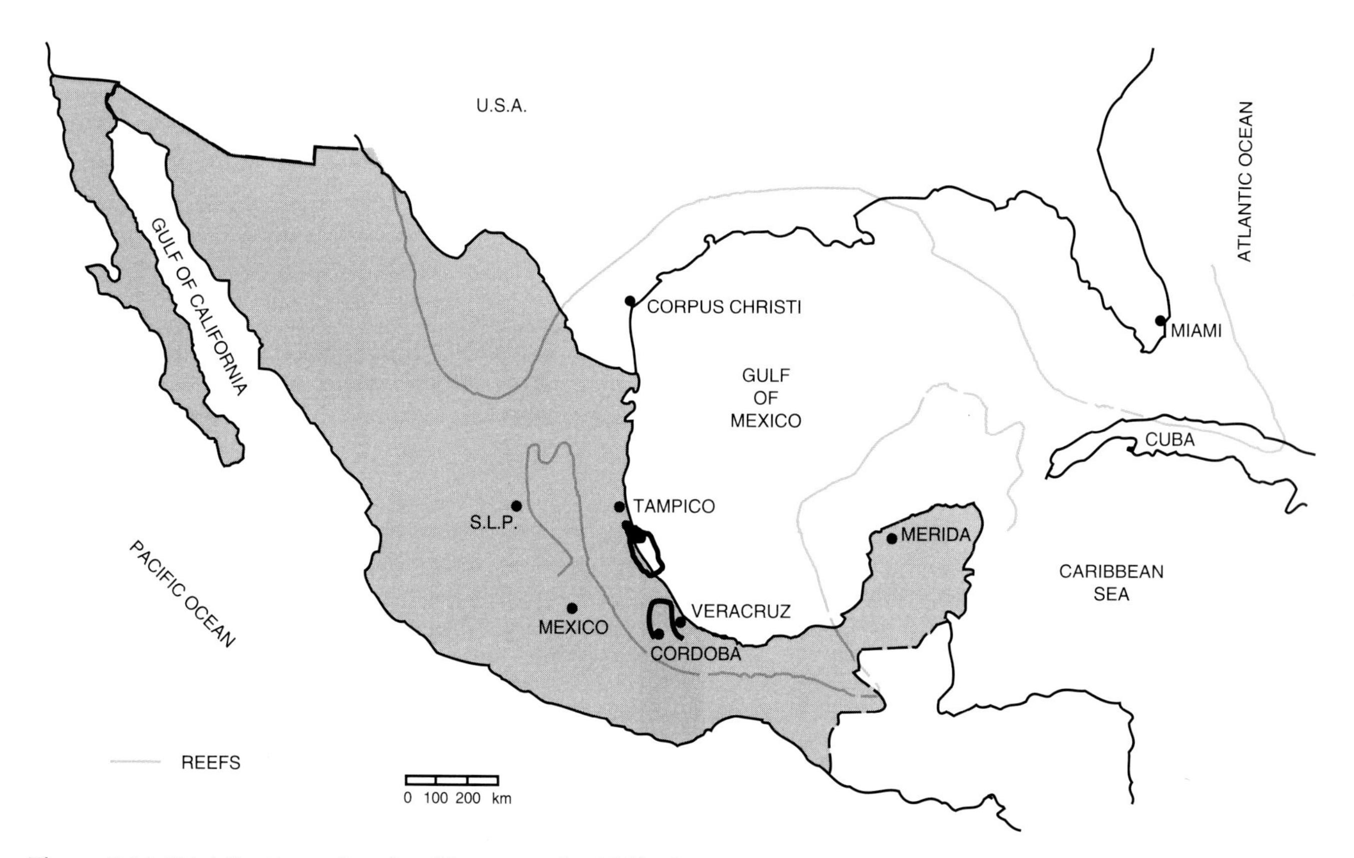

Figure 2.11. Distribution of reefs of Lower and middle Cretaceous around the margins of the Gulf of Mexico, after Carrillo-Bravo, 1971.

drilled by Petróleos Mexicanos (López-Ramos, 1979). In the Tampico-Misantla basin, thick marine sandy argillaceous sediments developed. This basin is seen to be limited to the following geographic borders, principally by orogenic structures from the early Cenozoic: to the north by the Sierra Tamualipas, to the east by the Sierra Madre Oriental and the Chicontepec foredeep, and to the south by the Tezuitlan massif. In the two basins, the Tertiary deposits occur within the setting of general eastward regression that left successive belts of outcrops parallel to the present coastline.

The plutonic and volcanic activity in the Sierra Madre Oriental and Gulf of Mexico coastal plain was very minor during the Cenozoic and is represented only by isolated plutons emplaced in the Mesozoic strata, and some flows in the areas of the Sierra Madre Oriental and the Neovolcanic axis. These are mineralogically similar to the alkaline province of eastern Mexico. The most important plutonic emplacements are located in the Sierra de San Carlos, in Tamaulipas, where nepheline syenites, gabbros, and monzonites disposed in laccoliths, dikes, and mantos are found (see Figure 2.12). These rocks constitute a southward continuation of the alkaline province that begins toward the north in the Big Bend area in Texas (Clark et al., 1980). Radiometric studies of intrusive rocks in the Tamaulipas area, published by Bloomfield and Cepeda (1973), reveal dates that vary between 28 and 30 million years. The flows of alkaline basalt located to the north of Tampico represent a later event attributable to a distension (Cantagrel and Robin, 1979).

Tectonic Summary

The characteristics of the Precambrian and Paleozoic basement, above which the great Mesozoic sequence of eastern Mexico evolved, are not clear because in general there are very scarce outcrops. The belts that form this basement should have been strongly dislocated by lateral and vertical movements during the first half of the Mesozoic. These tectonic movements prepared the paleogeographic distribution of basins and platforms that would control the sedimentation and the Laramide deformations during the end of the Mesozoic. In whatever structural form and orientation they might have, the Paleozoic outcrops of the Sierra Madre Oriental have been considered to be a prolongation of the Ouachita belt of the southeastern United States, since many writers have pointed out similarities with the rocks of this belt (De Cserna, 1956; Flawn, 1961; Denison et al., 1971; Ramírez-Ramírez, 1978). This belt was formed as a consequence of the closing of the proto-Atlantic Ocean.

During the Triassic this part of the country evolved in continental aspect with the development of distensive tectonics that caused the formation of troughs filled with thick continental sediments. In the Jurassic,

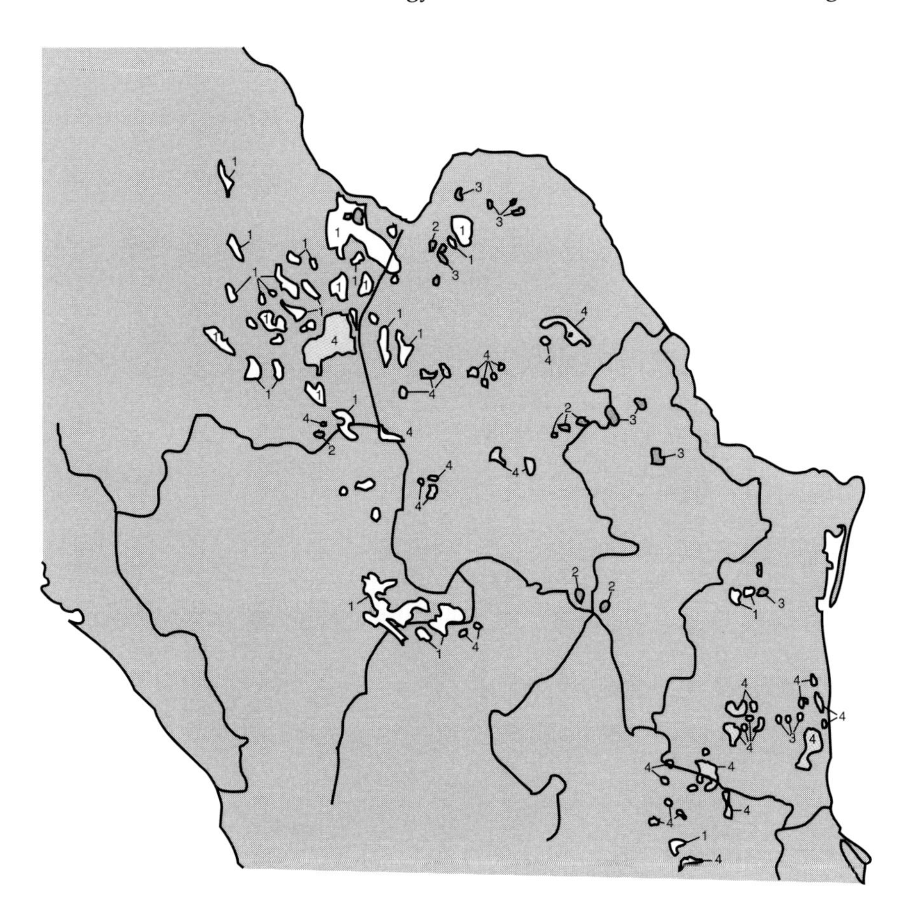

1 Oligocene ignimbrites

2 Lower Cenozoic granitic intrusives

3 Alkaline intrusive rocks associated with lower Cenozoic granites

4 Basaltic-alkaline rocks of the upper Tertiary and Quaternary

Figure 2.12. Distribution of outcropping igneous rocks in northeastern Mexico.

two important domains were established in Mexico as a result of the opening of the Atlantic and the Gulf of Mexico, as well as by the northwestward movement of North America. The first of these, located in western Mexico, was represented by a convergent margin and a zone of magmatic arc of Andean type, resulting from the subsidence of the Farallon plate under the North American continent. The second domain of geosynclinal type or of an aulacogenic system originated during the marine transgression over the eastern part of the country at the time of the opening of the Gulf of Mexico. This transgression caused a great thickness of calcareous deposits within a setting of intermittent subsidence and the presence of emergent cratonic elements and marine "highs."

At the beginning of the Late Cretaceous, a marked change occurred in the sedimentary regimen in this region, as a consequence of the uplift and deformation of the western area where there was continuously active subduction of the Farallon plate under the continental portion of Mexico. The detrital sediments that began to cover the calcareous sequence of the east were distributed widely and came to reach great thicknesses in the foredeeps of Parras (Upper Cretaceous) and Chicontepec (Paleocene). Their formation presages the orogenic activity that would affect all the region. In this way the western and eastern domains of Mexico, which had acted in a relatively independent manner, each with its own characteristics, are seen to be closely interrelated with the final Mesozoic deformations.

According to a model proposed by Coney (1983), the orogenic deformations at the end of the Cretaceous and beginning of the Tertiary coincide with a change in movement of the tectonic plates, since the North American and Farallon plates that converged obliquely in western Mexico began to face each other frontally and to move with greater velocity. De Cserna (1956) considers that the folds of the Mesozoic sequence increased in intensity from the Mesa Central toward the Sierra Madre Oriental because of the presence, in the time of deformation, of the massive cratons of the Coahuila platform and the Tamaulipas Peninsula. The forces coming from the southwest provoked the deformation of the sequence from the base of the Oxfordian evaporites on up. The evaporites served as a surface of sliding or decollement in the style of the Jura Mountains of Europe (De Cserna, 1979). Tardy et al. (1975) supposed the existence of a nappe with NNE direction, i.e., an overthrust of hundreds of kilometers that relocated the pelagic sequence of the internal basin (Mesa Central and High Chain of the Sierra Madre Oriental) over the Coahuila and Valles–San Luis Potosí platforms with reefal and subreefal sequences that form similar paleogeographical uplifts. The model of this author presumes the unbuckling of the internal basinal sequence at the level of the Oxfordian gypsum and establishes the possibility that the basement might have taken part in this tectonic phenomenon. According to the model of Padilla y Sánchez (1982), the distribution of the folds and overthrusts of northeast Mexico can be explained by movement of North America toward the northwest with respect to Mexico, more than by the action of compressive forces coaxial with a southwest-northeast orientation.

Economic Resources

The Gulf Coastal Plain and adjacent areas constitute a very important region for petroleum production, which has been obtained from both Mesozoic and Tertiary sequences.

The Faja de Oro (Golden Lane) has been traditionally a productive zone that years ago constituted the principal source of hydrocarbons in the nation. The productive unit is a reef that developed in the Early Cretaceous and that extends in a semicircular form out to the continental shelf at the latitude of Tuxpan. The Tamabra belt in the Poza Rica area that consists of an ancient forereef zone has also been an important source of hydrocarbons. In the Burgos basin zone to the north of Tamaulipas and east of Nuevo Leon, an important petroleum-producing Tertiary sequence is encountered. Furthermore, the Paleocene sequence of the Chicontepec area consists in actuality of an assemblage of very important reserves. Other zones with petroleum potentials, principally in Mesozoic rocks, are the Valles–San Luis Potosí platform, where important reefal development is present in the Lower Cretaceous and the Mesa Central has a significant marine sedimentary sequence of Jurassic and Cretaceous age.

In summary, the oil and gas productive districts of the region north of the Gulf Coastal Plain are Pánuco-Ebano, Faja de Oro, Poza Rica, and Veracruz (Díaz, 1980).

Mineral deposits include notable hydrothermal developments in the Tertiary of the Mesa Central area and on the western flank of the Sierra Madre Oriental. The most important recognized resources of lead, silver, and zinc are localized in the areas of Fresnillo, Zacatecas, Sierra de Catorce, Charcas, and Zimapán, in addition to the mineral district of Guanajuato, where the principal association is silver and gold. Likewise, of outstanding importance are the hydrothermal deposits of fluorite of the area of Las Cuevas and Río Verde. These constitute a southern continuation of the northwest–southeast-oriented belt that developed in Coahuila during the Tertiary, the time in which the magmatic arc reached its most eastern position. Finally, it is appropriate to indicate that the barite deposits that were developed on the eastern slopes of the Sierra Madre Oriental are also substantial and bear relationship to this episode of volcanic and subvolcanic activity (see Figure 2.13).

BIBLIOGRAPHY AND REFERENCES

Abbreviation UNAM is Universidad Nacional Autónoma de México

Alfonso-Zwanziger, J.A., 1978, Geología regional del sistema sedimentario Cupido: Boletín Asociación Mexicana Geólogos Petroleros, v. 30, n. 1 and 2.

Bloomfield, K., and D.L. Cepeda, 1973, Oligocene alkaline igneous activity in N.E. Mexico: Geological Magazine, v. 110, p. 561–559.

Burckhardt, C., 1930, Etude Synthétique sur de Mesozoique Mexicain: Memoire Société Paleontologique Suisse, v. 49–50, 280 p.

Cantagrel, J.M., and C. Robin, 1979, K-Ar dating on eastern Mexican volcanic rocks—relations between the andesitic and the alkaline provinces: Journal of Vulcanology Geotherm Research., v. 5, p. 99–114.

Carrillo-Bravo, J., 1961, Geología del Anticlinorio Huizachal-Peregrina al noroeste de Ciudad Victoria, Tamaulipas. Boletín Asociación Mexicana Geólogos Petroleros, v. 13, p. 1–98.

Carrillo-Bravo, J., 1965, Estudio de una parte del Anticlinorio de Huayacocotla: Boletín Asociación Mexicana Geólogos Petroleros, v. 17, p. 73–96.

Carrillo-Bravo, J., 1971, La plataforma Valles-San Luis Potosí: Boletin Asociación Mexicana Geólogos Petroleros, v. 23, nos. 1–6, p. 1–112.

Charleston, S., 1973, Stratigraphy, tectonics, and hydrocarbon potential of the Lower Cretaceous, Coahuila Series, Coahuila, Mexico. Doctoral Dissertation, University of Michigan, Ann Arbor, 268 p.

Clark, K., P. Damon, S. Shutter, and M. Shafiquillah, 1980, Magmatismo en el norte de México en relación con los yacimientos metalíferos: Revista, Geomimet, no. 106, p. 49–71.

Coney, P., 1983, Plate tectonics and the Laramide Orogeny: New Mexico Geological Society, Special Publication no. 6, p. 5–10.

Córdoba, D., 1970, Mesozoic stratigraphy of northeastern Chihuahua, Mexico: The Geologic Framework of the Chihuahua Tectonic Belt. Symposium West Texas Geological Society, in honor of Prof. R.K. DeFord, p. 91–96.

De Cserna, Z., 1956, Tectónica de la Sierra Madre Oriental de México, entre Torreón y Monterrey, XX Congreso Internacional, 87 p.

De Cserna, Z., 1979, Cuadro tectónico de la sedimentación y magmatismo en algunas regiones de México durante el Mesozoico: Programas y resúmenes del V Simposio sobre la Evolución Tectónica de México: Instituto de Geología, UNAM, p. 11–14.

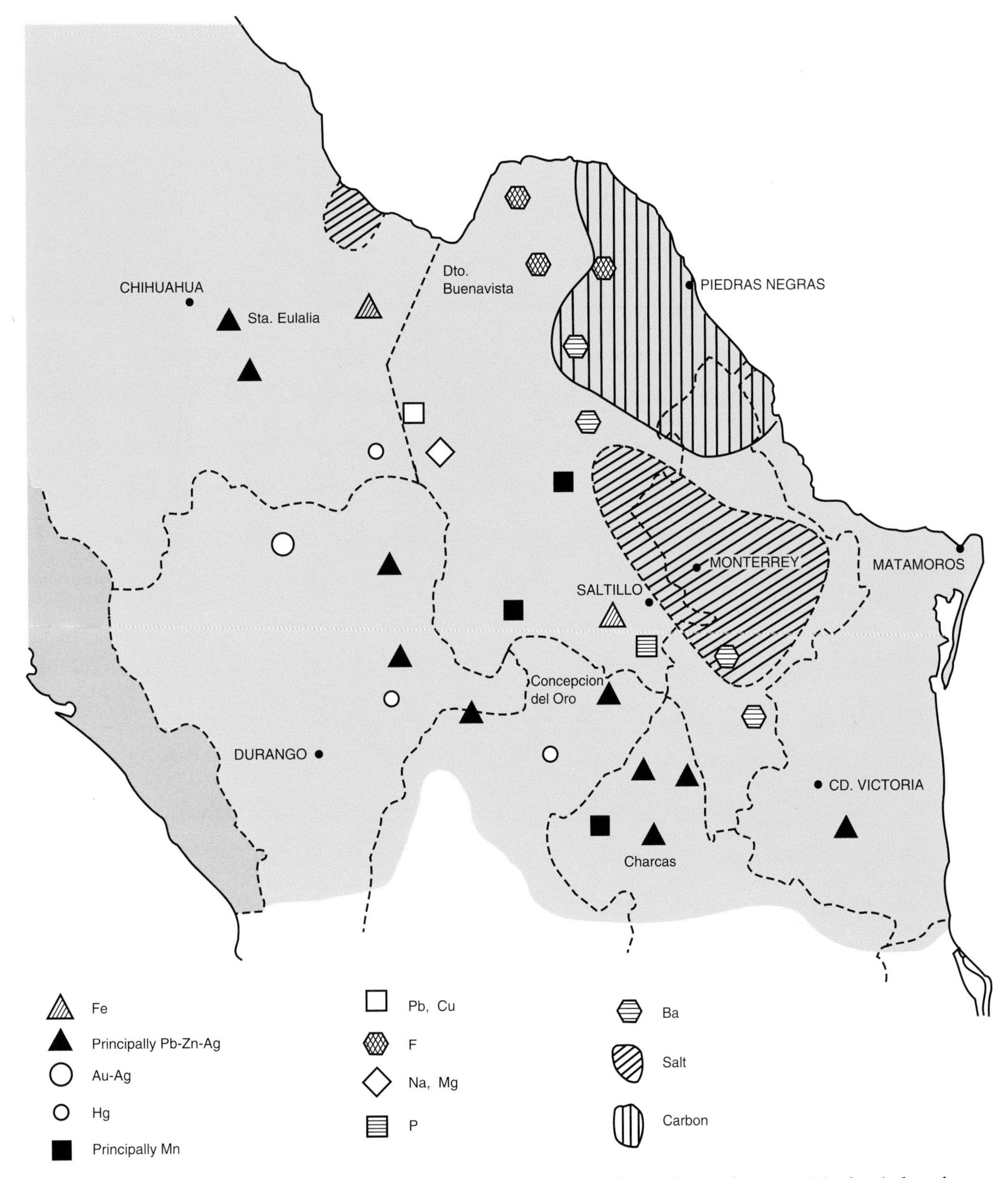

Figure 2.13. Distribution of the principal known mineral resources found in northeastern Mexico (taken from the Metallogenic Map of the Republic of Mexico, G.P. Salas, 1975).

De Cserna, Z., J.L. Graf, and F. Ortega-Gutiérrez, 1977, Alóctono del Paleozóico Inferior en la región de Ciudad Victoria, Estado de Tamaulipas: Revista del Instituto de Geología, UNAM, v. 1, p. 33–43.

DeFord, R.K., 1969, Some keys to the geology of northern Chihuahua, New Mexico Geological Society Guidebook, 20th Field Trip, p. 61–65.

Denison, R.E., et al., 1971, Basement rock framework of parts of Texas, southern New Mexico, and northern Mexico: Geologic Framework of the Chihuahua

Tectonic Belt, Midland: West Texas Geological Society, p. 3–14.
Diaz, J., 1980, ¿En qué consiste una reserva petrolera? El Petróleo en México y en el Mundo, CONACYT, 2nd edition, p. 221–223.
Flawn, P.T., 1961, Rocas metamórficas en el armazón tectónico de la parte septentrional de México. Boletín Asociación Mexicana Geólogos Petroleros, v. 13, p. 105–116.
Franco, M., 1978, Estratigrafía del Albiano–Cenomaniano en la región de Naica, Chihuahua: Revista, Instituto de Geología, UNAM, v. 2, no. 2, p. 132–149.
Fries, C., and O.C. Rincón, 1965, Nuevas aportaciones geocronológicas y tectónicas empleadas en el Laboratorio de Geocronometria: Boletín Instituto de Geología, UNAM, no. 73, p. 57–133.
González-Garcia, R., 1976, Bosquejo geologico de la Zona Noreste: Boletín Asociación Mexicana Geólogos Petroleros, v. 28, no. 1 and 2, p. 2–49.
Gries, C.J., and W.T. Haenggi, 1970, Structural evolution of the eastern Chihuahua tectonic belt: Geologic Framework of the Chihuahua Tectonic Belt: Symposium of West Texas Geological Society, in honor of R.K. DeFord, p. 119–137.
Imlay, R.W., 1938, Studies of the Mexican Geosyncline: Geological Society of America Bulletin, v. 50, p. 1–77.
López-Ramos, E., 1979, Geología de México, 2nd edition, México, D.F.: Scholastic edition, 3 volumes.
Mauger, R.L., F. McDowell, and J.C. Blount, 1983, Grenville Precambrian rocks of the Los Filtros near Aldama, Chihuahua, Mexico. Geology and Mineral Resources of north central Chihuahua: El Paso Geological Society, p. 165–168.
McDowell, F.W., and S.E. Clabaugh, 1979, Ignimbrites of the Sierra Madre Occidental and their relation to the tectonic history of western Mexico, *in* C.E. Chapin and W.E. Elston, eds., Ash Flow Tuffs: Geological Society of America Special Paper 180.
Navarro, A., and R. J. Tovar, 1975, Stratigraphy and tectonics of the State of Chihuahua: Exploration from the Mountains to the Basin: El Paso Geological Society, p. 23–27.
Padilla y Sánchez, R., 1982, Geologic evolution of the Sierra Madre Oriental between Linares, Concepción del Oro, Saltillo, and Monterrey, Mexico: Ph.D. Dissertation, University of Texas at Austin, 217 p.
Pilger, R.H., Jr., 1978, A closed Gulf of Mexico, pre-Atlantic Ocean plate reconstruction and the early rift history of the Gulf and the north Atlantic: Transactions Gulf Coast Association Geological Societies, v. 8, p. 385–593.
Quintero, O., and J. Guerrero, 1985, Una localidad de basamento Precámbrico de Chihuahua, en el área de Carrizalillo: Revista, Instituto de Geología, UNAM., v. 6, p. 98–99
Ramírez, J.C., and F.D. Acevedo, 1957, Notas sobre la geología de Chihuahua: Boletín Asociación Mexicana Geólogos Petroleros, v. 9, p. 583–766.
Ramírez-Ramírez, C., 1978, Reinterpretación tectónica del Esquisto Granjeno de Ciudad Victoria, Tamaulipas: Revista, Instituto de Geología, UNAM, v. 2, p. 31–36.
Rangin, C., and D.A. Córdoba, 1976, Extensión de la cuenca Cretácica Chihuahuense en Sonora septentrional y sus deformaciones: Memória del Tercer Congreso Latinoamericana de Geología, México, 14 p.
Rogers, L.C., et al., 1961, Reconocimiento geológico y depósitos de fosfatos del norte de Zacatecas y áreas adyacentes en Coahuila, Nuevo León, y San Luis Potosí: Boletín de Consejo de Recursos Naturales no Renovables, no. 56.
Salas, G.P., 1975, Mapa Metalogenético de la República Mexicana: Consejo de Recursos Minerales.
Schmidt-Effing, R., 1980, The Huayacocotla aulocogen in Mexico (Lower Jurassic) and the origin of the Gulf of Mexico, *in* R.H. Pilger, ed., Proceedings of a Symposium, The Origin of the Gulf of Mexico in the Early Opening of the central North Atlantic Ocean. Louisiana State University, Baton Rouge, Louisiana, p. 79–86.
Smith, C.I., 1970, Lower Cretaceous stratigraphy, northern Coahuila, Mexico: Bureau of Economic Geology, University of Texas, Report of Investigation no. 65. 101 p.
Tardy, M., 1980. La transversal de Guatemala y la Sierra Madre de México: J. Aubouin, Tratado de Geologia, v. III Tectónica y Tectonofísica, y Morfología—David Serret, translator, Barcelona, España Editorial Omega, p. 117–182.
Tardy, M., J. Sigal, and G. Gacon, 1974, Bosquejo sobre la estratigrafía y paleogeografía de los flysch Cretácicos del sector tranversal de Parras, Sierra Madre Oriental, México: Instituto de Geología, UNAM, Serie Divul, no. 2.
Tardy, M., et al., 1975, Observaciones generales sobre la estructuras de la Sierra Madre Oriental. La alóctonia del conjunto Cadena Alta-Altiplano Central, entre Torreon, Coahuila y San Luis Potosí, S.L.P., México: Revista, Instituto de Geologia, UNAM, v. 75, no. 1, p. 1–11.

3. Geology of the Central Region of Mexico

GENERAL CONSIDERATIONS

In describing the geology of the central area of Mexico, the following limits have been used: to the north, the northern edge of the Neovolcanic axis; to the west and south, the coastlines of the Pacific; and to the east, the shore of the Gulf of Mexico and the Isthmus of Tehuantepec.

The physiographic provinces of the Neovolcanic axis, the Sierra Madre del Sur, and the northern part of the Southern Gulf Coastal Plain are included in this region (see Figure 1.1). In terms of the division of geological provinces used by López-Ramos (1979), the provinces of the Veracruz basin (with the subprovince of Sierra de Juárez), the province of San Andres Tuxtla, the Tlaxiaco basin, the Sierra Madre del Sur, the Altiplano of Oaxaca, the Guerrero-Morelos basin, and the Neovolcanic axis are included.

The climate of the region is highly variable owing to the complex physiography. On the slopes of the Gulf of Mexico the climate changes from humid temperate in high parts of the Sierra Madre Oriental, to semi-hot and humid in the lowlands. On the Pacific slopes the climates vary from hot and subhumid on the southeast flank of the Sierra Madre del Sur and the banks of the Río Balsas to semi-arid, hot, and very hot in the Valley of Oaxaca and the major part of the Balsas basin. In the regions of basins within the Neovolcanic axis, the climate is in general subhumid and varies from temperate to semi-frigid and cold.

In the central Mexico region, sequences outcrop that attest to diverse domains of various stratigraphic levels, which in some regions are observed to be superimposed. This makes general descriptions rather fruitless. For this reason, this chapter treats each of the six domains of this region separately. This format facilitates description and synthesis, since within each one of the domains stratigraphic and tectonic conditions are more or less homogeneous with well-defined limits. These domains coincide in large part with the geological provinces proposed by López-Ramos (1979) for this region.

NEOVOLCANIC AXIS

The Transmexican Neovolcanic axis is composed of an upper Cenozoic belt that transversely crosses the Republic of Mexico at the 20th parallel (see Figure 3.1). It is formed by a large variety of volcanic rocks that were emitted along a significant number of volcanos, some of which constitute the highest peaks of the country. The volcanic activity in this belt has given rise to a large number of internal basins, with the consequent occurrence of lakes that give the geomorphic landscape a very characteristic appearance.

The principal volcanos located in this province are stratovolcanos of highly variable dimensions, such as El Pico de Orizaba, El Popocatépetl, El Iztaccíhuatl, El Nevado de Toluca, and El Nevado de Colima (see Figure 3.2). All of them were built by alternating pyroclastic emissions and lava flows. In addition, there exist vents of cinder cone type that are generally small, such as Paricutín, and rhyolitic volcanos such as are encountered southwest of Guadalajara. In addition to these types of centralized emissions, there is evidence of numerous fissure emissions and adventitious developments on the sides of the great stratovolcanos. There are also some calderas caused by both collapse and explosion; examples of the largest are La Primavera in the State of Jalisco and Los Húmeros in the State of Puebla.

According to Mooser (1972), the Neovolcanic axis has a zigzag pattern caused by the presence of a fun-

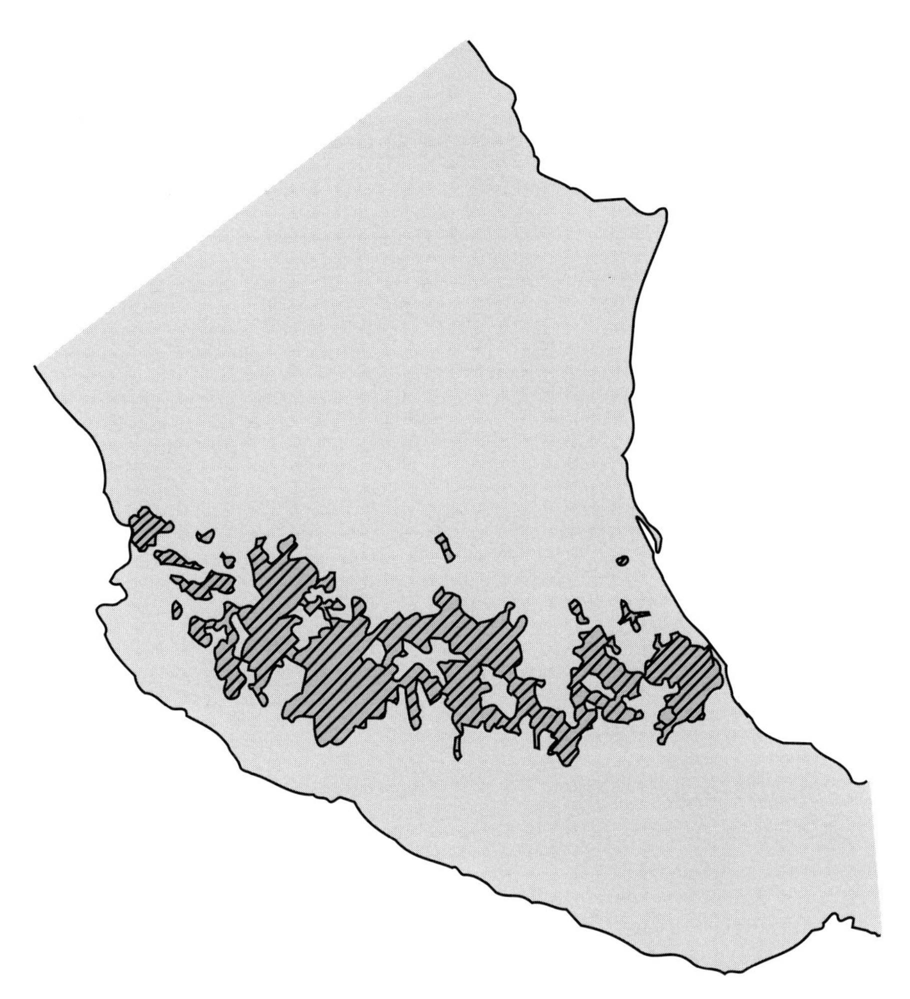

Figure 3.1. Distribution of rocks forming the Transmexican Neovolcanic axis.

damental system of orthogonal fragmentation with a northwest and northeast direction of the fractures. The latter orientation seems to be related to transcurrent movements principally in the eastern and central portions of the belt. This imparts the zigzag aspect to the axis. The great stratovolcanos, as at Tancítaro, Nevado de Toluca, Popocatépetl, and Nevado de Colima, would be situated on the southern apices of this system, while the great mining centers of the region, such as Guanajuato and Pachuca, would remain situated on the northern apices.

Demant (1978) considers that the Neovolcanic axis, more than just forming a continuous belt of volcanic rocks, constitutes a group of five principal focal points of activity with distinct orientation and characteristics. Within these five principal centers it is possible to recognize two types of volcanic structures: (1) those represented by great stratovolcanos in north-south alignment, and (2) those represented by numerous small volcanos aligned in a northeast-southwest trend, developed over tensional fractures.

The first volcanic manifestations in the area of the Valley of Mexico, in the upper Oligocene, are observed to be principally associated with fractures of west-northwest and east-southeast orientation and with influence of fractures with northeast-southwest orientation. In contrast, the last volcanic episodes of the Pleistocene and Quaternary in this portion of the axis seem to be related to a system of fractures with east-west orientation as in the Sierra de Chichinautzin (Mooser et al., 1974). In the central portion of the axis, seven phases of vulcanism have been recognized (Table 3.1) that have occurred since the Oligocene. The most important of these is the fifth, which occurred at the end of the Miocene and that gave rise to the mountain ranges of Las Cruces, Río Frío, and Nevada. During the sixth phase the cones and domes of Iztaccíhuatl and the active cone of Popocatépetl were developed. The last phase, equal to the former, was developed in the Quaternary and is responsible for the volcanic activity that cut off the drainage of the basin of Mexico toward the basin of the Río Balsas and caused the enclosed interior drainage of the former feature (Mooser et al., 1974).

In its western part, the Neovolcanic axis is bordered by the tectonic troughs of Tepic-Chapala and Colima. The first has a northwest-southeast orientation and is associated with the volcanos of San Juan, Sanganguey, Ceboruco, and Tequila. The second possesses a north-south orientation and is associated with

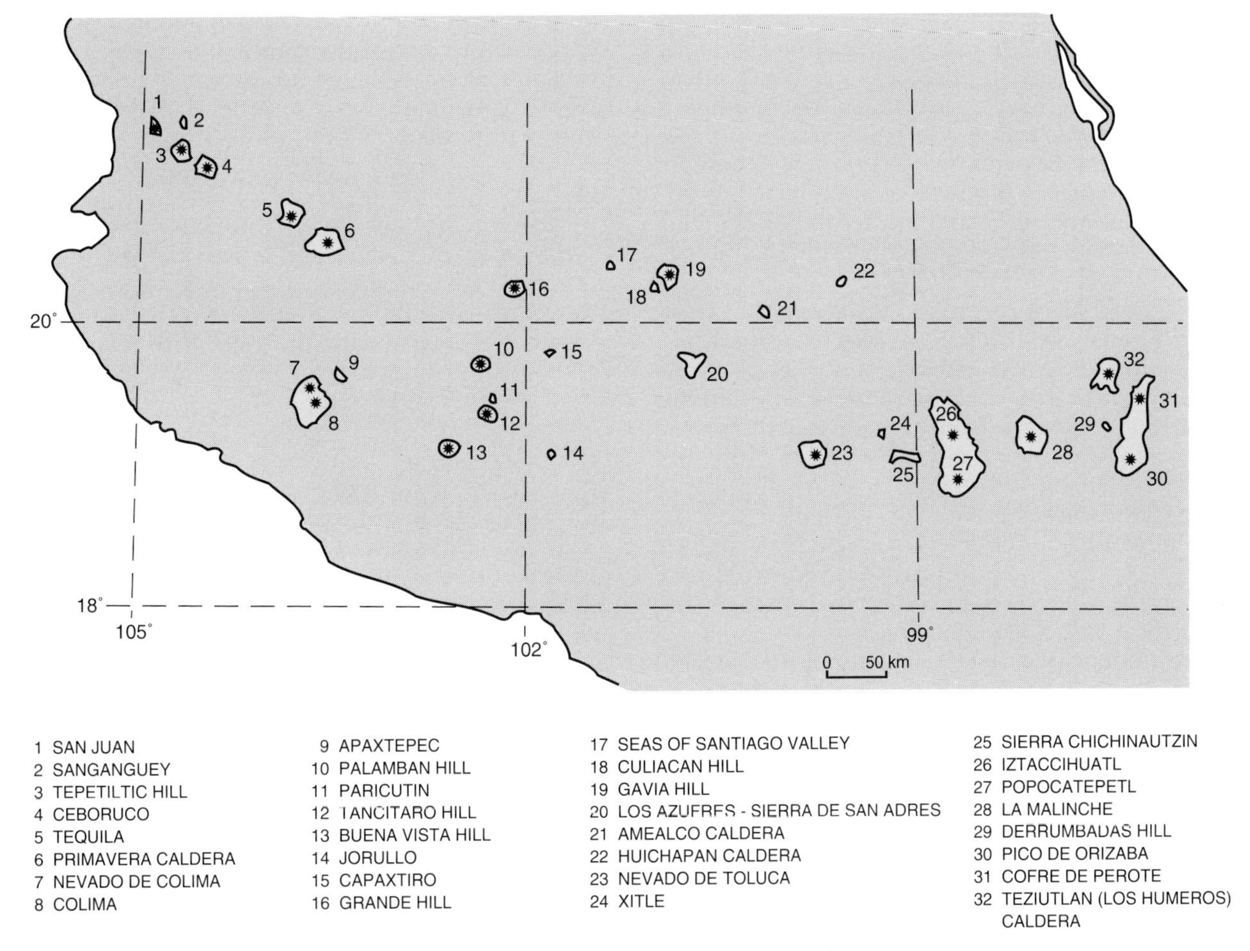

Figure 3.2. Distribution of the principal vents in the Mexican Neovolcanic axis.

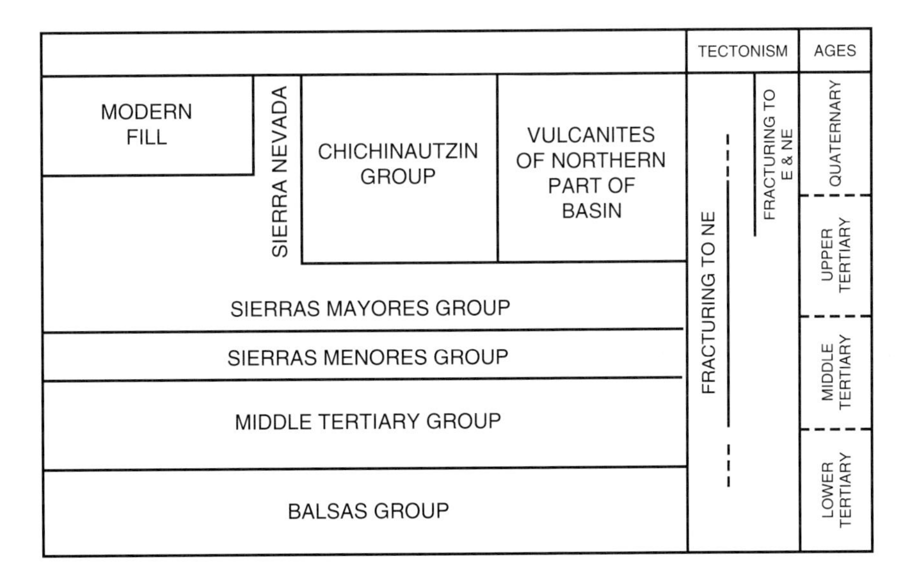

Table 3.1. Sequences of volcanic groups and tectonic events of the basin of Mexico.

the Nevado de Colima and the Volcán de Fuego (Colima Volcano). The latter vent constitutes, in the judgment of Demant (1978), the most dangerous volcano of the Neovolcanic axis, since it is a vent of the Mt. Pelee type with a plug of dacite lava that might cause the development of Nues Ardentes.

Toward the east the axis is bordered by the volcanic rocks of the San Andrés–Tuxtla region, although

Demant and Robin (1975) consider the rocks of this region to belong to the Eastern Alkaline Province since they fix the eastern limit of the province at the latitude of Pico de Orizaba and at Cofre de Perote.

The petrographic composition of the rocks forming the Transmexican Neovolcanic axis is highly variable. Flows and pyroclastic products of andesitic composition are abundant, although numerous dacite and rhyolite units also exist. Some units traditionally recognized as basalts, such as the Chichinautzin Group, have been considered recently to be andesitic in view of chemical analyses of rock samples (Mooser et al., 1974). In addition, local isolated occurrences of recent rhyolitic volcanics exist, such as those localized in the domes of the Primavera Caldera in Jalisco, in the area of Azufres in Michoacán, as well as in Los Húmeros and Puebla (Demant, 1978). From a chemical point of view, the Transmexican Neovolcanic axis is considered by numerous authors as a calcalkaline province characterized by its abundance of andesites and dacites and by the ratio maintained by content of silica and sodium and potassium oxides.

Most authors agree that the activity of the Neovolcanic axis began in the Oligocene and that it has continued up to the Recent (Mooser et al., 1974; Negendank, 1972; Bloomfield, 1975). Within this activity two principal cycles have been recognized: (1) Oligocene–Miocene and (2) Pliocene–Quaternary. Demant (1978) considers that the vulcanism of the axis is solely Pliocene–Quaternary, since the lower cycle of the Oligocene–Miocene constitutes the southern prolongation of the volcanic system of the Sierra Madre Occidental. This author indicates that the andesites of the Oligocene are folded, as in the Sierra de Mil Cumbres in the Lake Chapala region and in the Tzitzio-Huetamo anticlinorium. In contrast, he notes that in the eastern segment of the axis, outcrops of these andesites are very rare. This author does not clearly establish the relationship of these intermediate rocks with the Oligocene ignimbrites of the Sierra Madre Occidental, where the real andesitic activity had ceased by the end of the Eocene, about 40 million years ago (McDowell and Clabaugh, 1979).

The origin of the Neovolcanic axis has been related chiefly to the subduction of the Cocos plate beneath the continental crust of Mexico, which at the level of the asthenosphere induced partial fusion and originated the magmas of the axis (Mooser, 1975; Urrutia-Fucugauchi and Del Castillo, 1977; Demant, 1978) (Figure 3.3).

The calcalkaline character of this province supports the above hypothesis, although the oblique position of the axis with respect to the Acapulco trench does not result in a feature typical of this type of phenomenon. Urrutia-Fucugauchi and Del Castillo (1977) explain this lack of parallelism by means of a model which demonstrates that the direction of movement of the Cocos and American plates is not perpendicular to the Acapulco trench and that the northwest and southeast extremes of the Cocos plate become more dense, less warm, and older as well as of greater thickness and rigidity. All this is responsible for a gradual decrease in the angle of subduction toward the southeast end of the trench and causes a horizontal angle of 20° between the Acapulco trench and the Neovolcanic axis. According to Demant (1978), the subduction of the Cocos plate along the Acapulco trench commenced to develop progressively in the Oligocene, in the form of a zone of left-lateral displacement between the American plate and the Caribbean plate, which still is active along the system of the Polichic-Montagua-Cayman trench. The lateral movement within this system reflects the rotation of North America toward the west with respect to the Caribbean plate, which includes the continental portion of Central America.

Negendank (1972) supposes, based on the chemical characteristics of the rocks of the Neovolcanic axis, that this calcalkaline province originated as a result of the partial fusion of materials from the lower crust, more than from partial fusion of the Cocos plate at the level of the asthenosphere.

Some authors have indicated that the Neovolcanic axis coincides with a zone of lateral slippage that was active during the past. According to a model of Gastil and Jensky (1973), important right-lateral displacements occurred in the axis in the Late Cretaceous and in the early Tertiary, in concordance with the movements observed in the western United States. Nevertheless, Urrutia-Fucugauchi (personal communication) considers that the movement has been left-lateral, and he calls attention to the available paleomagnetic data. This author believes that the zone of lateral displacement indicated above could have operated as a structural control for the exit of the magmas produced by the subduction of the Cocos plate under the American plate. Mooser (1975) considers that the Neovolcanic axis could have coincided with the scar (geosuture) that marks the union between two ancient cratonal masses and whose zigzag arrangement would reflect the fact that the Cocos plate, after foundering in the Acapulco trench, would have been divided into slightly overlapping zigzagging fragments.

THE MORELOS-GUERRERO PLATFORM

The area of the Morelos-Guerrero platform, in which important marine Mesozoic deposits are developed, is located for the most part within the State of Morelos and in small portions of the northeastern State of Guerrero and southeastern State of Mexico. The marine sedimentary sequence exposed in this region covers a chronostratigraphic range from the Upper Jurassic to the Upper Cretaceous. This sequence rests on Precambrian metamorphic basement represented apparently by the Taxco Schist (Fries, 1960; De Cserna et al., 1975), which in a similar area underlies a lightly metamorphosed andesite unit that Fries (1960) termed the Old Taxco Greenstone. Campa (1978) indicates much similarity between the Taxco Schist rocks described by Fries and the volcano-sedimentary rocks of the Lower Cretaceous that crop out to the west of Teloloapan. This would indicate that the age of the Taxco Schist is not Precambrian,

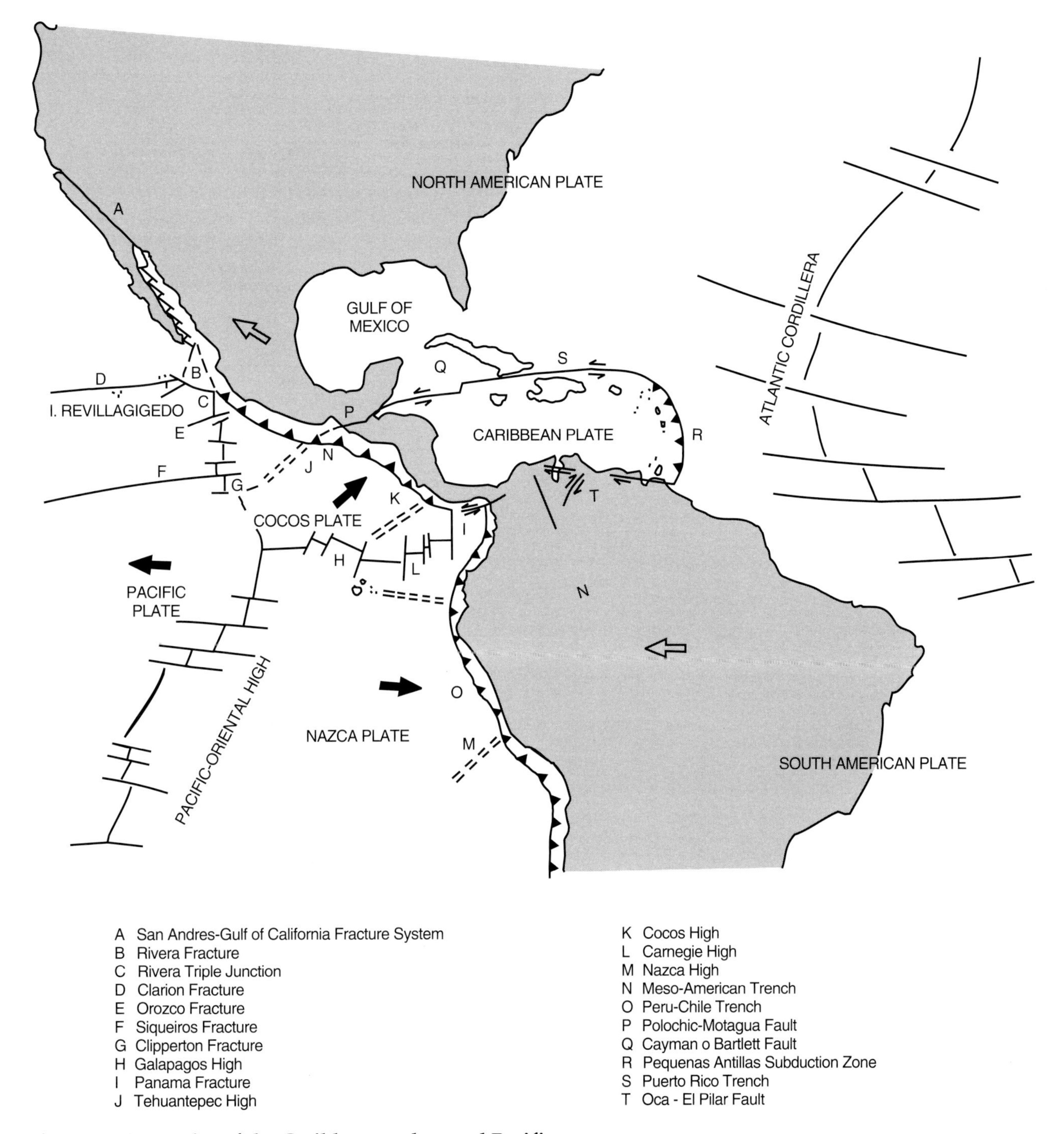

Figure 3.3. Tectonics of the Caribbean and central Pacific.

and in that case this unit would not form part of the of the metamorphic basement above which evolved the Mesozoic sedimentary sequence of the Morelos-Guerrero platform. Toward the borders of Guerrero and Oaxaca, the sedimentary marine sequence of the Morelos-Guerrero platform rests over a metamorphic basement of Paleozoic strata represented by the Acatlán Complex.

The marine sedimentary units of this region are covered discordantly by Cenozoic continental deposits and volcanic rocks of the Neovolcanic axis, as well as by some Oligocene remnants of rhyolitic volcanism.

The base of the Mesozoic marine package is represented by the Acahuitzotla Formation of Late Jurassic age (Fries, 1956), which is formed by calcareous and argillaceous sediments that crop out in isolated localities. This formation underlies with erosional discordance calcareous shales of the Neocomian Acuitlapan Formation. Both formations show the effect of weak dynamic metamorphism. The Xochicalco Formation of Aptian age, also in rare outcrops, is formed by a sequence of thin limestone beds that rest upon the Acuitlapan Formation. After deposition of the Xochicalco Formation an uplift occurred in the region that gave rise to the paleopeninsula of Taxco (Fries, 1956) and a period of erosion marked by the presence of an unconformity that places the Xochicalco Formation in contact with various parts of the Morelos Formation.

This latter formation consists of a calcareous unit that accounts for the most extensive outcrops of the region. Its name has been applied to sequences of limestone that extend toward Michoacán, Jalisco, and Colima, although its characteristics are not always the same. It is formed of thick beds of limestone and dolomite that in one sequence reach up to 900 m thick and that have in the base an anhydrite member some meters thick. The lithologic characteristics and the fauna reveal that this unit formed from shallow water marine deposition during the Albian–Cenomanian interval.

At the end of the Cenomanian an emersion occurred in the area with the emplacement of various stocks of granite and with differential erosion of the top of the Morelos Formation (Fries, 1956).

During the Turonian an invasion of the sea was repeated and calcareous sedimentation was reestablished with the development of a calcareous bank toward the west of a line trending from Cuernavaca to Huitzuco.

The end of the Turonian marked a drastic change in the sedimentation of the Morelos-Guerrero platform resulting from uplift of a major part of the volcano-sedimentary areas located in the western region of this part of Mexico. The deposits of shale, siltstone, sandstone, and conglomerate came to form a sequence more than 1200 m thick developed in the Turonian–Campanian interval. At the end of the Cretaceous and beginning of the Tertiary, compressional deformation occurred that resulted in the formation of anticlinal and synclinal folds.

In the Oligocene–Eocene interval, intense normal faulting occurred, accompanied by continental clastic sedimentation over the low parts of the newly created topography. This clastic continental sedimentation was initiated in the middle of the Cretaceous in the areas located west of this region. Deposition of conglomeratic materials was contemporaneous with some basaltic lava flows that gave rise to the lithostratigraphic assemblage termed the Balsas Group (Fries, 1960). These deposits were followed by important siliceous volcanic emissions that formed the ignimbrite cover of the Taxco area termed the Tilzapotla Rhyolite and by volcanics and volcanoclastic deposits of the Tepoztlán Formation. According to Campa (1978), this region suffered considerable warping during the Miocene that is evidenced by the dipping beds of the Balsas Group and by the abnormal elevation of the Oligocene ignimbrites.

The upper Tertiary and Quaternary are characterized in this region by the influence of volcanic activity of the Neovolcanic axis and by the development of tectonic trenches that caused the deposition of continental clastic sediments of the Cuernavaca Formation.

METAMORPHIC REGION OF ACATLÁN

The region that includes the higher part of the Balsas basin, drained by the Mixteco and Acateco rivers, is characterized by extensive outcrops of metamorphic rocks of various types that form a complex of early Paleozoic age (Ortega-Gutiérrez, 1978) (see Figure 3.4).

This metamorphic unit was termed originally the Acatlán Schist by Salas (1949). Later, Fries and Rincón (1965) defined it as the Acatlán Formation. Recently, Ortega-Gutiérrez (1978) elevated this unit to the rank of a complex, pointing to its varied lithology and structure. This author divided the Acatlán Complex into two subgroups termed Petlancingo and Acateco. In the lithostratigraphic division that Ortega-Gutiérrez introduced at the formational level, he employed some names that had already been utilized by Rodríguez (1970) in an informal subdivision that included in the Acateco Group the formations Esperanza, Acatlán, Salado, and Tecomate.

The formation that constitutes the structurally lower part of the Acatlán Complex is the Magdalena Migmatite, a classic migmatite derived from sedimentary rocks. The Chazumba Formation is formed principally by biotite schists with intervals of quartzite, differentiated metagabbro, and pelitic schist. The Cosoltepec Formation, which together with the two above units makes up the structurally lower Petlancingo Subgroup, is composed of psammitic and pelitic schists with the presence of greenstones, talc schists, calcareous schists, metamorphosed chert, and manganiferous rocks (Ortega-Gutiérrez, 1978).

The Acateco Subgroup is composed of the Xayacatlan, Tecomate, and Esperanza granitoid formations as well as the Totoltepec stock and the San Miguel dikes. The first formation is composed of greenschists, amphibolite, metagabbro, eclogite, serpentinite, mylonite, pelitic schists, and quartzite in an assemblage that, according to Ortega-Gutiérrez (1978), possibly makes up an ophiolite complex and is of great importance, since this is the first time in Mexico where the presence of eclogite rocks has been reported. The Tecomate Formation is composed of metarenite, pelites, and semipelites partially of tuffaceous origin, as well as metamorphosed limestone and metaconglomerate. The Esperanza granitoids are formed by granitic, aplitic, and pegmatitic rocks that are cataclastic and metamorphosed and in certain areas have been considered by Rodríguez (1970) as part of the Oaxacan Complex. The Totoltepec stock is an intrusive of trondhjemitic composition with slight foliation and could have resulted from the differentiation of a tholeiitic gabbro (Ortega-Gutiérrez, 1978). Fries et al. (1970) indicated an age of 440 ± 50 Ma for this intrusive, which would be in the Ordovician. The name "San Miguel dikes" has been applied to a series of tabular intrusive bodies of granitic and tonalitic composition that affect some units of the Acatlán Complex.

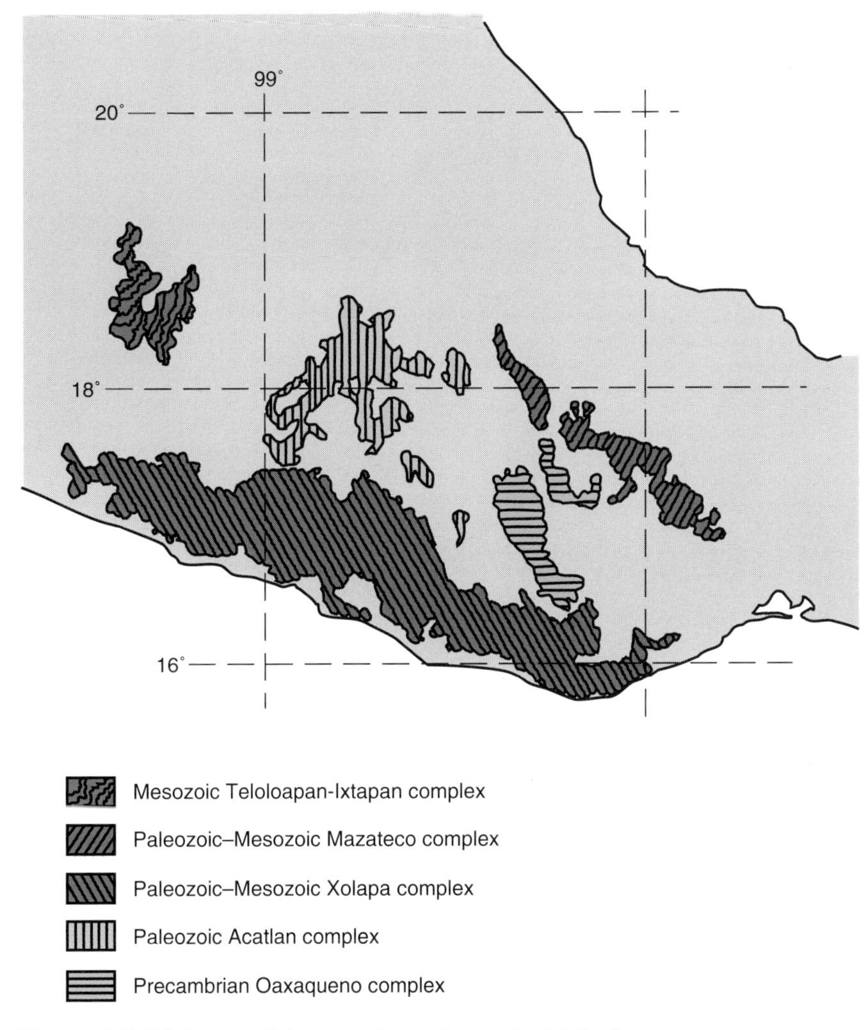

Figure 3.4. Metamorphic complexes in central Mexico.

The assemblage of the Acatlán Complex is found covered discordantly by numerous igneous and sedimentary units that include an age range which varies from the late Paleozoic to Quaternary and constitutes the basement of an extensive region that includes parts of the states of Puebla, Oaxaca, Guerrero, and Morelos. This group correlates with the Chacús Group of Guatemala and with the metamorphic rocks of the Sierra de Omoa in Honduras. It seems to have no similar relationship with the Xolapa Complex of the Sierra Madre del Sur, nor with the Oaxacan Complex (see Figure 3.4).

According to Ortega-Gutiérrez (1978), the character of the Acatlán Complex leads to the supposition that it is an ancient marine eugeosynclinal deposit with a style of tectonic deformation and metamorphism resembling that of the internal or deep zones of an alpine orogenic belt.

SIERRA MADRE DEL SUR AND ADJACENT AREAS

The Sierra Madre del Sur, from Colima to Oaxaca and contiguous areas of northwestern Guerrero, Michoacán, and the State of Mexico, makes up a region of high structural complexity that contains various juxtaposed tectonic domains (Figure 3.5).

The most northern segment of the Sierra Madre del Sur is formed by outcrops of Mesozoic sequences, both platform sediments and volcanic rock sediments of island arc type. Areas found in northwestern Guerrero, west of the State of Mexico and south of Michoacán, form a region with partially metamorphosed volcano-sedimentary rocks of Jurassic and Cretaceous age. These are covered by Cenozoic continental volcanic and sedimentary rocks. This region borders on the east the area of the Cretaceous Morelos-Guerrero platform at the latitude of the lineament of Ixtapan de la Sal-Taxco-Iguala. The southern segment of the Sierra Madre del Sur is formed by extensive outcrops of metamorphic rocks that have a geochronologic range varying from Paleozoic to Mesozoic and that are seen to be affected by batholithic emplacements of late Mesozoic and even Cenozoic age. The Pacific area of the Sierra Madre del Sur includes the States of Colima, Michoacán, and northern Guerrero and contains outcrops of andesitic volcanic rocks interstratified with silty redbeds, volcanic

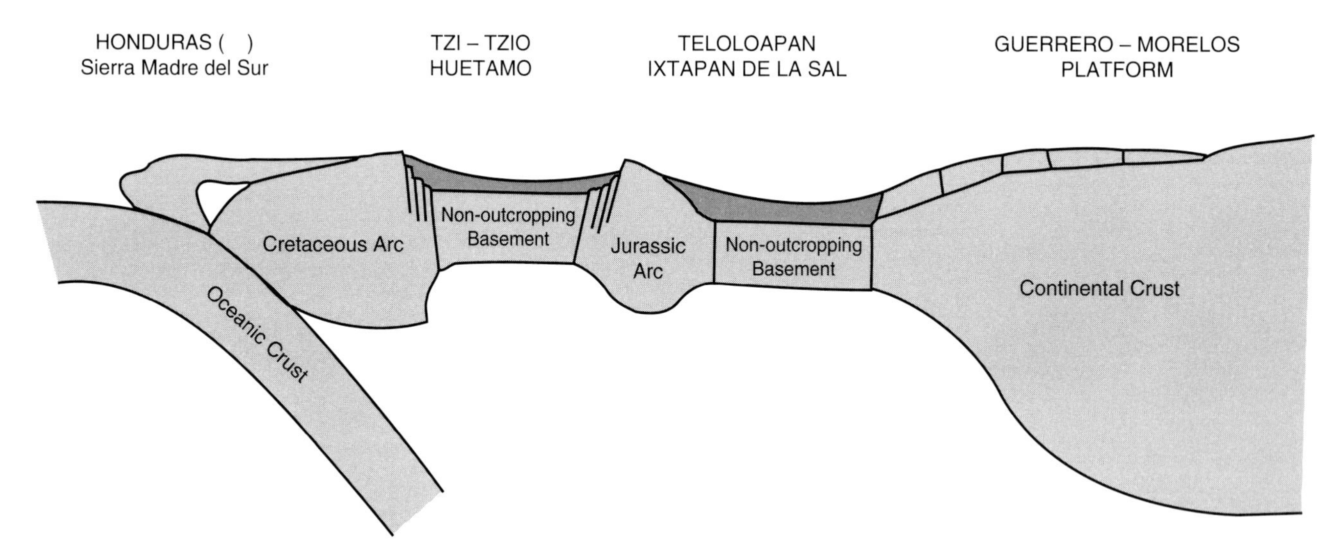

Figure 3.5. Schematic tectonic model of the Sierra Madre del Sur.

conglomerate, and subreef (slope) limestone beds that contain an Albian fauna. These outcrops form part of what Vidal et al. (1980) have called the Petrotectonic Assemblage of Zihuatanejo, Guerrero, Coalcomán, Michoacán. There exist, furthermore, in this northern portion of the Sierra, extensive outcrops of sedimentary sequences of platform limestones with Albian fauna and rhythmic sequences of terrigenous sandy muds. In areas situated in the neighborhood of Colima City, the platform limestones contain great thicknesses of intercalated evaporites. These underlie, in apparently transitional contact, continental terrigenous Upper Cretaceous sediments. In a large part of the Sierra Madre del Sur, from the northern tributaries to the area near Zihuatanejo, Campa and Ramírez (1979) have reported the existence of numerous mountains formed by andesitic materials interstratified with some beds of limestone and terrigenous clastics disseminated in small areas in the Sierra. This Mesozoic vulcanism continues toward the north bordering the Pacific coast until it becomes blurred with similar areas of the North American Pacific Cordillera (Campa and Ramírez, 1979).

Ferrusquía and co-workers (1978) have reported the presence, in the area of Playa Azul, Michoacán, of a transitional volcanic-sedimentary sequence with dinosaur footprints that indicate perhaps a Middle Jurassic to Early Cretaceous age. In addition they indicate that this is the first record of dinosaur tracks in Mexico and constitutes the southernmost trace of dinosaurs in North America.

Most authors have reported the volcanic-sedimentary sequences of this Pacific region of Mexico as being of Mesozoic age. Nevertheless, De Cserna at al. (1978a) obtained a Rb-Sr radiometric age of 311 ± 30 million years for intrusive rocks strictly related to volcanic rocks belonging to the metavolcanic complex of Zapotillo, east of Zihuatanejo.

Campa and Ramírez (1979) as well as Vidal and co-workers (1980) consider that the Mesozoic volcano-sedimentary sequences of a major part of the Sierra Madre del Sur are the result of magmatic activity from convergent edges of plates developed in this part of Mexico during the Early Cretaceous.

The southern half of the Sierra Madre del Sur is formed from metamorphic rocks that constitute the Xolapa Complex (De Cserna, 1965). This is found to be intruded by batholiths of granite (see Figure 3.4). De Cserna (1965) reported the Xolapa Complex on the highway from Chilpancingo to Acapulco as an assemblage of metasedimentary rocks formed of biotite schists and gneiss with some quartzite and cipolin marble horizons, and including the presence of pegmatites. Nevertheless, Guerrero and co-workers (1978) consider that in the major part of this region the complex is formed from quartz-feldspathic orthogneiss of granodiorite composition. In the most southern section of the Sierra Madre del Sur, corresponding to southern Guerrero and western Oaxaca, the Xolapa Complex has an ampholite facies derived from sedimentary rocks and orthogneiss with abundant migmatites.

De Cserna considers this metamorphic complex to be of Paleozoic age, given that it underlies the volcano-sedimentary sequence of the Chapolapa Formation, which is probably of Triassic age. As well, the complex is never seen in a locality where it underlies Paleozoic sedimentary rocks. Nonetheless, the stratigraphic range of this complex has not been precisely determined because the geochronologic studies have given very disparate radiometric ages indicating that thermal events occurred in the Paleozoic (Halpern et al., 1974), in the Mesozoic (Guerrero et al., 1978), and in the Tertiary (De Cserna, 1965). Guerrero et al. (1978) rely on the existence of a thermal event in the Tertiary (about 32 million years ago) in the area of the highway of Chilpancingo, and in their radiometric determinations, which failed to indicate Paleozoic or Precambrian ages as suggested by other authors. These authors recognized the oldest thermal event as Jurassic by means of uranium-lead (165 ± 3 million years) and rubidium-strontium (180 ± 84 million years).

In the Tierra Caliente region and adjacent areas of the western part of the State of Mexico and southeast Michoacán, extensive outcrops of partly metamorphosed volcano-sedimentary sequences exist that are juxtaposed against other extensive outcrops of marine Cretaceous platform sequences from the areas of Morelos and Huetamo-Coyuca, along the borders of Guerrero and Michoacán.

In the Teloloapan-Arcelia sector, a sequence of andesitic volcanic rocks as well as calcareous-argillaceous foliated sedimentary rocks and graywackes constitute deposits of an island volcanic arc and marginal seas developed in the Late Jurassic and Early Cretaceous (Campa and Ramírez, 1979) (Figure 3.5). These volcano-sedimentary sequences crop out in continuous pattern toward the north, up to the area of Tejupilco. From here the outcrops become isolated and less extensive. They may also be observed in the areas of Ixtapan de la Sal, Zitácuaro, and Tlalpujahua.

In the Huetamo-Coyuca section, a Jurassic–Cretaceous volcano-sedimentary sequence is exposed that gradually becomes more sedimentary toward its top. The base contains detrital sedimentary rocks interstratified with lavas and andesitic tuffs of the Jurassic that constitute the Angao Formation (Pantoja, 1959). Above this formation rest interbedded shales and sandstones with some tuffaceous horizons and with siltstones and reefal limestones deposited in the Lower Cretaceous (Neocomian–Aptian–lower Albian). These deposits make up the San Lucas Formation (Pantoja, 1959). Finally, the top of the sequence is formed by beds of argillaceous limestone attributed to the Morelos Formation of Albian age (Pantoja, 1959).

The Huetamo-Coyuca sector forms a transitional zone between the external Mesozoic domain represented by the Guerrero-Morelos platform and the Mesozoic island arc represented by the volcano-sedimentary outcrops of the Sierra Madre del Sur. The volcano-sedimentary sequences of Teloloapan and Ixtapan, situated to the east of Huetamo, would then be considered as tectonic allochthons transported over the platform of the external domain (Campa and Ramírez, 1979) and to have come from the western island arc domain. De Cserna (1978b) believes that the absence of platform limestone in the Morelos Formation to the west of Teloloapan is due to a facies change into a basin in this area during the Albian and Cenomanian. This author considers that the volcanic rocks of the Teloloapan-Arcelia area, which form the volcano-sedimentary sequence of the marginal sea and island arc proposed by Campa and Ramírez, belong to a stage of Cenomanian–Turonian vulcanism (Xochipala Formation) or could well be a basement of ancient volcanic rocks, all this in a model without major tectonic complications.

OAXACA AND ADJACENT ZONES

In the central region of Oaxaca and adjacent areas of southern Puebla and eastern Guerrero, an important sedimentary Mesozoic sequence crops out that attests to the development of a basin beginning in the Early Jurassic (Figures 3.6–3.8).

This region of Mesozoic outcrops is limited by various metamorphic complexes that are exposed in this part of the country. To the northeast, metamorphics of the Acatlán Complex are present, belonging to the lower Paleozoic and resulting from marine eugeosynclinal deposition (Ortega-Gutiérrez, 1978). Above this metamorphic complex rest sedimentary rocks of the Jurassic and Cretaceous and some unmetamorphosed units of the Paleozoic. To the west and south, the nonsedimentary Mesozoic exposures are bordered by the Xolapa Complex, composed of gneiss, migmatite, and biotite schists with amphibolite metamorphic facies (Ortega-Gutiérrez, 1976). The age of this complex is apparently Mesozoic, but thermal events of Paleozoic, Jurassic, and Tertiary have been reported (Halpern et al., 1974; Guerrero et al., 1978; De Cserna et al., 1962).

To the southeast, the Oaxacan Complex forms the limit of the basin. It is formed by banded gneiss metamorphosed from granulite facies to transitional granulite-amphibolite, including charnockite, anorthosite, and pegmatite. Fries and co-workers (1962) carried out radiometric studies of the Oaxacan Complex that resulted in age dates of 1100 ± 125, 920 ± 30, and 940 million years (Precambrian). Additionally, these authors indicate that the pegmatites and the last stage of metamorphism that affected the host rocks are equivalent to the Grenville metamorphic province of the eastern United States and Canada. The outcrops of this complex form a considerable part of the mountainous zone that is located to the west of the city of Oaxaca. Finally, to the northeast, the basin is found to be bordered by the metamorphic outcrops of the western flank of the Sierra de Juárez with a markedly rectilinear contact that forms the Oaxacan Ravine; this probably is a regional tectonic feature. These metamorphic rocks have been traditionally assigned to Precambrian (orthogneiss) and Paleozoic (phyllites and incipient meta-arkose) (López-Ramos, 1979). However, Charleston (1980) reported the existence of an ample metamorphic complex derived from eugeosynclinal deposition of sandstone, clays, and volcanic flows of Cretaceous age. Radiometric studies of these rocks gave ages for the metamorphism corresponding to Upper Cretaceous and lower Tertiary (Charleston, 1980). According to this author, this complex is formed by allochthonous blocks whose provenance is to the west and that have been thrust over miogeosynclinal sediments of Jurassic and Cretaceous during the Laramide Orogeny.

The Puebla, Oaxaca, Guerrero, and Morelos regions, underlain by the Acatlán and Oaxacan complexes, contain extensive outcrops of Mesozoic sedimentary units arranged in north-northeasterly folds. Under this Mesozoic sequence there have been reported, in isolated exposures, some Paleozoic sedimentary units resting discordantly above the metamorphic basement. Above the Acatlán Complex, Corona (1981) and Flores and Buitrón (1982) discovered in the Olinalá area a sequence of detrital and calcareous rocks with Pennsylvanian and Permian fossils. Also

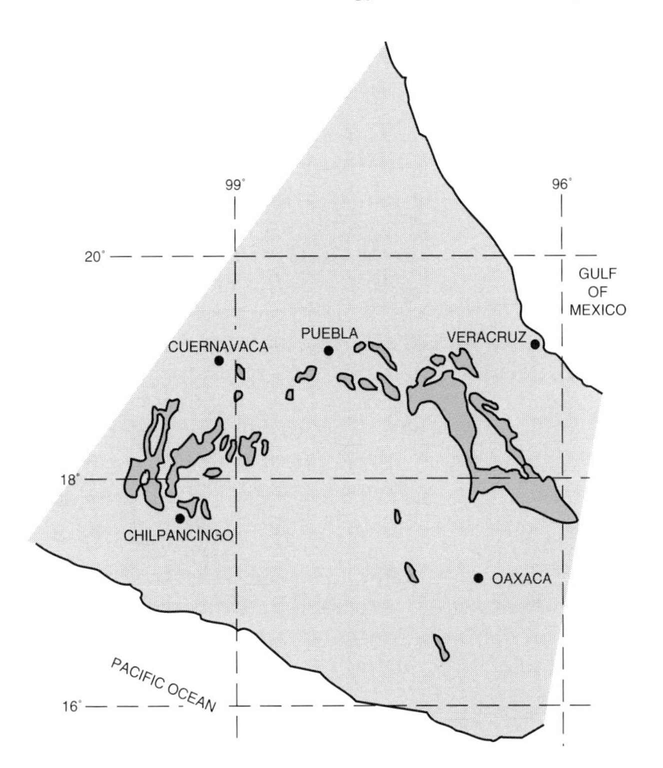

Figure 3.6. Sedimentary rocks of the Upper Cretaceous of the Guerrero-Morelos platform, Tlaxiaco basin, and southern section of the Sierra Madre Oriental.

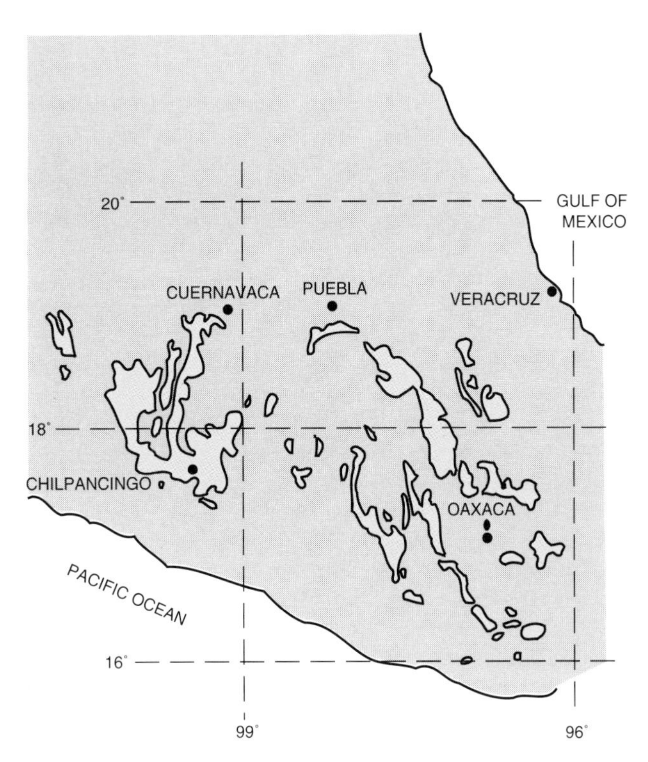

Figure 3.7. Sedimentary rocks of the Lower Cretaceous of the Guerrero-Morelos platform, Tlaxiaco basin, and southern sector of the Sierra Madre Oriental.

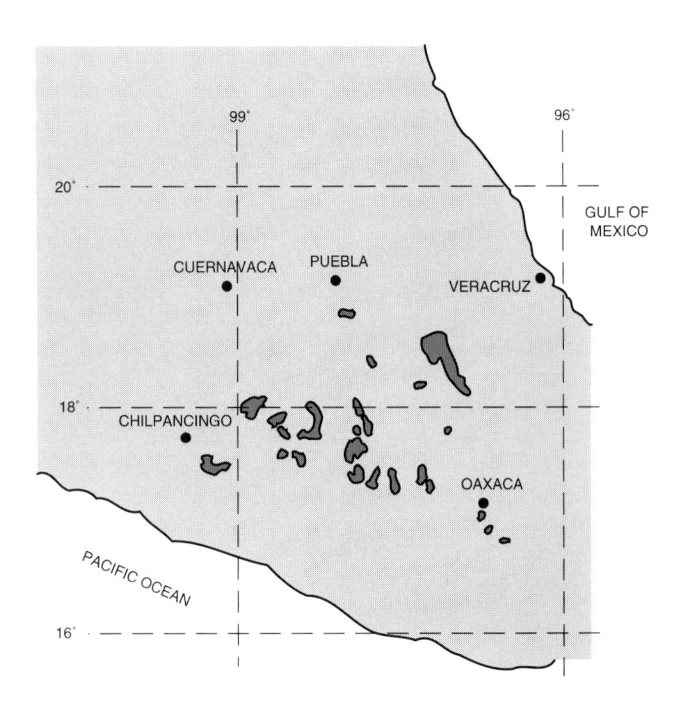

Figure 3.8. Sedimentary rocks of the Jurassic of the Guerrero-Morelos platform, Tlaxiaco basin, and southern sector of the Sierra Madre Oriental.

there have been reported above this complex discoveries of upper Paleozoic sedimentary rocks in Mixtepec, Oaxaca (Flores and Buitron, 1984), and in Tuxtepeque, Puebla (De la Vega, 1983). The Matzitzi Formation, with Pennsylvanian plant fossils (De Cserna, 1970), is found covering apparently both the Acatlán and Oaxacan complexes; its main outcrops occur to the southwest of Tehuacán.

In the Nochixtlán region above the Oaxacan Complex, Pantoja-Alor and Robison reported in 1967 the discovery of a marine sequence with Cambrian–Ordovician trilobites that was termed the Tiñú Formation. Above this unit a sequence made up of the Santiago, Ixtaltepec, and Yododeñe formations rests discordantly. These units form more than 1000 m of clastics belonging to the Mississippian, Pennsylvanian, and Permian systems (Pantoja-Alor, 1970).

The base of the Mesozoic sequence that is exposed in the Tlaxiaco basin is represented by detrital sediments of the lower part of the Rosario Formation, which is of continental origin and contains coal horizons (Erben, 1956). According to this author, the sediments of the Rosario Formation were deposited in a coal basin that developed during the Early Jurassic in northwestern Oaxaca, northeastern Guerrero, and southwestern Puebla. On the western and eastern borders of the basin, the lower strata of the formation were not deposited. Above the Rosario Formation rests the Cualac Conglomerate which, together with the middle and upper strata of the former formation, belongs to the Middle Jurassic. Both formations constitute the Consuelo Group, which underlies the Tecocoyunca Group whose formations crop out in various localities in the Tlaxiaco basin and also belong to the Middle Jurassic.

This group is composed of both detrital and carbonate sediments, both continental and marine, and contains plant fossils and ammonites that indicate several marine invasions and regressions. During the Late Jurassic, clearly marine sediments were deposited in some areas of the basin, such as the Cidaris Limestone in the Mixtepec-Tlaxiaco area (Erben, 1956) and the Chimeco and Mapache del Sur de Puebla, formations formed of limestone, argillaceous limestone, and calcareous shales (Pérez et al., 1965). The Teposcolula Limestone, considered originally Jurassic by Salas (1949) and later by Erben (1956), has been recently assigned to the Albian–Cenomanian (Ferrusquía, 1970) on account of its faunal content. On the other hand, the Cidaris Limestone has been confirmed as belonging to the Late Jurassic because of its echinoid fauna of Oxfordian, Callovian, and Kimmeridgian age (Buitron, 1970). It should be noted that these Jurassic units that outcrop in the region of the Acatlán Complex are not reported to be represented by similar strata above the Oaxacan Complex.

The Upper Cretaceous also is represented by marine sediments; in some localities the Neocomian and Aptian are present. In the Tehuacán area a clastic-calcareous sequence with beds of limestone crops out and constitutes the Zapotitlán Formation. Above this unit lie 1300 m of both fine and coarse clastic-calcareous beds of the Aptian San Juan Raya Formation. The Neocomian and Aptian formations of the central area of Oaxaca and central and south of Puebla have been included within the group termed Puebla. However, in various localities this group is absent and Albian limestones rest discordantly above the Jurassic sequence. López-Ramos (1979) mentions that the wells Yacudá no. 1 and Teposcolula no. 1 cut a sequence of more than 2500 m of Upper Jurassic and Lower Cretaceous evaporites.

During the Albian–Cenomanian interval a sequence of thick-bedded limestones developed in a transgressive sea. These limestones have received different names in different areas. Calderón (1956) designated a widespread sequence of massive micritic and biomicritic limestone with chert nodules that crops out in the Tehuacán region as the Cipiapa Formation. Ferrusquía (1970) designated a massive biomicrite as the Teposcolula Limestone. This crops out in a similar area and was considered Jurassic by Salas (1949). Finally, Pérez and co-workers (1965) applied the name Morelos Formation to these limestones in the region of Acatla and related them to the Albian–Cenomanian strata that crop out on the Guerrero-Morelos platform.

Above the Albian–Cenomanian limestones lies a sequence of marly limestone designated by Ferrusquía (1976) as the Yucunama Formation. It contains fossils of the Coniacian–Maastrichtian stages and crops out northwest of Nochixtlán. It can be correlated with the Tilantongo Marls (Salas, 1949) that are exposed southeast of Nochixtlán and with the Mexcala Formation of the Guerrero-Morelos platform.

The folded Mesozoic sequence of the Tlaxiaco basin is covered with angular discordance by extensive outcrops of continental deposits. These are sandstone-conglomerate and argillaceous sandy beds of the Tertiary and include siliceous, intermediate, and mafic volcanic rocks.

The Tertiary continental deposits have been assigned to the Yanhuitlán and Huajuapan formations (Salas, 1949) that according to Erben (1956) are distinct facies of the same unit. The first is formed by clays with some intercalations of sandstone and volcanic ash, argillaceous sandstones, and beds of conglomerate and breccia. Ferrusquía (1976) mentions a radiometric age of ±49.0 million years for a tuff interstratified within the Yanhuitlán Formation of Sayultepec, dating this formation as late Paleocene–middle Eocene. This author indicates that the formation has a stratigraphic position similar to that of the Tehuacán Formation (Calderón, 1956) and the Balsas Group (Fries, 1960).

The Oligocene, in various localities in the state of Oaxaca, was developed in a period of active vulcanism that originated initially with the emission of siliceous and intermediate tuffs and later andesitic lava flows. The volcanic activity culminated with some basaltic flows in the Neogene

THE SECTOR SOUTH OF THE SIERRA MADRE ORIENTAL AND THE COASTAL PLAIN OF THE SOUTHERN GULF

On the eastern flank of the sector south of the Sierra Madre Oriental (Sierra de Juárez) a thick sequence of Mesozoic sedimentary rocks is exposed that rests on a metamorphic basement composed of schists, gneisses, and phyllites. These have been derived principally from sedimentary rocks and have been traditionally attributed to the Paleozoic and Precambrian. However, in a section located at the 18th parallel, Charleston (1980) recognized a thick sequence of schists and metavolcanic rocks that he attributed to the Lower Cretaceous.

The sedimentary sequence of the eastern flank of the sector, which forms folds asymmetric toward the east, has, in the Zongolica-Tehuacán sector, a basal unit of dark-colored slates with some intercalations of fine-grained sand and calcareous shales. This is widely exposed and has been tentatively attributed to the Middle Jurassic (López-Ramos, 1979). In the sector located south of the 18th parallel and down to the region of the Isthmus of Tehuantepec, the base of the Mesozoic is formed by the Todos Santos Formation, which is a sequence of continental redbeds with cross-stratified sandstones, conglomerate, and shale. This formation has, furthermore, been recognized in Chiapas and northern Central America where its lower part is considered Lower and Middle Jurassic (Mulleried, 1957). However, López-Ramos (1979) believes that it could extend to the Triassic.

The Upper Jurassic is exposed in the Zongolica area (Viniegra, 1965) in the form of marine sequences of bituminous limestones with intercalations of sandy-argillaceous limestones and with ammonites. However, in the southern sector of the eastern flank of the Sierra de Juárez, outcrops of this age have not been reported.

The marine Cretaceous sequence, which crops out in the northern portion of the Sierra de Juárez, is composed principally of calcareous rocks that have been recognized by Petróleos Mexicanos in both surface and subsurface studies. These rocks include the following formations: Tuxpanguillo (Neocomian), Capolucan (Aptian), Orizaba Limestone (Albian–Cenomanian), Maltrata Limestone (Turonian–Coniacian), Guzmantla Unit (Turonian–Senonian) as well as the Necoxtla and Atoyac formations of Senonian–Campanian and Campanian–Maastrichtian age (Viniegra, 1965). Additionally, the marine Cretaceous is represented in the area of the Isthmus of Tehuantepec by neritic fossiliferous limestones that López–Ramos (1979) included within the series of middle Cretaceous limestones of Nizanda-Lagunas.

In the portion of the Gulf Coastal Plain that borders the Sierra de Juárez, Petróleos Mexicanos has drilled exploratory wells that have afforded recognition of Mesozoic units in the subsurface. From these it has been possible to reconstruct a paleoplatform termed the Córdoba Platform, which formed the marine sea floor during the second half of the Mesozoic (Figures 3.9–3.10). The western half of the platform is exposed in the Sierra Madre Oriental and the eastern half is buried under the Coastal Plain of the Gulf. In addition, it is limited on the west by the Zongolica paleobasin and on the east by the Veracruz paleobasin (González–Alvarado, 1976). More than 5000 m of sediments accumulated in this latter basin. Petroleum production has been obtained from these strata, chiefly in fields located in its eastern portion (González–Alvarado, 1976).

During the Tertiary, in a setting of eastward marine regression, terrigenous sediments were deposited in the Gulf Coastal Plain. These are: Chicontepec-Velasco (Paleocene); Aragon, Guayabal, and Chapopote (Eocene); Horcones and La Laja (Oligocene); Depósito, Encanto, Concepción, Filisola, and Paraje (Miocene). These deposits began to form at the inception of the orogenic deformation of the Sierra Madre Oriental during the beginning of the Cenozoic.

The igneous activity of the southern sector of the Sierra Madre Oriental, which is manifested in the form of granitic intrusions at the end of the Mesozoic and beginning of the Cenozoic, is restricted to alkaline basaltic emissions in the area of Tuxtlas in the upper Tertiary and Quaternary. Demant (1978) related this volcanic zone with the alkaline province of the Gulf of Mexico, more than with the eastern extreme of the Neovolcanic axis as some other authors have indicated.

TECTONIC SUMMARY

The complicated structural and stratigraphic setting of the central-southern portion of Mexico makes difficult a paleogeographic and tectonic reconstruction that permits a clear explanation of the origin of features in this part of the nation.

Recently the structure of the region has been interpreted in terms of a mosaic of tectonostratigraphic terranes (see Figure 3.11) that were accreted in different episodes during the tectonic evolution of this part of Mexico (Campa et al., 1981; Campa and Coney, 1983). Each terrane contains a distinctly different basement and their limits have been generally interpreted as tectonic boundaries.

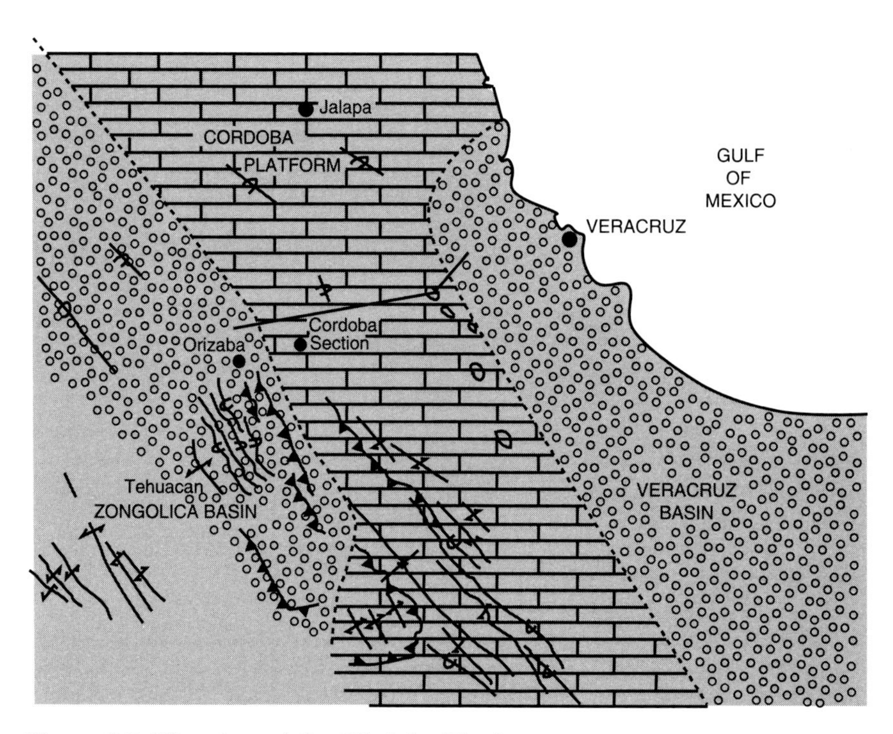

Figure 3.9. Situation of the Córdoba Platform.

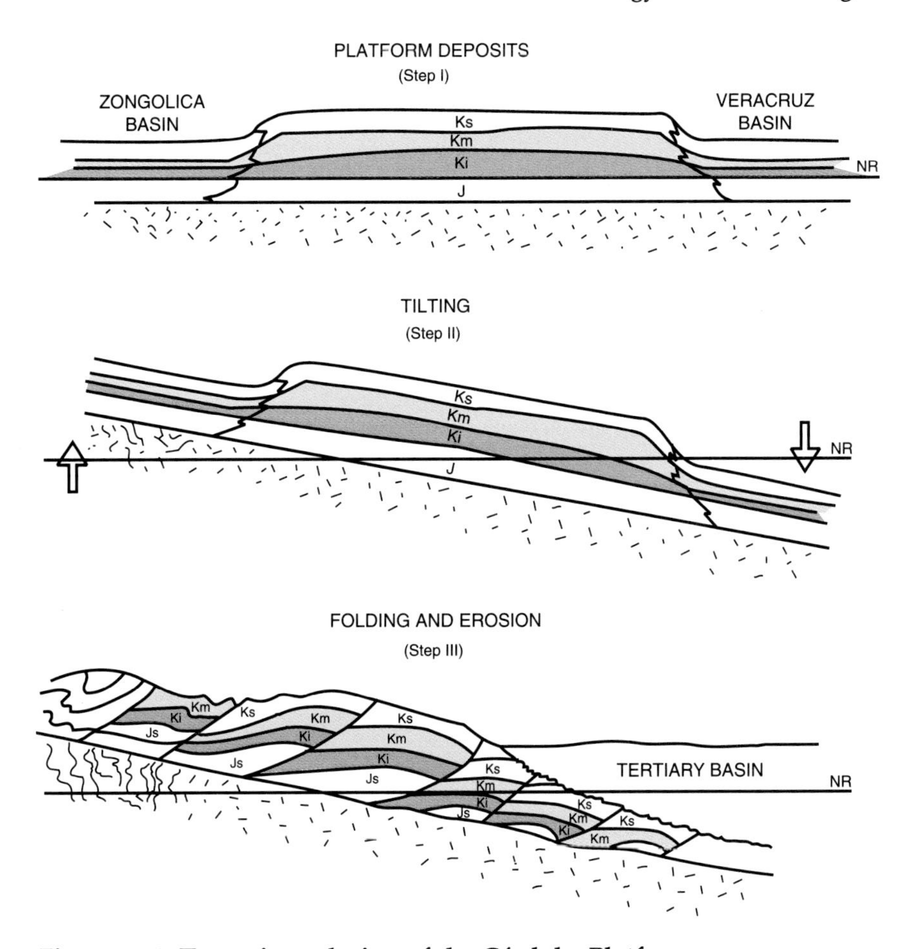

Figure 3.10. Tectonic evolution of the Córdoba Platform.

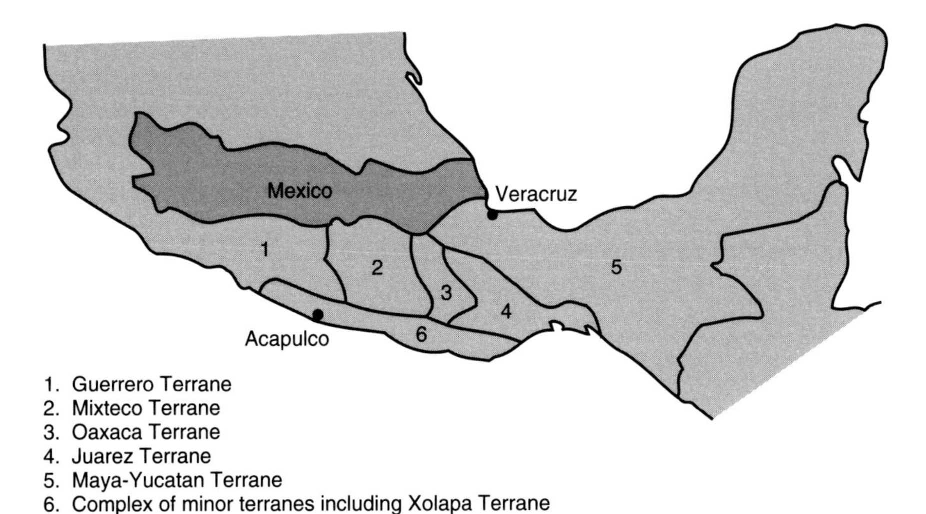

1. Guerrero Terrane
2. Mixteco Terrane
3. Oaxaca Terrane
4. Juarez Terrane
5. Maya-Yucatan Terrane
6. Complex of minor terranes including Xolapa Terrane

Figure 3.11. Tectonostratigraphic terranes from southern Mexico according to the divisions of Campa and Coney (1983).

The terrane with the oldest basement is the Oaxacan, which occupies part of the state of that name and contains unmetamorphosed sequences of Cambrian–Ordovician and Mississippian–Pennsylvanian. The metamorphic basement, formed in the Oaxacan Complex of Precambrian age (900–1100 million years), has been interpreted as the result of the evolution of a rift with sedimentation on ancient continental crust and later metamorphism to the granulite facies. This resulted from an ensialic evolution or from continental collision (Ortega-Gutiérrez, 1981). This complex is considered to be a southern continuation of the Grenvillian belt (Fries et al., 1962). However, the trilobite fauna of the Cambrian–

Ordovician cover show more affinity with the fauna of Europe and South America than with the fauna of North America (Whittington and Hughes, 1974). Bazán (1984) does not discard the idea of the existence of arc-type rocks in this complex, based on the interpretation of greenstone belts in Precambrian shield areas.

To the west of the Oaxacan Terrane lies the Mixtecan Terrane, which has the lower Paleozoic Acatlán Complex as basement (Campa and Coney, 1983) and which, in contrast to contemporaneous rocks in the Oaxacan Terrane, contains diverse grades of metamorphic rocks. The boundary between these two terranes has been interpreted to be tectonic (Ortega-Gutiérrez, 1981). The time of its accretion has not yet been confirmed, but it has been suggested to be Devonian (Ortega-Gutiérrez, 1981). As well, the time of accretion has been placed in the Late Jurassic–Early Cretaceous interval (Ramírez, 1984). The first paleomagnetic dates for Permian units in both terranes indicate similar directions of primary magnetism. This does not discount totally a later accretion (of the blocks) along the same magnetic paleolatitude (Urrutia-Fucugauchi and Morán-Zenteno, 1984).

The Acatlán Complex has been interpreted as an aggregation of petrotectonic assemblages resulting from the opening and closure of an ocean basin (Ortega-Gutiérrez, 1981). The Petlancingo Subgroup would constitute a sequence of an autochthonous passive margin and the Acateco Subgroup would form the allochthonous assemblage, including the Xayacatlán Formation as the vestige of an ancient oceanic lithosphere consumed in the subduction process.

To the southwest, the Mixteco and Oaxacan terranes are bordered in tectonic contact by the Xolapa Complex, whose age and time of accretion to the tectonic mosaic of southern Mexico are not well known, but whose characteristics identify them as roots of a mountain range from an ancient magmatic arc (Halpern et al., 1974).

In the extreme east of the central-southern portion of Mexico, deformed Mesozoic marine sequences are recognized that reveal a paleogeographic setting of interspersed deep and shallow marine substrates developed over the Paleozoic basement, which has traditionally been considered related to Appalachian deformation. These assemblages form part of the Maya Terrane, which extends into south and southeast Mexico (Campa and Coney, 1983). Separating the Maya and Oaxacan terranes, a belt of apparently Mesozoic strata has been recognized. These are marine beds that include calcareous, detrital, and volcanic rocks. They are highly deformed and have a general eastward vergence. The western boundary of this belt forms a mylonitic band that separates it from the Oaxacan Terrane.

In the central-southern portion of Mexico, two principal Mesozoic domains with clearly distinct characteristics are recognized. In the west an andesitic island arc was developed, associated with the subduction of oceanic lithosphere (Campa and Ramírez, 1979). This is a feature that is common to a major part of western North America and that originated during the initial breakup of Pangea. Additionally, in the east an external zone with marine sedimentation evolved over the Guerrero-Morelos platform, the Tlaxiaco basin, and the area of the east flank of the Sierra de Juárez, the coastal plain and the platform of the Gulf of Mexico—all developed on continental crust. Marine sedimentation of this external zone was initiated with the opening of the Gulf of Mexico and the marine transgression over this part of Mexico. The partly metamorphosed volcanic and sedimentary assemblages of the Sierra de Juárez alter the homogeneity of this domain, and their presence is not clearly understood. Carfantan (1983) has suggested that this petrotectonic assemblage is the result of the opening and closing of an ocean basin occurring between Portlandian and Turonian and caused by development of a rift that was connected to a triple junction over a ridge located between Yucatán and South America.

Two alternative models have been postulated in order to explain the development of a volcanic island arc in the western domain of the central-southern portion of Mexico. One of them proposes the accretion by obduction of an island arc system developed in the Pacific and displaced in the direction of its collision with the Mexican continental crust (Urrutia-Fucugauchi, 1980; Coney, 1983). In the other model, the development of an arc domain in the vicinity of the continental crust of Mexico is proposed, limited to the southwest by an eastward subduction (Campa and Ramírez, 1979). Preliminary paleomagnetic data of the volcano-sedimentary sequence of Ixtapan-Telolapan (Urrutia-Fucugauchi and Valencio, 1986) seem to point to the first hypothesis, although no report exists of assemblages of oceanic affinity that would indicate a suture.

According to Campa and Ramírez (1979), in the northwest region of Guerrero and adjoining regions of other states, five phases of deformation can be recognized that were active in Mesozoic and Cenozoic time. The first of these occurred at the end of the Jurassic, affected the Jurassic volcano-sedimentary deposits, and manifests itself by the presence of folds refolded in two generations with a relatively increasing metamorphism in some zones. The second phase occurred in the Cenomanian and is manifested in the Teloapan-Ixtapan area by metamorphism that folded and foliated the volcano-sedimentary sequence. This phase, in the Sierra Madre del Sur, caused the emergence of the island arc terranes and marginal seas. During this time marine sedimentation continued in the Guerrero-Morelos platform, and to its east, contemporaneously with a major introduction of terrigenous sediment coming from the western emergent region. The next phase occurred in the Paleocene and deformed the whole Mesozoic blanket of the two domains and is responsible for the folds in the external zone as well as the overthrusting of the internal domain over the external zone.

Campa (1978) has proposed two alternative models to explain the presence of the volcano-sedimentary assemblage of Ixtapan-Teloloapan between the Guerrero-Morelos and Huetamo platforms. In one of these it is suggested that the Ixtapan-Teloloapan assemblage is the result of the evolution of an arc between the two platforms, but this does not explain the metamorphism of this assemblage between the unmetamorphosed sequences of the two platforms and the absence of facies changes from these platforms to the volcanic arc. In the other model, the author suggests that the Guerrero-Morelos and Huetamo sequences, belonging to the Albian–Cenomanian, could be part of a single platform and that the Ixtapan-Teloloapan assemblage would be a tectonic allochthon of the compressional phase of Paleocene age.

At the end of the Miocene there occurred a phase of deformation that resulted in a warping that is observed in the Arcelia-Altamirano region and evidenced by the abnormally elevated metamorphic sequences and the pre-Miocene lithostratigraphic units. The origin of the great structural anticline of Tzitzio-Tiquicheo of southeastern Michoacán is attributed to this phase because of the consideration that the continental sequence on the flanks of the structure is correlative with the lower Tertiary Balsas Group. Campos (1984) has attributed the folding to the Paleocene compressional phase since he considers that the continental sequence on the flanks of the structure belongs to the Upper Cretaceous and not to the Tertiary.

Campa and co-workers (1980) believe that in the western part of the central-southern portion of Mexico one can recognize tectonostratigraphic terranes that are characterized by homogeneity and continuity of internal stratigraphy, but that have an obscure and poorly understood relationship between themselves. The borders of each terrane separate sequences that have different physical and temporal characteristics. The discontinuities of these borders cannot be explained clearly by conventional facies changes or unconformities. These authors have recognized in this region the following fundamental terranes: the Assemblage of Guerrero-Morelos platform, Assemblage of Teloloapan, Assemblage of Huetamo-Cutzamala, Assemblage of Zihuatenjo, and Assemblage of Taxco and Taxco Viejo, all of them integrated into the Guerrero composite terrane.

In the Pliocene–Quaternary interval, the central-southern region of Mexico has been affected by normal faulting and lateral displacement within a setting of general uplift and very great geodynamic activity.

ECONOMIC RESOURCES

The principal mineral resources known in the central-southern region of Mexico are the sulfides of lead, zinc, and silver in a central belt, as well as iron oxides localized chiefly in the Sierra Madre del Sur (Figure 3.12). To the first category belong the mineral deposits of the Pachuca mining district, which is located at the northern edge of the Neovolcanic axis and has been one of the principal silver producers in the world. To the south of the Neovolcanic axis, mineral districts of hydrothermal sulfides appear along a belt with north-northwest and south-southeast orientation in the states of México, Guerrero, and Michoacán. The band includes the field areas of Taxco, Xitinga, Zacualpan, Temascaltepec, Angangueo, and Tlapujahua. Within this belt, the mercury deposits of Huitzuco and Huahuaxtla also are developed. These hydrothermal deposits are attributed by Campa and Ramírez (1979) to the end of the Miocene period, contemporaneous with the

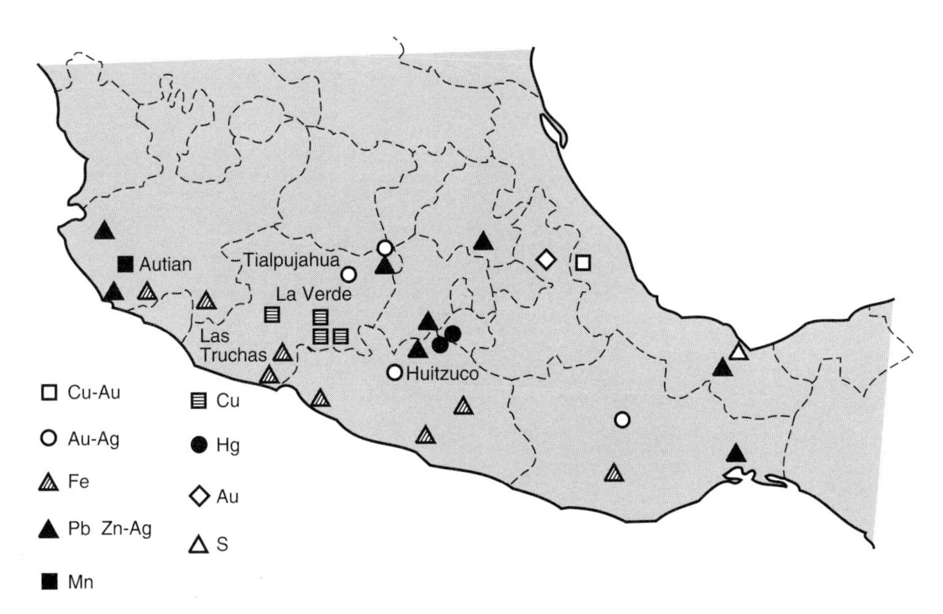

Figure 3.12. Distribution of the principal mineral deposits known in the central part of Mexico.

warping that affected pre-Miocene strata. Another group of sulfide mineral deposits exists in this region. Their origin has been attributed to volcanogenic processes that do not have a preferred orientation but are encountered associated with volcano-sedimentary Upper Jurassic and Lower Cretaceous rocks. To this group belong the fields of Pinzán Morado, Tlapehuala–Las Fraguas, Campo Morado–La Suriana, Rey de la Plata, Teloloapan, and Cuetzalán del Progreso, as well as the volcano field in the north of Michoacán. The deposits mentioned are considered to be contemporaneous with the volcanic activity that occurred in the island arc zone formed during the Mesozoic in this part of Mexico (Gaytán et al., 1979; Campa and Ramírez, 1979).

In a belt situated along the Sierra Madre del Sur, numerous deposits of iron are located; these make up the major reserves of the country. The origin of these deposits is attributed to processes of contact metasomatism unleashed by the effect of silicic and intermediate intrusions of the lower Cenozoic on the Cretaceous limestones (Gómez, 1961; Mapes, 1959; Pineda et al., 1969; Zamora et al., 1975). Among the more important iron deposits that are known in this region are those of Peña Colorada, in Colima; Pihuamo, in Jalisco; Las Truchas in Michoacán; and El Violín and Tiber in Guerrero. Also, deposits of copper such as those of Inguarán and La Verde in Michoacán exist in this belt.

Furthermore, the zone of major petroleum interest is in the coastal plain of the Gulf where petroleum has been extracted in fields located along the eastern edge of the Córdoba Platform in sedimentary rocks of the Cretaceous and where there exist good prospects in sediments deeper than the Upper Jurassic (González, 1976).

In considering possibilities of obtaining geothermal energy, the Mexican Neovolcanic axis constitutes the geologic province with the major manifestations and potentials in the country, owing to its contemporaneous igneous activity. The principal thermal manifestations are related to acid igneous activity. Some of these are located in the areas of La Primavera, Jalisco; Ixtlán de los Hervores, Negritos, and Lago de Cuitzeo, Michoacán; Los Húmeros, Puebla; and San Bártolo de los Baños, Queretaro (see Figure 3.13 and Table 3.2).

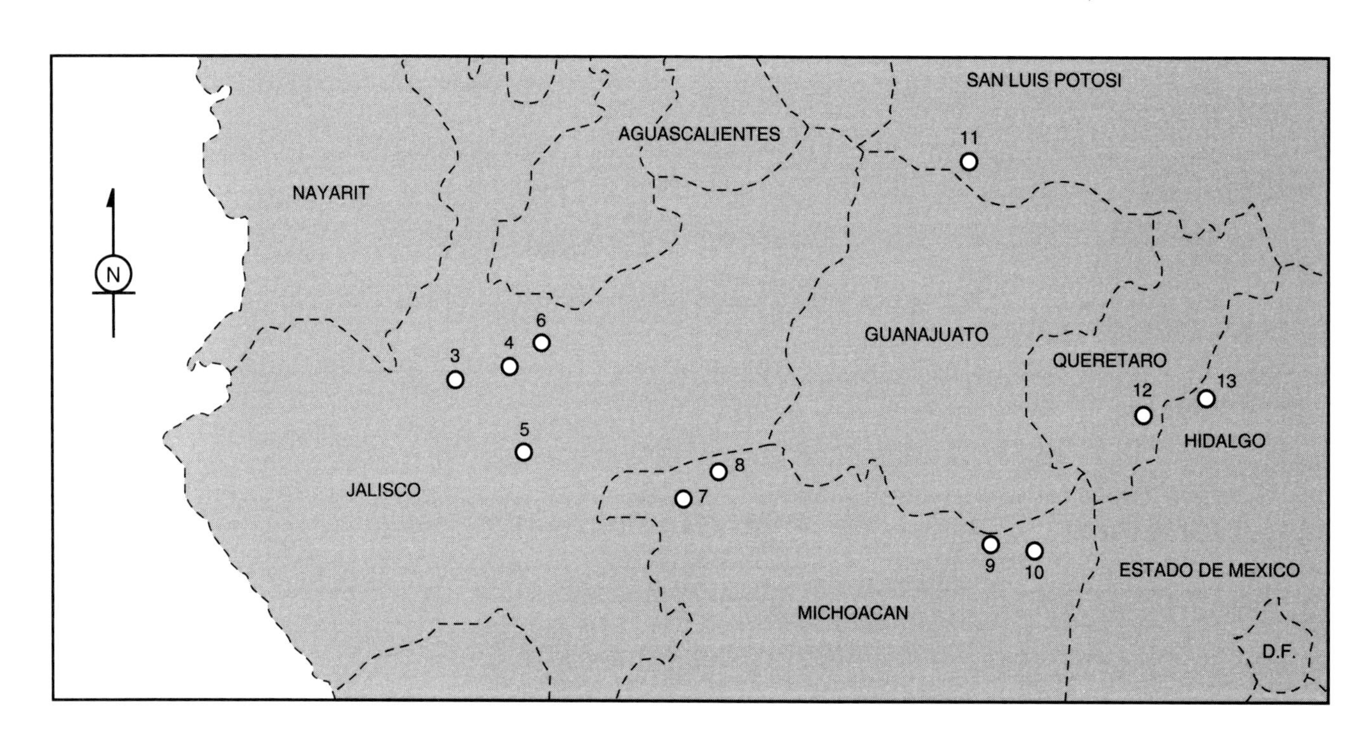

1. CERRO PRIETO, B.C.N.
2. DESIERTO DE ALTAR, SON.
3. HERVORES DE LA VEGA, JAL.
4. LA PRIMAVERA, JAL.
5. SAN MARCOS, JAL.
6. LA SOLEDAD, JAL.
7. LOS NEGRITOS, MICH.
8. IXTLAN DE LOS HERVORES, MICH.
9. LAGO DE CUITZEO Y ARARO, MICH.
10. LOS AZUFRES, MICH.
11. EL GOGORRON, S.L.P.
12. SAN BARTOLO, QRO.
13. PATHE, HGO.
14. EL CHICHONAL, CHIS.
15. TOLIMAN, CHIS.
16. LOS HUMEROS, PUE.

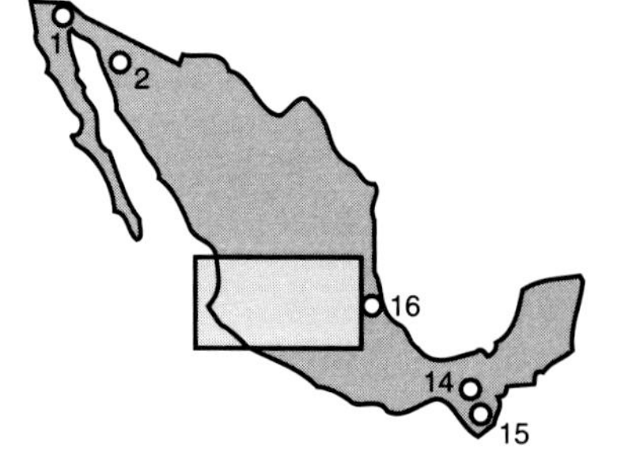

Figure 3.13. Location of the most important geothermal fields in the Republic of Mexico.

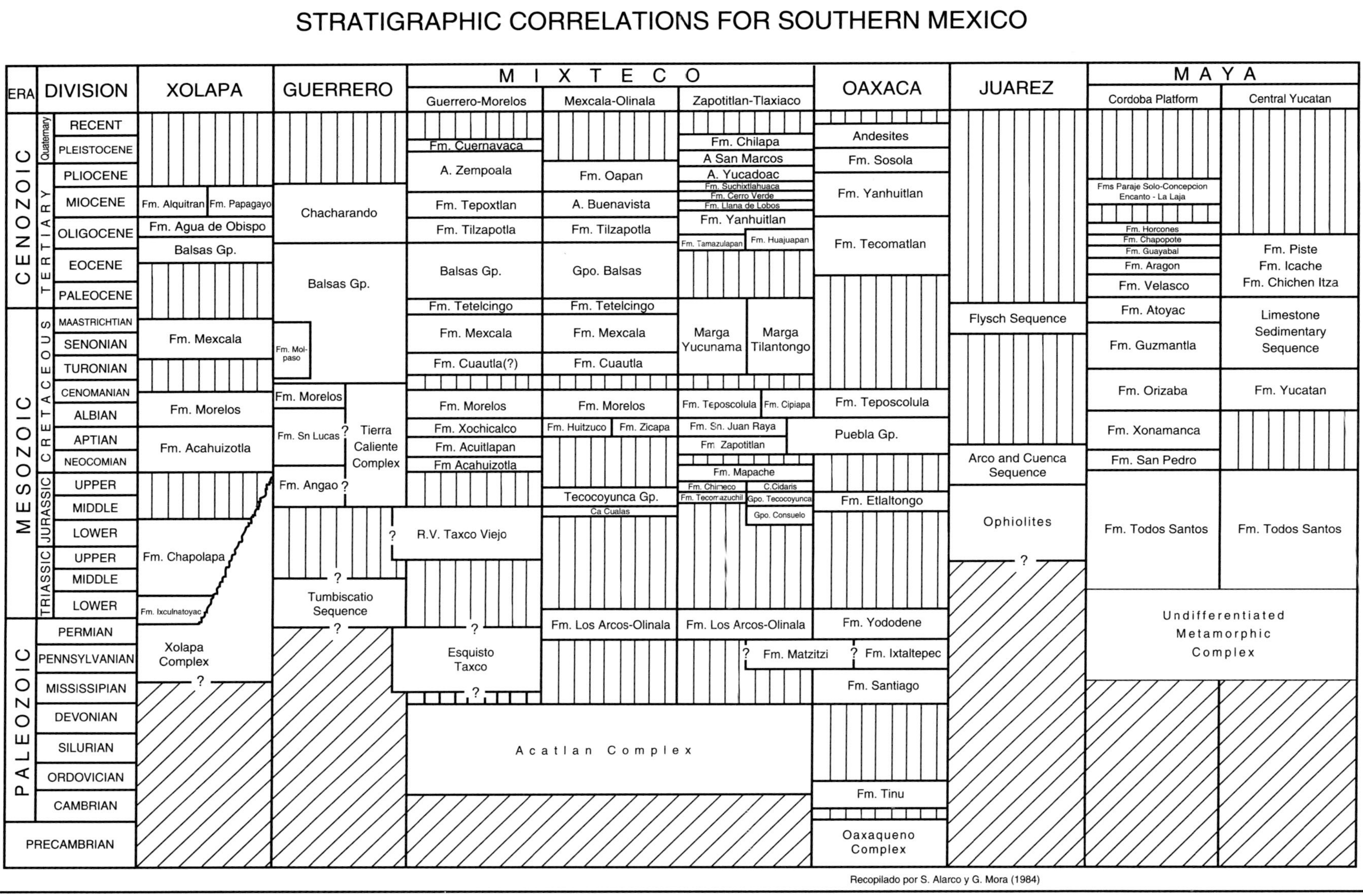

STRATIGRAPHIC CORRELATIONS FOR SOUTHERN MEXICO

ERA		DIVISION	XOLAPA	GUERRERO	MIXTECO: Guerrero-Morelos	MIXTECO: Mexcala-Olinala	MIXTECO: Zapotitlan-Tlaxiaco	OAXACA	JUAREZ	MAYA: Cordoba Platform	MAYA: Central Yucatan
CENOZOIC	Quaternary	RECENT									
CENOZOIC	Quaternary	PLEISTOCENE			Fm. Cuernavaca		Fm. Chilapa	Andesites			
CENOZOIC	TERTIARY	PLIOCENE			A. Zempoala	Fm. Oapan	A San Marcos A. Yucadoac	Fm. Sosola			
CENOZOIC	TERTIARY	MIOCENE	Fm. Alquitran Fm. Papagayo	Chacharando	Fm. Tepoxtlan	A. Buenavista	Fm. Suchixtlahuaca Fm. Cerro Verde Fm. Llana de Lobos	Fm. Yanhuitlan		Fms Paraje Solo-Concepcion Encanto - La Laja	
CENOZOIC	TERTIARY	OLIGOCENE	Fm. Agua de Obispo		Fm. Tilzapotla	Fm. Tilzapotla	Fm. Yanhuitlan	Fm. Tecomatlan		Fm. Horcones Fm. Chapopote Fm. Guayabal	
CENOZOIC	TERTIARY	EOCENE	Balsas Gp.	Balsas Gp.	Balsas Gp.	Gpo. Balsas	Fm. Tamazulapan Fm. Huajuapan			Fm. Aragon	Fm. Piste Fm. Icache Fm. Chichen Itza
CENOZOIC	TERTIARY	PALEOCENE			Fm. Tetelcingo	Fm. Tetelcingo				Fm. Velasco	
MESOZOIC	CRETACEOUS	MAASTRICHTIAN							Flysch Sequence	Fm. Atoyac	
MESOZOIC	CRETACEOUS	SENONIAN	Fm. Mexcala	Fm. Mol-paso	Fm. Mexcala	Fm. Mexcala	Marga Yucunama Marga Tilantongo			Fm. Guzmantla	Limestone Sedimentary Sequence
MESOZOIC	CRETACEOUS	TURONIAN			Fm. Cuautla(?)	Fm. Cuautla					
MESOZOIC	CRETACEOUS	CENOMANIAN		Fm. Morelos			Fm. Teposcolula Fm. Cipiapa	Fm. Teposcolula		Fm. Orizaba	Fm. Yucatan
MESOZOIC	CRETACEOUS	ALBIAN	Fm. Morelos		Fm. Morelos	Fm. Morelos	Fm. Sn. Juan Raya				
MESOZOIC	CRETACEOUS	APTIAN	Fm. Acahuizotla	Fm. Sn Lucas ? Tierra Caliente Complex	Fm. Xochicalco Fm. Acuitlapan	Fm. Huitzuco Fm. Zicapa	Fm. Zapotitlan	Puebla Gp.		Fm. Xonamanca	
MESOZOIC	CRETACEOUS	NEOCOMIAN			Fm Acahuizotla		Fm. Mapache		Arco and Cuenca Sequence	Fm. San Pedro	
MESOZOIC	JURASSIC	UPPER		Fm. Angao ?		Tecocoyunca Gp.	Fm. Chimeco Fm. Tecomazuchil C.Cidaris Gpo. Tecocoyunca	Fm. Etlaltongo			
MESOZOIC	JURASSIC	MIDDLE				Ca Cualas	Gpo. Consuelo				
MESOZOIC	JURASSIC	LOWER			? R.V. Taxco Viejo				Ophiolites	Fm. Todos Santos	Fm. Todos Santos
MESOZOIC	TRIASSIC	UPPER	Fm. Chapolapa						?		
MESOZOIC	TRIASSIC	MIDDLE		?							
MESOZOIC	TRIASSIC	LOWER	Fm. Ixculnatoyac	Tumbiscatio Sequence							
PALEOZOIC		PERMIAN		?	?	Fm. Los Arcos-Olinala	Fm. Los Arcos-Olinala	Fm. Yododene		Undifferentiated Metamorphic Complex	
PALEOZOIC		PENNSYLVANIAN	Xolapa Complex		Esquisto Taxco		? Fm. Matzitzi	? Fm. Ixtaltepec			
PALEOZOIC		MISSISSIPPIAN	?		?			Fm. Santiago			
PALEOZOIC		DEVONIAN									
PALEOZOIC		SILURIAN			Acatlan Complex						
PALEOZOIC		ORDOVICIAN									
PALEOZOIC		CAMBRIAN						Fm. Tinu			
PRECAMBRIAN								Oaxaqueno Complex			

Recopilado por S. Alarco y G. Mora (1984)

Table 3.2. Stratigraphic correlations for southern Mexico.

BIBLIOGRAPHY AND REFERENCES

Abbreviation UNAM is Universidad Nacional Autónoma de México

Bazán, B.S., 1981, Distribución y metalogénesis de la provincia uranífera del Mesozóico de México: Geomimet 3a. epoca Julio/Agosto, n. 112, p. 65–96.

Bazán, B.S., 1984, Litoestratigrafía y rasgos estructurales del Complejo Oaxaqueño, Mixteca Alta, Oaxaca: Geomimet, no. 129, p. 35–63.

Bloomfield, K., 1975, A late Quaternary monogenetic field in Central Mexico: Geologische Rundschau, v. 64, n. 2, p. 476–497.

Buitrón, B.S., 1970, Equinóides del Jurásico Superior y del Cretácico Inferior de Tlaxiaco, Oaxaca: Libro Guia de la Excursión México-Oaxaca de la Sociedad Geológica Mexicana, p. 154–163.

Bullard, E., 1969, El orígen de los océanos *in* J.T. Wilson, 1976, Deriva Continental y Tectónia de Placas. Selections of Scientific American, C. Martin-E. and A. González-U., translators, 2nd edition, Madrid: H. Blume editions, p. 98–109.

Calderón, A., 1956, Bosquejo geológico de la región de San Juan Raya, Puebla: XX Congreso Geológico Internacional, México. Libro Guia de la Excursion A-11, p. 9–33.

Campa, M.F., 1978, La evolución tectónica de Tierra Caliente. Bulletin Geological Society of Mexico, v. 39, n. 2.

Campa, M.F., and P.J. Coney, 1983, Tectono-stratigraphic terranes and mineral resource distributions in Mexico: Canadian Journal of Earth Science, v. 20, p. 1040–1051.

Campa, M.F., and J. Ramírez, 1979, La evolución geológica y la metalogénesis del noroccidente de Guerrero. Série técnico-científica de la Universidad Autónoma de Guerrero, n. 1, 102 p.

Campa, M.F., et al., 1977, La evolución tectónica y la mineralización en la región de Valle de Bravo, México, e Iguala, Gro: Asoc. Ing. Min. Met.Geol. Méx. Memoria de la XII Convención Nacional, p. 143–170.

Campa, M.F., J. Ramírez, R. Flores, and P. Coney, 1980, Conjuntos estratotectónicos del occidente de Guerrero y oriente de Michoacán: Resúmenes de la V Convención Geológica Nacional, México, D.F., p. 106–107.

Campa, M.F., et al., 1981, Terrenos tectono-estratigráficos de la Sierra Madre del Sur, región comprendida entre los estados de Guerrero, Michoacán, México, y. Morelos: Série técnico-científica de la Universidad Autónoma de Guerrero, n. 10, 28 p.

Campos, E., 1984, Estudio geológico regional del área de Valle de Bravo-Tzitzio, estados de México y Michoacán. Tesis Profesional Facultad de Ingenieria, UNAM.

Carfantan, J.C., 1983, Les ensembles géologiques du Mexique meridional. Evolution géodynamique durante le Mesozóique et le Cénozique: Geofísica Internacional v. 22, n. 1, p. 39–56.

Charleston, S., 1980, Stratigraphy and tectonics of the Rio Santo Domingo area, State of Oaxaca, Mexico: 26th Congrés Géologique International (Paris), Abstracts, v. 1, sections 1–5, 324 p.

Coney, P., 1983, Un modelo tectónico de México y sus relaciones con América del Norte, América del Sur y el Caribe: Revista, Instituto Mexicano del Petróleo, v. 15, n. l, p. 6–15.

Corona, J.R., 1981 (delayed until 1983), Estratigrafía de la región Olinalá-Tecocoyunca, noreste del Estado de Guerrero. Revista del Instituto de Geología, UNAM, v. 5, n. l, p. 17–24.

De Cserna, Z., 1965, Reconocimiento geológico en la Sierra Madre del Sur de México, entre Chilpancingo y Acapulco, Estado de Guerrero: Boletín del Instituto de Geología, UNAM, n. 62, 77 p.

De Cserna, Z., 1970, Reflexiones sobre algunas problemas de la geologia de la parte central-meridional de México. Libro Guia de la excursión México-Oaxaca de la Sociedad Geología Mexicana, p. 37–50.

De Cserna, Z., 1979, Cuadro tectónico de la sedimentación y magmatismo en algunas regiones de México durante el Mesozóico: Programas y Resúmenes del V Simposio sobre la Evolución Tectónica de México: Instituito de Geología, UNAM, p. 11–14.

De Cserna, Z., et al., 1962, Edades isotópicas de rocas metamórficas del centro y sur de Guerrero y de una monzonita cuarcífera del norte de Sinaloa: Boletín del Instituto de Geología, UNAM, n. 64, p. 71–84.

De Cserna, Z., et al., 1975, Edad Precámbrica Tardia del Esquisto Taxco, Estado de Guerrero: Boletín Asociación Mexicana de Geólogos Petroleros, v. 26, p. 183–193.

De Cserna, Z., et al., 1978a, Rocas metavolcánicas e intrusivos relacionados Paleozóicos de la región Petatlán, Estado de Guerrero: Revista, Instituto de Geología, UNAM, v. 2, n. 1, p. 1–7.

De Cserna, Z., et al., 1978b. Relaciones de facies de las rocas Cretácicas en el noroeste de Guerrero y en las áreas colindantes de México y Michoacán: Revista, Instituto de Geología, UNAM, v. 2, n. 1, p. 8–18.

De la Vega, E., 1983, Una nueva localidad Pérmica en México fechada con fusilinidos, porción meridional del Estado de Puebla: Resúmenes de la VII Convención Geológica Nacional Sociedad Geológica Mexicana, p. 51.

Demant, A., 1978. Características del Eje Neovolcánico Transmexicano y sus problemas de interpretación. Revista, Instituto de Geología, UNAM, v. 2, n. 2, p. 172–187.

Demant, A., and C. Robin, 1975, Las fases del volcanismo en México; una síntesis en relación con la evolución geodinámica desde el Cretácico: Revista, Instituto de Geología, UNAM, v. 75, n. 1, p. 70–83.

Dietz, R.S., and J.C. Holden, 1970, La disgregación de la Pangea, *in* J.T. Wilson, 1976, Deriva Continental y Tectonica de Placas. Selecciones de Scientific American, C. Martin–E. and A. Gonzalez-U., translators, 2nd edition, Madrid, H. Blume Ediciones, p. 154–167.

Erben, H.K., 1956, El Jurásico Medio y el Calloviana de México: XX Congreso Geológico Internacional, México: Contribution to the Congress by the Instituto de Geología de UNAM, 140 p.

Ferrusquía, I., 1970, Geología del área Tamazulapan-Teposcolula-Yanhuitlán. Mixteca Alta, Estado de Oaxaca: Libro Guia de la excursión México-Oaxaca de la Sociedad Geológica Mexicana, p. 97–119.

Ferrusquía, I., 1976, Estudios geológico-paleontológicos en la Región Mixteca, Part 1: Geología del área Tamazulapan-Teposcolula-Yanhuitlán, Mixteca Alta, Estado de Oaxaca, Mexico: Boletín del Instituto de Geología, UNAM, n. 97, 106 p.

Ferrusquía, I., S.P. Appelgate, and L. Espinosa. 1978, Rocas volcanosedimentarias Mesozóicas y huellas de dinosaurios en la región suroccidental Pacífica de México: Revista, Instituto de Geología, UNAM, v. 2, n. 2, p. 150–162.

Flores, L.A., and B.E. Buitrón, 1982, Revisión y aportes de la estratigrafia de la montaña de Guerrero: Serie técnico-científica de la Universidad Autónoma de Guerrero, n. 12, 28 p.

Flores, L.A., and B.E. Buitrón, 1984, Una nueva localidad del Paleozoico Superior de la región de la Mixteca Oaxaqueña: Resúmenes de la VII Convención Geológica Nacional Sociedad Geológica Mexicana, p. 207.

Fries, C., 1956, Bosquejo geológico de la región entre Mexico D.F. y Acapulco, Gro: Excursions A-9 and C-12, XX Congreso Geológico Internacional México, p. 7–53.

Fries, C., 1960, Geología del Estado de Morelos y de partes adyacentes de México y Guerrero, región central meridional de México: Boletín del Instituto de Geología, UNAM, n. 60, 236 p.

Fries, C., 1966, Hoja Cuernavaca 14 Q-h (8), Estado de Morelos, Carta Geológica de México: Instituto de Geología, UNAM, Serie 1/100,000 mapa con texto.

Fries, C., and O.C. Rincón, 1965, Nuevas aportaciones geocronológicas y técnicas empleadas en el Laboratorio de Gecronometria: Boletín del Instituto de Geología, UNAM, n. 73, p. 57–133.

Fries, C., et al., 1962, Rocas Precámbricas de edad Grenvilliana, de la parte central de Oaxaca en el sur de México. Boletín del Instituto de Geología, UNAM, n. 64, part 3, p. 45–53.

Fries, C., et al. 1970, Una edad radiométrica Ordovícica de Totoltepec, Estado de Puebla: Libro Guia de la excursión México-Oaxaca de la Sociedad Geológica Mexicana, p. 164–166.

Gastil, G.R., and W. Jensky, 1973, Evidence for strikeslip displacement beneath the Trans-Mexican volcanic belt: Stanford University, Publications in Geologic Science, v. 13, p. 171–180.

Gaytán, J.E., E. Garza-V de la Arévalo, and A. Rosas, 1979, Descubrimiento, geología, y génesis del Yacimiento Vulcano, Michoacán: Memória de la XIII Convención Nacional de la Asociación de Ingeníeros de Minas, Metalogénesis, y Geología de México, Acapulco, Gro., p. 58–113.

Gómez, D., 1961, Inventario de los yacimientos ferríferos de México, Consejo de Recursos Naturales no Renovables, Publication 3E.

González-Alvarado, J., 1976, Resultados obtenidos en la exploración de la Plataforma de Córdoba y principales campos productores: Boletín, Sociedad Geológica Mexicana. v. 37, n. 2, p. 53–60.

Guerrero, J.C., L.T. Silver, and T.H. Anderson, 1978, Estudios geocronológicos en el Complejo Xolapa: Boletín, Sociedad de Geología Mexicana, Resúmenes de la Convención Geológica Nacional, v. 39, p. 22–23.

Halpern, M., J.C. Guerrero, and M. Ruiz-Castellanos, 1974, Rb-Sr dates of igneous and metamorphic rocks from southeastern and central Mexico, a progress report: Unión Geofísica Mexicana. Reunión Anual, Resúmenes, p. 30–31.

Lopez-Ramos, E., 1979, Geología de México, 2nd edition, Mexico D.F., Scholastic Edition, 3 volumes.

Mapes, E., 1959, Los yacimientos ferríferos de Las Truchas, Michoacán: Consejo de Recursos no Renovables. Boletín 46.

Mauvois, 1977, Cabalgamiento Miocénico (?) en la parte centro-meridional de México. Revista, Instituto de Geología, UNAM, v. 1, n. 1, p. 48–63.

McDowell, F.W., and S.E. Clabaugh, 1979, Ignimbrites of the Sierra Madre Occidental and their relation to the tectonic history of western Mexico, *in* C.E. Chapin and W.E. Elston, eds., Ash-Flow Tuffs, Geological Society of America Special Paper 180.

Mooser, F., 1972, El Eje Neovolcánico Mexicano, debilidad cortical pre-Paleozóica reactivada en el Terciario: Memória de la II Convención Nacional de la Sociedad Geológica Mexicana, Mazatlán, Sinaloa, p. 186–187.

Mooser, F., 1975, Historia geológica de la Cuenca de México: Memoria de las Obras del Sistema de Drenaje Profundo del Distrito Federal, v. 1, DDF, p. 7–38.

Mooser, F., A.E. Nairn, and J.F. Negendank, 1974, Paleomagnetic investigations of the Tertiary and Quaternary igneous rocks, VII, a paleomagnetic and petrologic study of volcanics of the Valley of Mexico: Geologische Rundschau, v. 63, n. 2, p. 452–483.

Mulleried, F.K.G., 1957, La Geología de Chiapas: Publicación del gobierno del Estado de Chiapas.

Negendank, J.F.W., 1972, Volcanics of the Valley of Mexico: Neues Jahrbuch Mineralogische Abhandlungen, v. 116, p. 308–320.

Ortega-Gutiérrez, F., 1974, Nota preliminar sobre las eclogitas de Acatlán, Puebla: Boletin, Sociedad Geológica Mexicana, v. 35, p. 1–6.

Ortega-Gutiérrez, F., 1976, Los complejos metamórficos del sur de México y su significado tectónico: Resúmenes del III Congreso Latinoamericano de Geología, México.

Ortega-Gutiérrez, F., 1978, Estratigrafia del Complejo Acatlán en la Mixteca Baja, estados de Puebla y Oaxaca: Revista, Instituto de Geología, UNAM v. 2, n. 2, p. 112–131.

Ortega-Gutiérrez, F., 1979, La evolución tectónica pre-Misisípica del sur de México: V Simposio Evolución Tectónica de México, Programas y Resúmenes, p. 27–29.

Ortega-Gutiérrez, F., 1981, Metamorphic belts of southern Mexico and their tectonic significance: Geofísica Internacional, v. 20, n. 3, p. 177–202.

Pantoja-Alor, J., 1959, Estudio geológico de reconocimiento de la región de Huetamo, Estado de Michoacán: Consejo Recursos Naturales no Renovables, Boletín 50, 36 p.

Pantoja-Alor, J., 1970, Rocas sedimentárias Paleozóicas de la región centro-septentrional de Oaxaca, México. Libro Guia de la Excursión México-Oaxaca de la Sociedad Geológica Mexicana, p. 67–84.

Pantoja-Alor, J., and R.A. Robison, 1967, Paleozoic sedimentary rocks in Oaxaca, Mexico: Science, v. 157, p. 1033–1035.

Pérez, J.M., A. Hokuto, and Z. de Cserna, 1965, Reconocimiento geológico del área de Petlalcingo-Santa Cruz, municipio de Acatlán, Estado de Puebla: Instituto de Geología, Paleontología Mexicana, UNAM, n. 21, part 1, 22 p.

Pineda, A., H. López, and A. Peña, 1969, Estudio geológico-magnetométrico de los yacimientos ferríferos de Peña Colorada, municipio de Minatitlán, Colima: Consejo de Recursos Naturales no Renovables, Boletín 77.

Ramírez, J., 1984, La acreción de los terrenos Mixteca y Oaxaca durante el Cretácico Inferior, Sierra Madre del Sur: Resúmenes de la VII Convención Geológica Nacional Sociedad Geológica Mexicana, p. 59.

Rodríguez, R., 1970, Geología metamórfica del áreas de Acatlán, Estado de Puebla: Libro Guia de la Excursión México-Oaxaca de la Sociedad Geológica Mexicana, p. 51–66.

Salas, G.P., 1949, Bosquejo geológico de la cuenca sedimentaria de Oaxaca: Boletín Asociación Mexicanos Geólogos Petroleros, v. 1, p. 79–156.

Urrutia-Fucugauchi, J., 1980, Paleomagnetic studies of Mexican rocks: Ph.D. Dissertation, University Newcastle upon Tyne, England, 689 p.

Urrutia-Fucugauchi, J., ed., 1981, Paleomagnetism and tectonics of Middle America and adjacent regions, Part 1: Geofísica Internacional, v. 20, n. 3, p. 139–270.

Urrutia-Fucugauchi, J., ed., 1983a, Paleomagnetism and tectonics of Middle America and adjacent regions, Part 2: Geofísica Internacional, v. 22, p. 87–110.

Urrutia-Fucugauchi, J., 1983b, On the tectonic evolution of Mexico: Paleomagnetic constraints: American Geophysical Union, Geodynamics Series, v. 12, p. 29–47.

Urrutia-Fucugauchi, J., and L. del Castillo, 1977, Un modelo del Eje Volcánico Mexicano: Boletín de la Sociedad Geológica Mexicana, v. 38, p. 18–28.

Urrutia-Fucugauchi, J., and D. J. Morán-Zenteno, 1984, Resultados preliminares paleomagnéticos para el sur de México y sus implicaciones tectónicas: Resúmenes de la VII Convención Geológica Nacional de Sociedad Geológica Mexicana, p. 5.

Urrutia-Fucugauchi, J., and D.A. Valencio, 1986, Paleomagnetic study of Mesozoic rocks from Ixtapán de la Sal, México: Geofísica Internacional, v. 25, p. 485–502.

Vidal, R., et al., 1980, El Conjunto petrotectónico de Zihuatanejo, Guerrero-Coalcomán, Michoacán. Sociedad Geológica Mexicana: Resúmenes de la V Convención Geológica Nacional, p. 111–112.

Viniegra-Osorio, F., 1965, Geologia del Macizo de Teziutlán y la Cuenca Cenozoica de Veracruz: Boletin Asociación Mexicana de Geólogos Petroleros, v. 17, p. 103–163.

Whittington, H.B., and C.P. Hughes, 1974, Geography and faunal provinces in the Tremadoc Epoch, *in* C.A. Ross, ed., Paleogeographic Provinces and Provinciality: SEPM Special Publication 21, p. 203–218.

Zamora, S., et al., 1975, Los yacimientos en el Cerro del Violín, municipio de Mochitlán, Gro, Geomimet n. 78.

4. Geology of the Southeastern Region of Mexico

GENERAL CONSIDERATIONS

For the description of the region of southeastern Mexico, the following limits have been selected: on the west, the Isthmus of Tehuantepec; to the north, the shores of the Gulf of Mexico; and to the south, the Pacific coast. The region includes the physiographic provinces of the Chiapas Mountains, the Central American Cordillera, the Yucatán Peninsula and the eastern extreme of the coastal plain of the southern Gulf (see Figure 1.1).

The climate of this region varies from temperate and semi-arid in the high parts of the Sierra de Soconusco and Sierra de Chiapas to hot in the coastal plain of the Gulf and Pacific as well as in the central depression of Chiapas. In this last area the climates are subhumid and different from the coastal plains where they are generally humid. In the Yucatán Peninsula the climate is typically hot and subhumid. In all places in southeastern Mexico the rainy season is in the summer except in some areas of the Gulf Coastal Plain where rains occur all year.

CHIAPAS AND TABASCO

In the region that includes the states of Chiapas and Tabasco, a great sequence of Mesozoic and Cenozoic rocks crops out. It consists principally of marine sedimentary rocks that are folded and faulted. This sequence rests discordantly on a crystalline basement of Precambrian and Paleozoic rocks that crops out to the southwest of the region, where the crystalline rocks of these Eras form an outlying metamorphic and plutonic complex constituting the nucleus of the Sierra de Soconusco.

Mulleried (1957) considers that a large part of the Sierra de Soconusco is formed by Precambrian igneous and metamorphic rocks. However, the majority of radiometric ages obtained from intrusive rock samples reveal Paleozoic dates for the principal events of igneous intrusion. Castro and co-workers (1975) report an age of 242 ± 9 million years for a diorite (analysis of biotite by K/Ar method) that forms part of the batholithic complex of the Sierra de Soconusco and that was discovered in the base of a section located at the borders of the states of Oaxaca and Chiapas.

Damon and co-workers (1981) report dates from 17 samples from eight areas of the batholithic complex that were studied by the K/Ar and Rb/Sr methods. After analyzing ten samples of the complex, these authors recognized an isochron of apparent age at 256 ± 10 million years, which indicates that these intrusions originated from the same Permian magma that was isotopically homogeneous and perhaps derived from the mantle. These authors mention unpublished dates for the eastern part of the Sierra Madre del Sur in Chiapas that indicate Carboniferous plutonic activity in this area. Furthermore, they consider that the batholithic emplacement in Chiapas would have been associated with the closing of the Proto-Atlantic ocean at the end of the Paleozoic, at the time of the Appalachian Orogeny.

Carfantan (1977) believes that the emplacement of the batholith should have occurred in the Appalachian phase of deformation and considers that the metamorphic rocks affected by this intrusion ought to have originated in the Grenville phase of the Precambrian. This is in accord with radiometric dates from gneiss samples in Chiapas as well as from various samples from the base of the complex in Oaxaca that have been correlated with these rocks (De Cserna, 1967, 1971).

In the extreme southeast of the Sierra de Soconusco, a sedimentary sequence of late Paleozoic age crops out that has been recognized in the area of Chicomuselo (Hernández-García, 1973) and that extends toward Guatemala. Strata in the base of this sequence comprise the Santa Rosa Formation and form a lower member consisting of a sequence of slates with some metaquartzite intercalations. The upper member is formed of slates, sandstone, and some beds of fossiliferous limestone. The formation appears to be partially metamorphosed and has been assigned an age of Mississippian–Pennsylvanian based on the reported fossils (Hernández-García, 1973).

A sequence of shales and limestones of the Grupera Formation that contains Lower Permian fusulinids rests unconformably on the Santa Rosa Formation (Gutiérrez, 1956). The Vainilla Limestone overlies this formation. It contains crinoids, brachiopods, and various fusulinid species and is covered discordantly by the Paso Hondo Formation. This latter unit is composed of massive limestones with fusulinids of the Middle Permian and basal Upper Permian (Gutiérrez, 1956). In a large part of the northeastern edge of the Sierra de Soconusco, an important continental sequence is exposed that consists of red conglomerates, sandstones, silts, and clay. These outcrops reach to the Isthmus of Tehuantepec and even to the eastern edge of the southern sector of the Sierra Madre Oriental. This sequence has been named the Todos Santos Formation and constitutes the base of the Mesozoic, which crops out chiefly in Chiapas. Most authors have assigned this formation to a stratigraphic interval that varies from Triassic to Jurassic (Mulleried, 1957; Gutiérrez, 1956; Castro et al., 1975; López-Ramos, 1979).

In central Chiapas, above the Todos Santos Formation, an Upper Jurassic sedimentary marine sequence occurs. This is formed by limestones of shallow water facies with some intercalated continental beds. The Tithonian sediments indicate an open platform environment with a pelagic fauna over the whole area where the states of Chiapas, Oaxaca, and Veracrúz converge, but to the southeast of Chiapas the facies become more sandy (Castro et al., 1975).

Viniegra-Osorio (1981) has interpreted the existence of a saline Oxfordian basin that occupied a major part of the present Sierra de Chiapas, the coastal plain of the southern Gulf, and the continental platform of Tabasco (see Figure 4.1). These saline deposits played a very important role in the deformation of the later Mesozoic sequence and in the development of petroleum traps. At present these bodies of salt form two great uplifts, which Viniegra (1981) termed the Campeche dome and the Jalpa dome (see Figure 4.1). In the petroleum areas of Tabasco and Campeche, Pemex has drilled Upper Jurassic sequences, principally of platform facies, and has obtained petroleum production from them (Figure 4.2).

Above the Upper Jurassic sediments there rests a Neocomian sequence that gives evidence of the existence of marginal marine and continental deposits in northwestern Chiapas and eastern Veracrúz.

In the Yucatán Peninsula and a large part of the State of Chiapas, a major calcareous bank was created by the marine transgression initiated in the Cretaceous. This resulted in the deposition of carbonate and anhydrite in these regions as well as sedimentation of slope deposits in a belt that bordered the Great Calcareous Bank (Viniegra-Osorio, 1981). This belt is located in the subsurface of the eastern half of the State of Tabasco and in parts of northeastern Chiapas and the marine platform of Campeche where these sediment types are important producers of hydrocarbons (see Figure 4.3).

In the area of Cintalapa, the Neocomian sequence has been termed the San Ricardo Formation (Richard, 1963) and is composed of sandy shales, red sandstones, intercalations of limestone and dolomite, and some horizons of gypsum. The Barremian–Aptian interval seems to be absent in the immediate vicinity of Sierra de Soconusco because rocks of this age have not been identified; there is a resulting discordance between the lower Neocomian units and the Albian–Cenomanian sequence. According to Castro et al. (1975), this discordance is accentuated toward the west with the disappearance of units corresponding to the Lower Cretaceous and Upper Jurassic, owing to probable erosion occurring at the end of the Aptian.

Limestones crop out extensively in a central belt in the State of Chiapas and reveal shallow water bank environments belonging to the Albian and Cenomanian Stages (Figure 4.4). During this time, seas transgressed numerous areas that had been eroded during the Barremian–Aptian. They extended to cover up the crystalline rocks of the Sierra de Soconusco. These Lower Cretaceous sequences that crop out across Chiapas disappear under the Tertiary deposits in the areas of Tabasco and Campeche but have been recognized in petroleum-producing wells. In the subsurface of this last-mentioned region, sequences reported by Petróleos Mexicano show slope facies that follow a persistent sedimentary pattern during Neocomian and Aptian and the continued existence of the Great Calcareous Bank of Yucatán during Albian and Cenomanian time.

In central Chiapas abundant calcareous sediments of the Upper Cretaceous are exposed. These show bank facies with peri-reefal limestones and rudist fragments (Castro et al., 1975) (Figure 4.5). In the

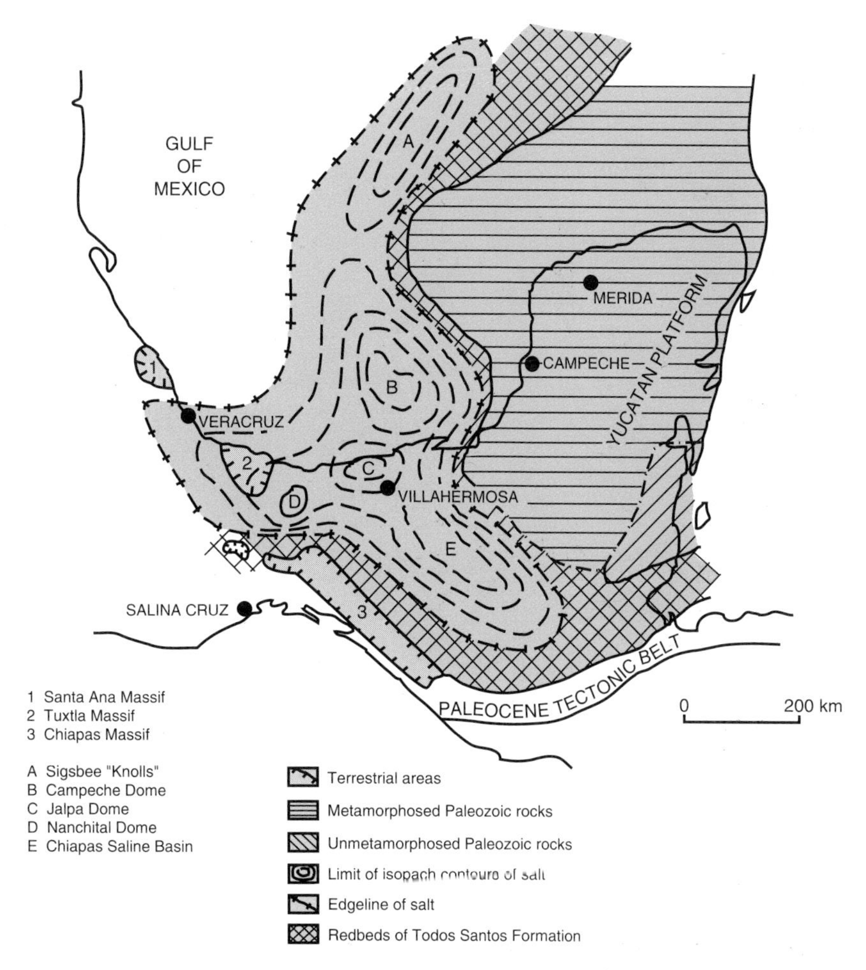

Figure 4.1. The large Saline Basin of Campeche during the Callovian–Oxfordian.

Reforma area the edges of the Great Calcareous Bank were exposed and eroded during the Upper Cretaceous because some Pemex wells in this area encounter Paleocene overlying Albian–Cenomanian sediments (Viniegra-Osorio, 1981). In offshore wells, recognition of the Upper Cretaceous has not been possible owing to dolomitization that has affected the Mesozoic sequence in this portion of the marine platform (Viniegra-Osorio, 1981).

During the Tertiary, in most of Chiapas and Tabasco, marine terrigenous sedimentation was initiated (Figure 4.6). These clastics are products of uplift of western Mexico and the folding of the Sierra Madre Oriental. At this same time, deposition of carbonates was continuing in the Yucatán Peninsula with the gradual emersion of its central part. Two basins of Tertiary age were developed in the Gulf of Mexico coastal plain (Comacalco and Macuspana). These are separated by a high formed by the "Villahermosa Horst," a result of normal faulting at the nose of the Chiapas anticlinorium. This anticlinorium is divided in sections by normal faults at the foot of the Sierra. The faulting has induced its subsidence into the Gulf Coastal Plain.

YUCATÁN PENINSULA

A Tertiary calcareous sequence crops out in a large part of Yucatán. The strata have no significant deformation and are horizontal. Both the Cretaceous sequence recognized in the subsurface and the Cenozoic sequence show no major structural perturbation and overlie a crystalline mass that has remained stable from the Paleozoic on.

The Cretaceous recognized in the Pemex wells is composed principally of anhydrites, limestones, dolomites, and intercalations of bentonites and some pyroclastic materials. Especially toward the base, the section consists of the Yucatán Evaporites (López-Ramos, 1979). All the Cretaceous sediments that have been encountered in the Pemex wells belong to the middle and upper parts of this system.

During the second half of the Cretaceous and a large part of the Cenozoic, the Yucatán Peninsula and its marine platform formed a calcareous bank. This was a marine high that extended to Chiapas and to the south of Veracrúz. A shelf margin developed that has been the principal petroleum objective in Tabasco and on the Campeche marine platform.

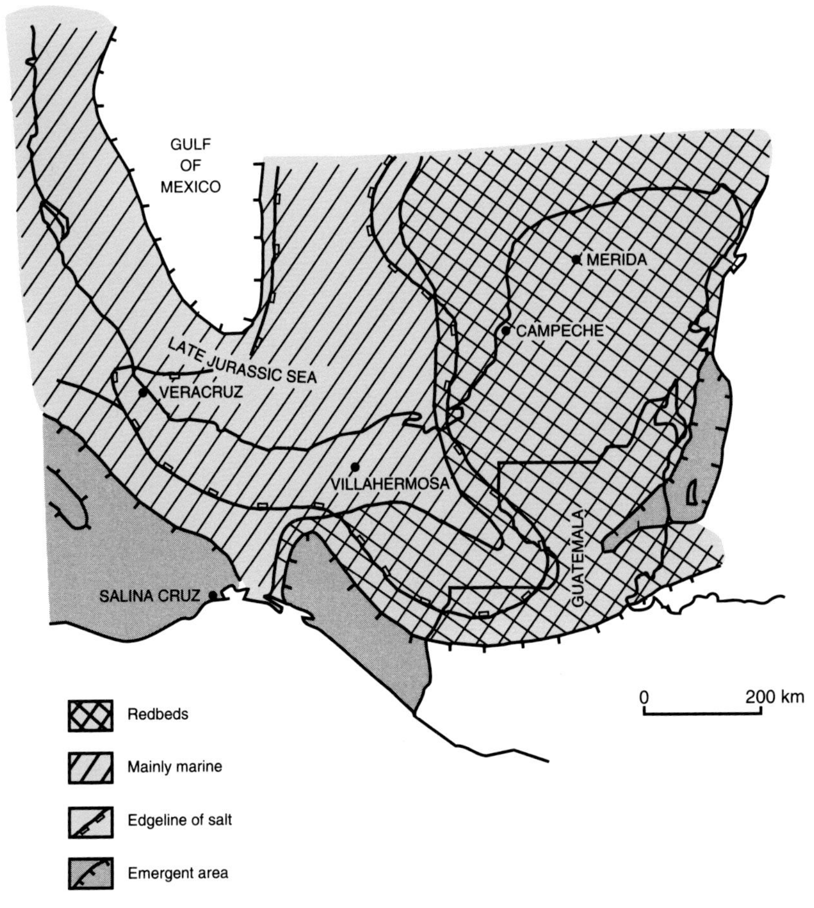

Figure 4.2. Map showing facies distribution in southeastern Mexico of Upper Jurassic facies.

Under this Cretaceous sequence, the wells Yucatán No. 1 and 4 cut through siltstones and sandstones with some intercalations of quartzose sand and gravel as well as green bentonite and dolomitic limestone. López-Ramos (1979) originally considered these as belonging to the Jurassic–Cretaceous interval.

These redbeds rest above a crystalline basement that was reached by the well Yucatán No. 1, at 3200 m depth (López-Ramos, 1979). From a sample of rhyolite porphyry from this well, a Rb/Sr date of 410 million years (Silurian) was obtained. This porphyry seems to have intruded a quartz and chlorite schist (López-Ramos, 1979). The Yucatán No. 4 well cut 8 m of slightly metamorphosed quartzite that underlies the Triassic–Jurassic redbeds (López-Ramos, 1979).

The Cenozoic deposits of the Yucatán Peninsula are represented principally by calcareous and dolomitic sequences with evaporite intercalations. Butterlin and Bonet (1963) formulated a column that extends from Paleocene to Quaternary. This column includes in ascending order: the Chichén Itzá and Icaiche formations (Paleocene–Eocene); the Bacalar, Estera, Franco, and Carillo Puerto formations (upper Miocene and Pliocene); and molluscan limestones of Pleistocene–Holocene. The Oligocene has not been recognized on the surface but was cut in the exploration wells of Chicxulub No. 1 and Cacapuc No. 1 (Butterlin and Bonet, 1963). The surface distribution of the Cenozoic units clearly shows a gradual retreat of the seas toward the present coast line, and only in the Eocene did the seas transgress and cover almost completely the Yucatán Peninsula (Butterlin and Bonet, 1963).

TECTONIC SUMMARY

The metamorphic rocks that crop out in the Sierra de Soconusco have been related to a metamorphic event contemporaneous with the Grenvillian deformation, which is well known in the eastern United States (Carfantan, 1977), and they have also been correlated with the metamorphic events that formed the Oaxaca Complex (Fries et al., 1962).

An important phase of deformation occurred at the end of the Paleozoic. This affected the Mississippian and Pennsylvanian sedimentary sequences of southeast Chiapas, and the chief plutonic activity began in the present-day Sierra de Soconusco. This phenomenon was followed by a prolonged interval of continental environments during which the lower beds of the Todos Santos Formation were deposited. Damon and co-workers (1981) relate the emplacement of the Sierra de Soconusco batholith to the closing of the Proto-Atlantic ocean and the unification of South

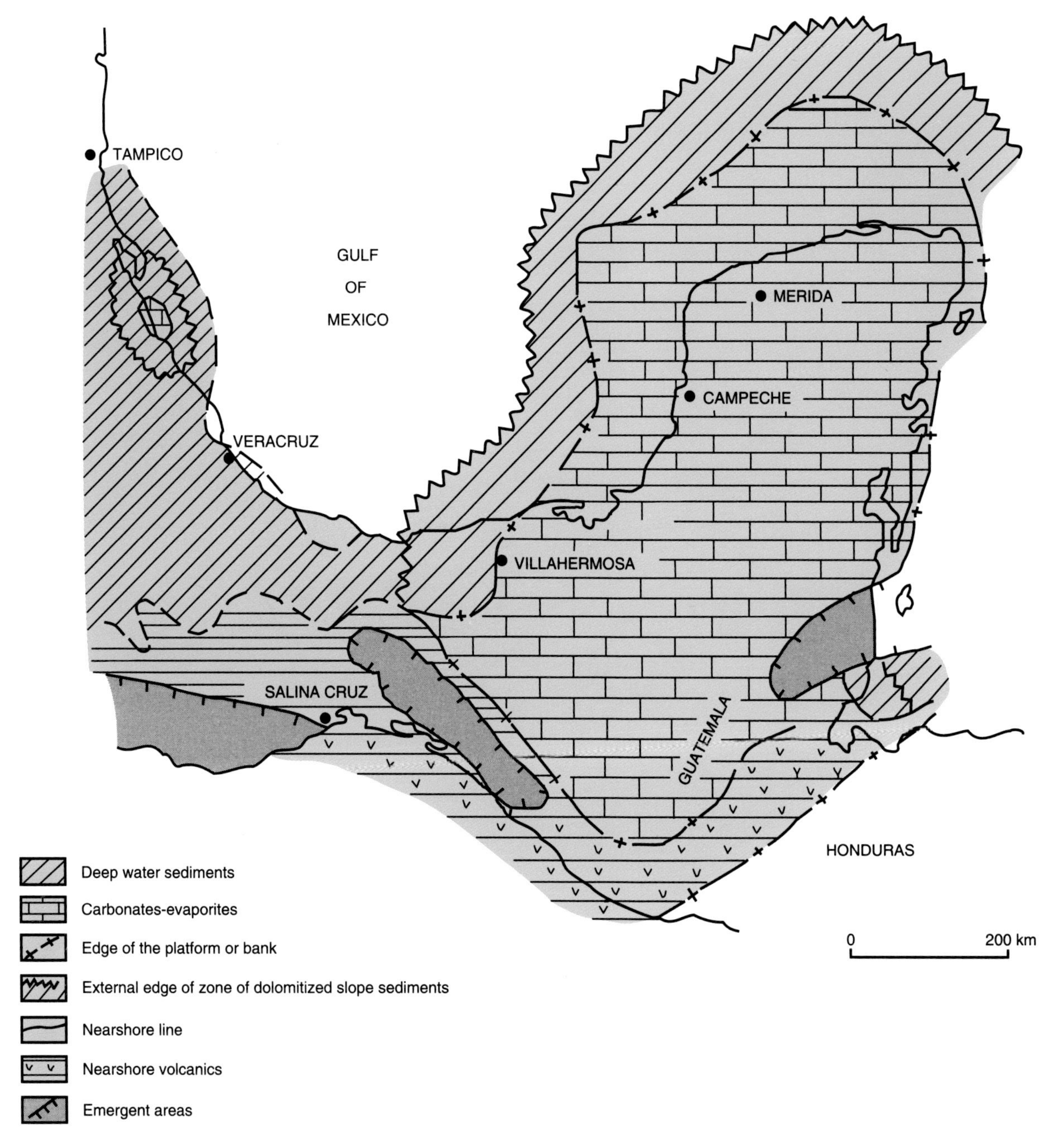

Figure 4.3. Map showing facies distribution of Neocomian–Aptian facies in southeastern Mexico.

America and Africa with North America, an action that culminated in the Appalachian Orogeny at the end of the Paleozoic. During the Late Jurassic a transgression occurred that gave rise to marine sedimentation, especially in the localities near the Gulf Coast in Tabasco and Veracrúz. In the Sierra Madre Oriental and other regions in the east of Mexico, a Jurassic transgression has been related to the opening of the western extreme of the Tethys (Tardy et al., 1975; Tardy, 1980; Campa and Ramírez, 1979) during the disintegration of Pangea.

In the Cretaceous there was general marine sedimentation that, in a major part of the state of Chiapas, is represented by the platform Sierra Madre Limestone. The area of the Yucatán Peninsula remained stable but submerged and had shallow water deposition, forming the Great Calcareous Bank that extended toward Chiapas and south of Veracrúz. Viniegra-Osorio (1981) believes that the Great Calcareous Bank of Yucatán tilted southwestward during its evolution. This interpretation is sustained by the fact that in the Pemex wells the basement was encountered at increasing depths from east to west across the marine platform of Campeche and finally reaches depths greater than 6500 m, with still greater thickness of the whole Mesozoic and Tertiary sequences. Dengo (1968) recognizes a partial deformation of the Mesozoic sequence at the end of the Albian

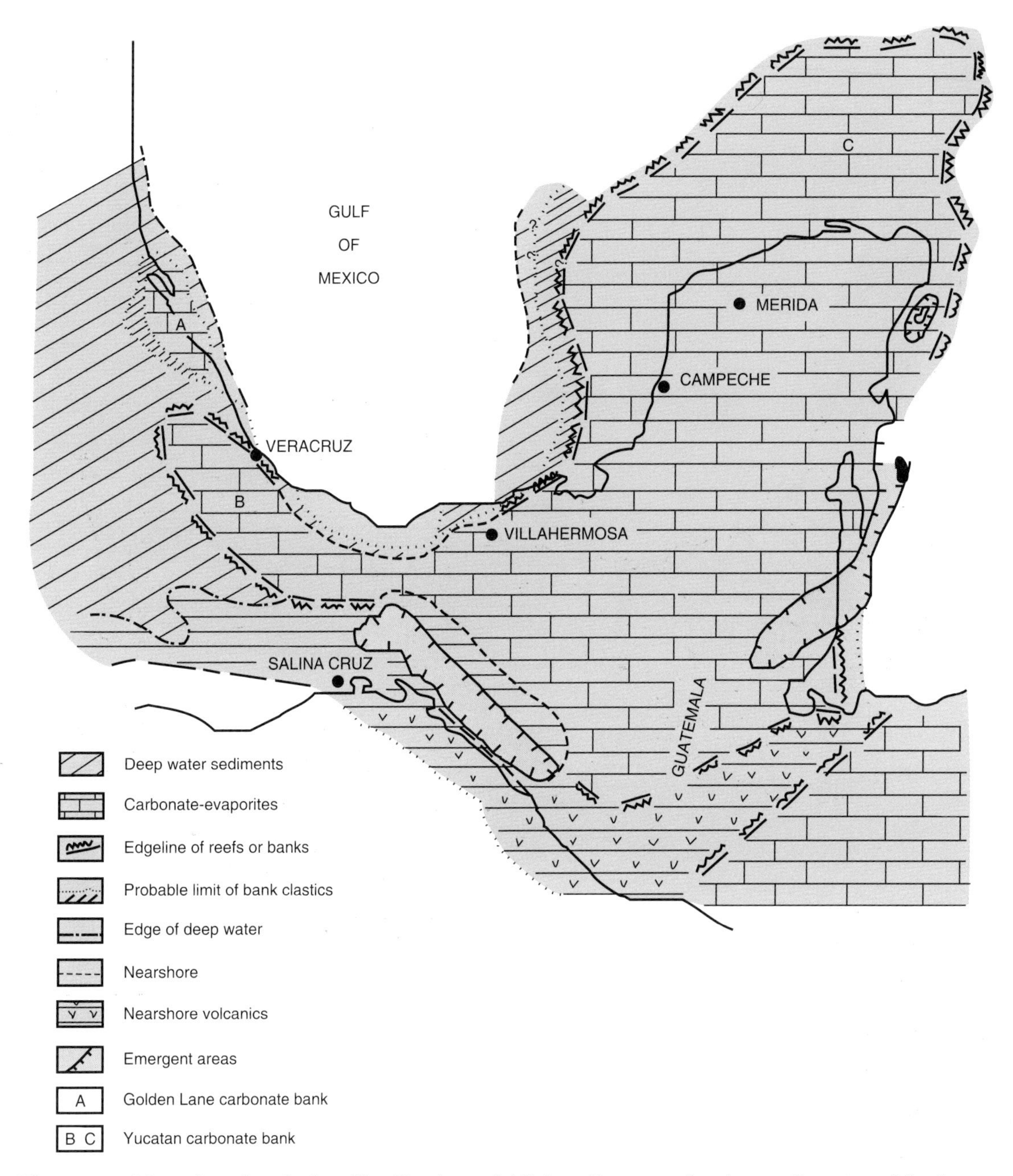

Figure 4.4. Map showing facies distribution of Albian–Cenomanian in southeastern Mexico.

that was accompanied by granitic intrusions that extend to the Sierra Madre del Sur in Chiapas and central Guatemala.

Carfantan (1977) mentioned a phase of Cenomanian deformation that placed a volcanic-plutonic complex of probable Mesozoic age in allochthonous position over the eroded Chiapas platform. This is located in the area of Motozintla. This complex corresponds to a volcanic arc similar to those recognized in the northwest and western parts of Mexico.

After this deformation, and during the Late Cretaceous, Paleocene, and Eocene, the Mesozoic sequence was affected by Laramide orogenic deformation. At this time an elongated marine basin was developed as a foredeep with flysch deposition of the Ocozocuautla Formation (Dengo, 1968).

Seemingly, the saline sediments at the base of the Mesozoic played a very important role in these deformations, since they served as plastic material during the development of the decollement in which time the Mesozoic and Cenozoic sequences were folded (Viniegra-Osorio, 1981). In the Reforma-Campeche belt, the origin of the domal and pillow-like structural system is related to vertical movement impelled by the subjacent salt.

During the Cenozoic, the Chiapas region was apparently caught in tectonism involving normal and strike-slip faulting, which complicates the structural relations of the Mesozoic and Cenozoic sequences.

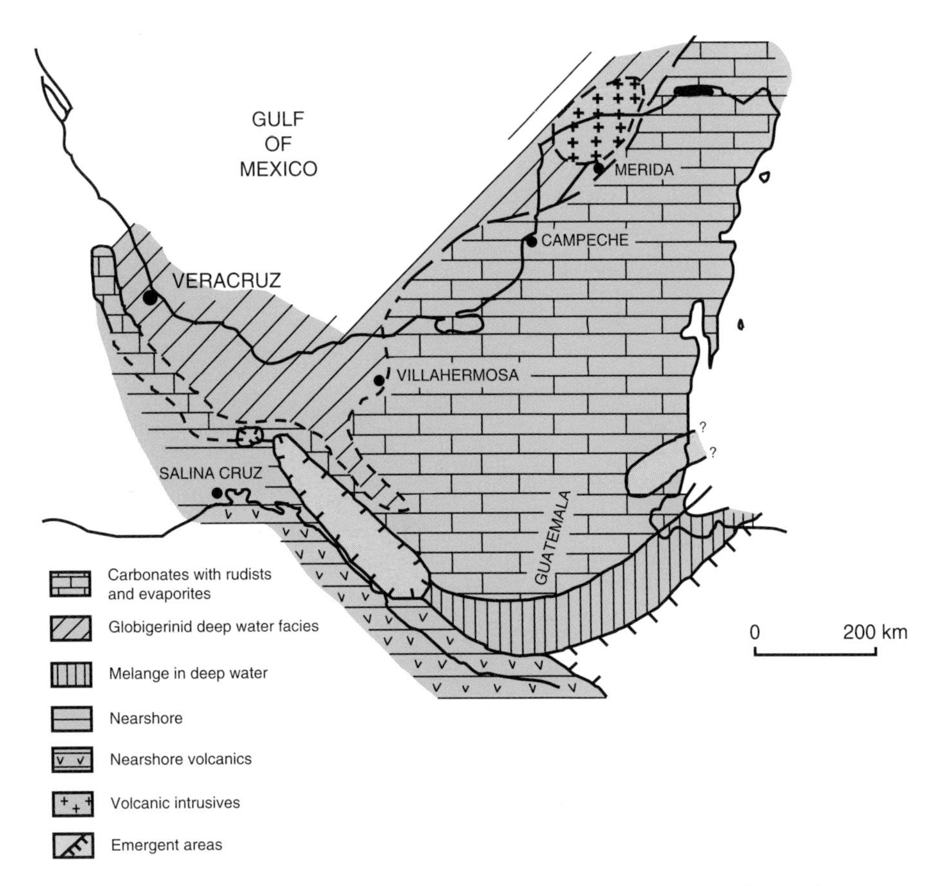

Figure 4.5. Map showing distribution of Upper Cretaceous facies in southeastern Mexico.

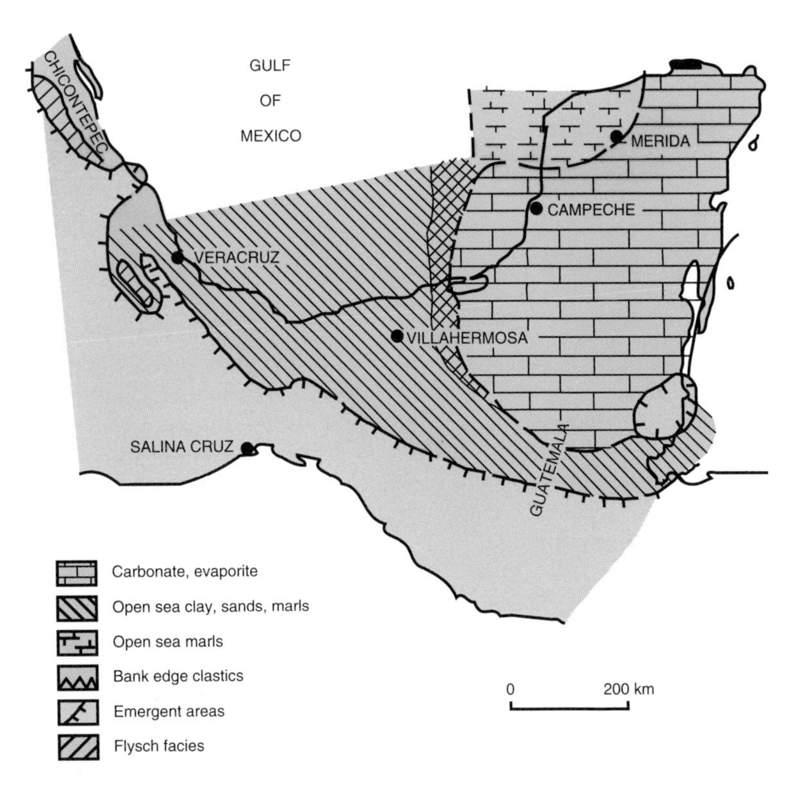

Figure 4.6. Map showing distribution of Paleocene facies in southeastern Mexico.

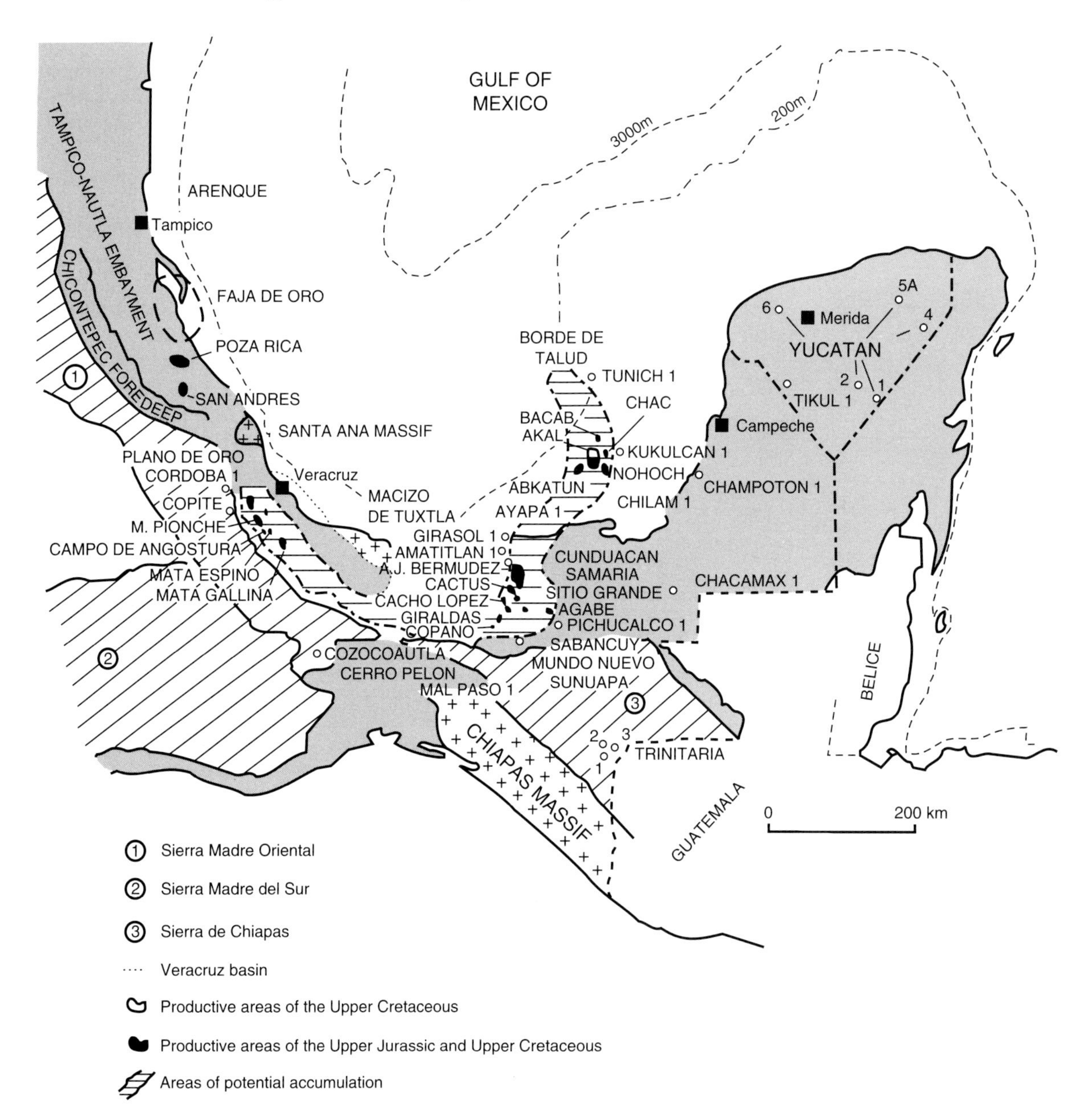

Figure 4.7. Map showing producing areas and oil wells in southeastern Mexico.

The directions of the faults of this period seem to be associated with the northwestern movement of North America in respect to the Caribbean plate along the Polochic-Montagua fault system of Guatemala and the southern border.

ECONOMIC RESOURCES

The principal petroleum reserves of the nation are located in the subsurface of the Reforma area at the Chiapas and Tabasco border, as well as on the marine Campeche platform (Figure 4.7). Most of the production comes from rocks of Late Jurassic and Cretaceous age as well as from the base of the Paleocene. In these areas the sequences have slope facies composed of fractured and dolomitized detrital material. The belt containing the slope where these sequences were deposited developed during the Cretaceous along the edge of the Great Calcareous Yucatán bank, which extends to Chiapas and Veracrúz.

It has long been considered that the source rocks of these hydrocarbons are Jurassic and that the reservoirs were developed in many varied traps resulting from a complex stratigraphic and structural evolution.

In the southeast region of Mexico, there exist some mineral deposits of known hydrothermal origin that show the association of silver-lead-zinc-gold-copper. In general they are small and localized generally in the south of Chiapas, in the localities of Pijijapan, Nueva Morelia, Lajeria, Payacal, and Almagres. In addition, metasomatic deposits of iron exist in the localities of Ventosa, Niltepec, and Fololapilla. Iron and copper are found in Arriaga and copper, lead, and zinc in Ixtapa.

BIBLIOGRAPHY AND REFERENCES

Abbreviation UNAM is Universidad Nacional Autónoma de México

Butterlin, J., and F. Bonet, 1963, Mapas geológicos de la Peninsula de Yucatán: Ingeniería Hidráulica en México.

Campa, M.F., and J. Ramírez, 1979, La evolución geológica y la metalogénesis del noroccidente de Guerrero: Serie técnico-científica de la Universidad Autónoma de Guerrero, n. 1, 102 p.

Carfantan, C.J., 1977, La cobijadura de Motozintla—un paleoarco volcánico en Chiapas: Revista, Instituto de Geología, UNAM, v. 1, n. 1, p. 133–137.

Castro, J., C.J. Schlaepfer, and E. Martínez, 1975, Estratigrafía y microfacies del Mesozoico de la Sierra Madre del Sur, Chiapas: Boletín Asociación Mexicana de Geólogos Petroleros, v. 27, n. 1–3, p. 1–103.

Damon, P.R., M. Shafiquillah, and K.F. Clark, 1981, Age trends of igneous activity in relation to metallogenesis in southern Cordillera, *in* W.R. Dickinson and D. Payne, eds.: Arizona Geological Society Digest, v. 14, p. 137–154.

De Cserna, Z., 1967 (1969), Tectonic framework of southern Mexico and its bearing on the problem of continental drift: Boletín de la Sociedad Geológica Mexicana, v. 30, p. 159–168.

De Cserna, Z., 1971, Precambrian sedimentaton, tectonics and magmatism in Mexico: Geologische Rundschau, v. 60, p. 1488–1513.

Dengo, G., 1968, Estructura geológica, história tectónica y morfología de América Central: Guatemala Instituto Centroamericano de Investigación y Tecnología Industrial: Centro Regional de Ayuda Técnica, Agencia para el Desarrollo Internacional, 45 p.

Fries, C., et al., 1962, Rocas Precámbricas de edad Grenvilliana de la parte central de Oaxaca en el sur de México: Boletín, Instituto de Geología, UNAM, n. 64, parte 3, p. 45–53.

Gutiérrez, R., 1956, Bosquejo geológico del Estado de Chiapas: XX Congreso Geológico Internacional, México: Excursion C-15, Geologia del Mesozóico y Estratgrafía Pérmica del Estado de Chiapas.

Hernández-Garcia, R., 1973, Paleogeografía del Paleozóico de Chiapas: Boletin, Asociación Mexicana Geólogos Petroleros, v. 25, p. 79–113.

López-Ramos, E., 1979, Geologia de México, 2nd edition: Scholastic Edition, v. III, 446 p.

Mulleried, F.K.G., 1957, La Geología de Chiapas: Publicación del gobierno del Estado de Chiapas.

Richard, H.G., 1963, Stratigraphy of Early Mesozoic sediments in southwest Mexico and western Guatemala: AAPG Bulletin, v. 47, p. 1861–1970.

Tardy, M., 1980, La transversal de Guatemala y las Sierra Madre de México, *in* J. Auboin, R. Brousse, J.P. Lehman, Tratado de Geologia, v. III, Tectónica, Tectonofísica, y Morfolgía. D. Serrat Translation, Barcelona, España: Editorial Omega, p. 117–182.

Tardy, M., et al., 1975, Observaciones generales sobre la estructura de la Sierra Madre Oriental. La aloctonia del conjunto cadena alta-altaplano central, entre Torreon, Coah. y San Luis Potosi, S.L.P., Mexico: Revista del Instituto de Geología, UNAM, v. 75, p. 1–11.

Viniegra-Osorio, F., 1981, El gran banco calcáreo yucateco: Revista Ingenieria n. 1, p. 20–44.

Section II
The Bibliography of Mexican Geology, 1983–1993

Introduction to the Bibliography of Mexican Geology, 1983–1993

Compiled by
James Lee Wilson
Luis Sanchez-Barreda
Della Moore Wilson

The date of the latest INEGI publication of the *Geology of the Mexican Republic* by Dante Morán Zenteno is 1984. For this reason, the translators believed that an up-to-date bibliography of the geology of Mexico could be very useful, and such is published with this translation. Many significant studies on Mexican geology have been reported in the literature in the last ten years and many people have contributed to the present compilation of references. A few important papers written prior to 1983, extending back as far as the 1970s, are also included. In all, there are about 1500 references. The list is reasonably comprehensive but not absolutely complete through 1993.

An unpublished bibliography by Norris W. Jones, who has worked for many years in northern Mexico with James McKee and Tom H. Anderson, was used as a nucleus for the reference list. A very comprehensive bibliography from the recently published tectonostratigraphic study of Mexico by R. L. Sedlock, R.C. Speed, and F. Gutierrez-Ortega (Geological Society of America Special Paper No. 278, 1993) was also used. The extensive references compiled by Amos Salvador and others in major papers of the DNAG Vol. J, "The Geology of the Gulf of Mexico," were also employed, as well as the research files of the translators, J.L. Wilson and Luis Sanchez-Barreda. Lastly, the bibliographies of the Gulf Coast Association of Geological Societies (volumes 5 and 6) derived from A.G.I. Georef files were checked, as well as the *Revista* of the Instituto Mexicano del Petróleo.

Assistance of Jon F. Blickwede (Amoco Production Co., Houston) in compiling some references is gratefully acknowledged. Della M. Wilson did most of the word processing.

The references are arranged alphabetically within four categories. Individual authors may appear in more than one category, but the references are not repeated and occur only in a single category.

Papers whose subject matter overlaps the categories should be searched for in more than a single category list.

The four categories are as follows:

Part 1. Stratigraphic and paleontologic studies, including reports on structural controls on sedimentation.

Part 2. Plate tectonic, paleomagnetic, and regional structural studies.

Part 3. Mineral resources, studies of igneous-metamorphic rocks, neovolcanic investigations, basement studies, and age dating.

Part 4. Specific studies of petroleum and gas, reports on structure and mapping of specific areas, engineering and environmental geology, hydrology, and remote sensing.

1. Stratigraphic and Paleontologic Studies, Structural Controls of Sedimentation

Abbou, P.L., and T.E. Smith, 1989, Sonora, Mexico, source for the Eocene Poway Conglomerate of southern California: Geology, v. 17, p. 329–332.

Adatte, T., W. Stinnesbeck, H. Hubberten, and J. Remane, 1992, Nuevos datos sobre el límite Jurásico-Cretácico en el noreste y en el centro este de México: XI Convención Geológica Nacional, Sociedad Geológica Mexicana, Libro de Resúmenes, p. 1–4.

Aguayo-Camargo, J.E., 1978, Sedimentary environments and diagenesis of a Cretaceous reef complex, eastern Mexico: Anales Centro Ciencias del Mar y Limnologia., Universidad Nacional Autónoma, México, 5, p. 83–140.

Aguayo-Camargo, J.E., 1982, Sedimentary environments and diagenesis of the Novillo Formation (Upper Jurassic), northeastern Mexico, *in* D.G. Bebout et al., eds., The Jurassic of the Gulf Rim: Annual Research Conference, Gulf Coast Section SEPM, Program and Abstracts, 3, p. 9–10.

Aguayo-Camargo, J.E., 1984, Estúdio de los sedimentos terrígenos de la Cuenca de Guaymas, Golfo de California, noroeste de México: Revista del Instituto Mexicano del Petróleo, v. 16, p. 5–19.

Aguayo-Camargo, J.E., and K. Kanamori, 1976, The Tamuín Member of the Méndez shale along the eastern flank of the Sierra de El Abra, San Luis Potosí, eastern Mexico: Boletín de la Sociedad Geológica Mexicana, v. 37, p. 11–17.

Aguayo-Camargo, J.E., et al., 1985, Tectonic evolution and carbonate sedimentary environments during the Mesozoic at Reforma-Jalpa area, southeast Mexico, *in* P.D. Crevello et al., eds., Deep-water Carbonates; Buildups, Turbidites, Debris Flows and Chalks: SEPM Core Workshop 6, p. 249–265.

Aguilar-Pina, M., 1992, Aspectos bioestratigráficos y paleoecológicos de los ostrácodos y moluscos de los pozos Xicalango 1010 y Pol 1 de la planicie costera del Golfo de México: XI Convención Geológica Nacional, Sociedad Geológica Mexicana, Libro de Resúmenes, p. 4.

Aguilera, J.G., 1986, Sinopsis de la geología Mexicana: Instituto de Geología de México Boletín, v. 4–6, p. 189–250.

Aguilera-F., N., and S. Franco-N., 1992, El genero *Saccocoma* Agassiz, 1836 y su importancia estratigráfica en el Jurasico Superior, en pozos profundos del sureste de México y zona marina de Campeche: XI Convención Geológica Nacional, Sociedad Geológica Mexicana, Libro de Resúmenes, p. 6.

Aguilera-F., N., H. Alzaga-R., J.L. Macias-V., and D. Zamudio-A., 1992, Bioestratigrafía de la Formación Apango (Turoniano-Santoniano) en el Estado de Guerrero, al sureste de México: XI Convención Geológica Nacional, Sociedad Geológica Mexicana, Libro de Resúmenes, p. 5.

Ahr, W.M., 1986, Mixed carbonate-clastic sedimentation on the western Yucatan Platform during late Pleistocene and early Holocene times: Geological Society of America Abstracts with Programs, v. 18, p. 523.

Akers, W.H., 1984, Planktic foraminifera and calcareous nannoplankton biostratigraphy of the Neogene of Mexico; Part II, Lower Pliocene: Tulane Studies in Geology and Paleontology, v. 18, p. 21–36.

Alemán, A., R. Flores, and C.M. Tejeda, 1992, Las interrelaciones de las fases tectónicas y la maduración térmica de la materia orgánica dispersa del Jurásico Inferior, en el area del anticlinorio de Huayacocotla-Rio Pantepec, Ver., Mex: XI Convención Geológica Nacional, Sociedad Geológica Mexicana, Libro de Resúmenes, p. 8.

Alencaster, G., 1984, Late Jurassic–Cretaceous molluscan paleogeography of the southern half of Mexico, *in* G.E.G. Westermann, ed., Jurassic–Cretaceous Biochronology and Palegeography of North America: Geological Association of Canada Special Paper 27, p. 77–88.

Alencaster, G., and F. Michaud, 1990, Rudistas (Bivalvia-*Hippuritacea*) del Cretácico Superior de la region de Tuxtla Gutiérrez, Chiapas, (México): Actas de la Facultad de Ciencias de la Universidad Autónoma de Nuevo León, Linares, v. 4, p. 175–194.

Alencaster, G., and J. Pantoja-Alor, 1992, *Amphitriscoelus,* genero de rudista (Bivalvia *Hippuritacea*) en el Cretácico Inferior de la region de Huetamo, Michoacán: XI Convención Geológica Nacional,

Sociedad Geológica Mexicana, Libro de Resúmenes, p. 9–10.

Alfonso-Zwanziger, J., 1978, Geología regional del sistema sedimentario Cupido: Asociación Mexicana de Geológos Petroleros Boletin, v. 30, p. 1–55.

Alfonso-Zwanziger, J., 1987, Paleogeografía Chihuahua-Coahuila: Ingenieros Petroleros Mexicanos, v. 27, p. 9–18.

Alfonso-Zwanziger, J., and J. García-Esparza, 1990, Regionalizaciones estratigráfico-sedimentológicas de Chihuahua, Parte I, Paleozóico and Parte II Mesozóico: X Convención Geológica Nacional, Sociedad Geológica Mexicana, Libro de Resúmenes, p. 72.

Allen, W.W., Jr., 1976, Petrology of the Middle Jurassic (?) La Joya Formation, Sierra Madre Oriental, southwestern Tamaulipas, Mexico: Master's Thesis, Texas A&M University, College Station, Texas.

Almazán-Vázquez, E., 1988, Marco paleosedimentário y geodinámico de la Formación Alisitos in la Península de Baja California: Revista del Instituto de Geología, Universidad Nacional Autónoma México, v. 7, p. 41–51.

Almazán-Vásquez, E., 1989, El Cámbrico–Ordovícico de Arivechi, en la región centro-oriental del Estado de Sonora: Revista del Instituto Geología, Universidad Nacional Autonóma de México, v. 8, p. 58–66.

Almazán-Vásquez, E., 1990a, Fauna Aptiano–Albiana del Cerro las Conchas, Sonora centro-oriental: Actas de la Facultad de Ciéncias de la Universidad Autónoma de Nuevo León, Linares, v. 4, p. 153–174.

Almazán-Vásquez, E., 1990b, Pre-Mesozoic sequences of the Sonora mountainous region and their tectonic significance: Geological Society of America Abstracts with Programs, v. 22, n. 3.

Almazán-Vásquez, E., and M.A. Fernández-Aguirre, 1987, Los terrenos Paleozóicos de la region Serrana de Sonora: Universidad Autonoma de Chihuahua Gaceta Geológica, v. 1, n. 1, p. 97–107.

Almazán-Vásquez, E., A.R., Palmer, and R.A. Robison, 1986, Constraints on the southern margin of North America during the early Paleozoic: Geological Society of America Abstracts with Programs, v. 18, p. 525.

Almázan-Vásquez, E., et al., 1987, Stratigraphic framework of Mesozoic strata in northern Sonora, Mexico: Geological Society of America Abstracts with Programs, v. 19, p. 570.

Alvarez, W., et al., 1980, Franciscan Complex limestone deposited at 17º south paleolatitude: Geological Society of America Bulletin, v. 91, p. 476–484.

Alzaga-Ruiz, H., 1991, Estratigrafía y consideraciones paleogeográficas de las rocas del Jurásico Tardío–Cretácico Temprano, en el Area de Tomellín-Santiago Nacaltepec, Oaxaca, México: Revista del Instituto Mexicano del Petróleo, v. 23, n. 2, p. 17–27.

Alzaga-Ruiz, H., and A. Pano-Arciniega, 1989, Orígen de la Formación Chivillas y preséncia del Jurásico Tardío en la región de Tehuacán, Puebla, México: Revista del Instituto Mexicano del Petróleo, v. 21, p. 5–15.

Alzaga-Ruiz, H., and O.D. Santamaría, 1988, La Formación Tecamalucan, Estado de Veracruz, México: Revista del Instituto Mexicano del Petróleo, 20, p. 5–12.

Amaya-Martínez, R., et al., 1992, Nuevos datos estratigráficos de la Formación Baucarit en Sonora centro oriental: XI Convención Geológica Nacional, Sociedad Geológica Mexicana, Libro de Resúmenes, p. 14.

Amaya-Martínez, R., M. González-Carlos, and J. Roldán-Quintana, 1992, Nuevas consideraciónes estratigráficas de la Formación Tarahumara en la porción centro-oriental de Sonora: XI Convención Geológica Nacional, Sociedad Geológica Mexicana, Libro de Resúmenes, p. 15.

Anderson, P.V., 1984, Prebatholithic strata of San Felipe: the recognition of Cordilleran miogeosynclinal deposits in northeastern Baja California, Mexico: Pacific Section, SEPM, San Diego meeting, p. 29–30.

Anderson, T.H., J.L. Eells, and L.T. Silver, 1979, Precambrian and Paleozoic rocks of the Caborca region, Sonora, Mexico, *in* Geology of Northern Sonora, T.H. Anderson and J. Roldán-Quintana, eds., Geological Society of America Field Trip Guidebook, p. 1–22.

Anderson, T.H., J.W. McKee, and N.W. Jones, 1990, Jurassic (?) melange in north-central Mexico: Geological Society of America Abstracts with Programs, v. 22, p. 3.

Applegate, S.P., 1986, The El Cien Formation; strata of Oligocene and early Miocene age in Baja California Sur: Universidad Nacional Autónoma de México, Instituto de Geología, Revista, v. 6, p. 145–162.

Applegate, S.P., 1988, A new genus and species of a holostean belonging to the family Ophiopsidae, *Teoichthys kallistos*, from the Cretaceous, near Tepexi de Rodríguez, Puebla: Universidad Nacional Autónoma de México, Instituto de Geología, Revista, v. 7, p. 200–205.

Aranda-García, M., 1988, La Isla de Jimulco: un paleoelemento Hauteriviano, en el Sector Transversal de la Sierra Madre Oriental, Coahuila (abs): IX Convención Nacional, México, D.F., Sociedad Geológica Mexicana, p. 30.

Aranda-García, M., M.E. Gómez-Luna, and B. Contreras-Montero, 1987, El Jurásico Superior (Kimeridgiano-Titoniano) en el área de Santa Maria del Oro, Durango, México: Revista de la Sociedad Mexicana Paleontologia, v. 1, p. 75–87.

Aranda-Garcia, M., O. Quintero, and E. Martínez-Hernández, 1988, Palinomorfos del Jurásico Temprano de la Formación Gran Tesoro Santa María del Oro, Durango: Revista del Instituto de Geología, Universidad Nacional Autónoma de México, v. 7, p. 112–115.

Araujo-Mendieta, J., and R. Arenas-Partida, 1986, Estúdio tectónico-sedimentário, en el Mar Mexicano, estados de Chihuahua y Durango:

Sociedad Geológica Mexicana Boletín, v. 47, p. 43–87.

Araujo-Mendieta, J., and R. Caesar-González, 1987, Estratigrafía y sedimentología del Jurásico Superior en la Cuenca de Chihuahua, norte de México: Revista del Instituto Mexicano del Petróleo, v. 19, p. 6–29.

Araujo-Mendieta, J., and C.F. Estavillo-González, 1987, Evolución tectónica sedimentaria del Jurásico Superior y Cretácico Inferior en el NE de Sonora, México: Revista del Instituto Mexicano del Petróleo, v. 19, p. 4–67.

Araujo-Mendieta, J., and C. Estavillo-González, 1992, Orígen, evolución sedimentaria, e implicaciones económicas de la Cuenca de Ojinaga, Chihuahua: XI Convención Geológica Nacional, Sociedad Geológica Mexicana, Libro de Resúmenes, p. 17.

Archila, M., M.A. Carballo, J.S Cruz, J.C. Franco, L.F. López, and R. Matias, 1990, Facies hidrocarburiferas del Cretácico Superior en la Cuenca Petén de Guatemala: Actas de la Facultad de Ciéncias de la Universidad Autónoma de Nuevo León, Linares, v. 4, p. 61–98.

Arenas-Partida, R., 1992, Normalización de la nomenclatura estratigráfica del Triásico–Jurásico en la región centro-norte de Mexico: XI Convención Geológica Nacional, Sociedad Geológica Mexicana, Libro de Resúmenes, p. 18.

Armin, R. A., 1987, Sedimentology and tectonic significance of Wolfcampian (Lower Permian) conglomerates in the Pedregosa basin: Southeastern Arizona, southwestern New Mexico, and northern Mexico: Geological Society of America Bulletin, v. 99, p. 42–65.

Armin, R. A., 1990, Ouachita orogeny in northern Mexico: Geological Society of America Abstracts with Programs, v. 22, p. 4.

Austin, J.A., Jr., S.D. Locker, and R.T. Buffler, 1982, Pre-Cretaceous stratigraphy; Gulf of Mexico and Bahamas, *in* Abstracts, Eos, American Geophysical Union Transactions, v. 63, p. 4.

Avila-Angulo, R., 1990, Lower Jurassic volcanic and sedimentary rocks of the Sierra Lopez, west-central Sonora, Mexico: Geological Society of America Abstracts with Programs, v. 22, p. 4.

Banks, N., and M.A. Carballo, 1987, Petroleum evaluation of Petén basin: Oil and Gas Journal, p. 58–72.

Bañuelos-Mendieta, O., 1993, Roca almacen del Mioceno y Plio-Pleistoceno en la Sonda de Campeche: Asociación Mexicana Geólogos Petroleros I Simposio de Geología de Subsuelo, Ciudad del Carmen, Resúmenes, p. 84–91.

Barattolo, F., 1987, Remarks on *Neomeris cretacea* Steinmann (Dasycladales) from the Cretaceous of Orizaba, Mexico, *in* Abstracts: 4th International Symposium on Fossil Algae, 4, p. 20.

Barceló-Duarte, J., 1986, La secuéncia progradacional Albiana en el noroeste de Coahuila: Revista Instituto de Geología, Universidad Nacional Autónoma Mexico. Also Ph.D. Dissertation, University of Texas at Austin.

Barceló-Duarte, J.M. Martínez-Medrano, G. Hernández-Reyes, and V. González-Pacheco, 1992, Aspectos diagenéticos de la secuéncia Cretácica en la Cuenca Guerrero-Morelos: XI Convención Geológica Nacional, Sociedad Geológica Mexicana, Libro de Resúmenes, p. 24.

Barker, C.E., 1991, Implications for organic maturation studies of evidence for a geologically rapid increase and stabilization of vitrinite reflectance at peak temperature (abs.): AAPG Bulletin, v. 75, p. 1852.

Barnes, D.A., 1982, Basin analysis of volcanic arc-derived, Jura-Cretaceous sedimentary rocks, Vizcaino Peninsula, Baja California Sur, Mexico: Unpublished Ph.D. Dissertation, University of California, Santa Barbara, 240 p.

Barnes, D.A., and J.M. Mattinson, 1981, Late Triassic-Early Cretaceous age of eugeosynclinal terranes, western Vizcaino Peninsula, Baja California Sur, Mexico: Geological Society of America Abstracts with Programs, v. 13, p. 43.

Barrera, P.G., and J. Pantoja-Alor, 1991, Equinóides del Albiano Tardío de la Formación Mal Paso, de la region de Chumbitaro, estados de Guerrero y Michoacán, México: Revista de la Sociedad Mexicana de Paleontología, v. 4, p. 23–41.

Bartling, W.A., and P.L. Abbott, 1983, Upper Cretaceous sedimentation and tectonics with reference to the Eocene, San Miguel Island, and San Diego area, Calfornia, *in* D.K. Larue and R.J. Steel, eds., Cenozoic Marine Sedimentation, Pacific Margin, USA, Los Angeles Section SEPM, p. 133–150.

Bartok, P., 1989, The origin of the eastern Gulf of Mexico and proto-Caribbean: Geological Society of America Abstract with Programs, v. 21, p. 3.

Bartolini, C., and C. González-León, 1989, Paleotectonic and depositional setting of the Early Cretaceous in Sonora, Mexico: Geological Society of America Abstracts with Programs, v. 21, p. A129.

Bartolini, C., and J.H. Stewart, 1990, Stratigraphy and structure of Paleozoic oceanic strata in Sierra El Aliso, central Sonora, Mexico: Geological Society of America Abstracts with Programs, v. 22, p. 6.

Bartolini, C., M. Morales, C. Spinosa, and S. Finney, 1990, Paleozoic off-shelf successions unconformably overlain by Triassic strata in central Sonora, Mexico: Evidence for a Permo-Triassic tectonic event?: Geological Society of America Abstracts with Programs, v. 22, p. A114.

Bartolini, C., M. Morales, and S. Finney, 1992, Chronostratigraphic correlation of Ordovician deep-marine, massive quartzarenite, central Sonora, Mexico: Geological Society of America Abstracts with Programs, v. 24, p. A359.

Bartolini, C., P.E. Damon, and M. Morales, 1992, Desarrollo de secuéncias volcánico-sedimentarias Terciarias durante la evolución estructural de la provincia de sierras y valles (Basin and Range) Sonora, México: XI Convención Geológica Nacional, Sociedad Geológica Mexicana, Libro de Resúmenes, p. 25.

Basáñez-Loyola, M.A., and A. Ruiz-Violante, 1992, Implicaciones eustáticas a nivel local y regional durante el Cretácico en las plataformas carbonatadas del centro de Mexico: XI Convención Geológica Nacional, Sociedad Geológica Mexicana, Libro de Resúmenes, p. 25.

Basáñez-Loyola, M.A., R. Fernández-T., and C. Rosales-D., 1993, Cretaceous platform of Valles-San Luis Potosi, northeast central Mexico, *in* A. Simo et al., eds., Atlas of Cretaceous Platforms, AAPG Memoir 56, p. 51–59.

Bello-Montoya, R., and J. Guardado-Cabrera, 1992, Estúdio estratigráfico-sedimentológico-diagenético de las rocas del Mesozóico en el area Gaucho, Chiapas: XI Convención Geológica Nacional, Sociedad Geológica Mexicana, Libro de Resúmenes, p. 33.

Bello-Montoya, R., et al., 1992, Estudio estratigráfico-sedimentológico de las rocas del Jurásico Superior en la zona marina de Campeche: XI Convención Geológica Nacional, Sociedad Geológica Mexicana, Libro de Resúmenes, p. 34.

Bello-Navarro, M.A., 1983, Integración e interpretación geológico-geofísica de la Cuenca de Veracruz: Revista del Instituto Mexicano del Petróleo, v. 15, p. 15–33.

Benavides-García, L., 1983, Domos salinos del sureste de México: Boletín de la Asociación de Geólogos Petroleros, v. 35, p. 9–35.

Bernabé-Martínez, Ma. G., and V.M. Dávila-Alcocer, 1992, Microfacies de la Formación Las Trancas en el anticlinal de Bonanza, Estado de Hidalgo: XI Convención Geológica Nacional, Sociedad Geológica Mexicana, Libro de Resúmenes, p. 35.

Bertagne, A.J., 1984, Seismic stratigraphy of Veracruz Tongue, deep southwestern Gulf of Mexico: AAPG Bulletin, v. 68, p. 1894–1907.

Bilodeau, W.L., 1986, The Mesozoic Mogollon highlands, Arizona: an Early Cretaceous rift shoulder: Journal of Geology, v. 94, p. 724–735.

Bilodeau, W.L., and S.B. Keith, 1986, Lower Jurassic Navajo-Aztec-equivalent sandstones in southern Arizona and their paleographic significance: AAPG Bulletin, v. 70, p. 690–701.

Bishop, W.F., 1980, Petroleum geology of northern Central America: Journal of Petroleum Geology, v. 3, p. 3–39.

Bitter, M., 1993, Sedimentación y petrología de las areniscas de la Formación Chicontepec con implicaciones petrolíferas, Cuenca de Tampico-Misantla, México: Asociación Mexicana Geólogos Petroleros, I Simposio de Geología de Subsuelo, Ciudad del Carmen, Resúmenes, p. 94–99.

Blair, T.C., 1981, Alluvial fan deposits of the Todos Santos Formation, central Chiapas, Mexico: Unpublished Master's Thesis, University of Texas at Arlington, 134 p.

Blair, T.C., 1986, Paleoenvironments, tectonics, and eustatic controls of sedimentation, regional stratigraphic correlation, and plate tectonic significance of the Jurassic–lowermost Cretaceous Todos Santos and San Ricardo Formations, Chiapas, Mexico: Ph.D. Thesis, University of Colorado, Boulder, 265 p.

Blair, T.C., 1987, Tectonic and hydrologic controls on cyclic alluvial fan, fluvial, and lacustrine rift-basin sedimentation, Jurassic–lowermost Cretaceous Todos Santos Formation, Chiapas, Mexico: Journal of Sedimentary Petrology, v. 57, p. 845–862.

Blair, T.C., 1988, Mixed siliciclastic-carbonate marine and continental syn-rift sedimentation, Upper Jurassic–lowermost Cretaceous Todos Santos and San Ricardo formations, western Chiapas, Mexico: Journal of Sedimentary Petrology, v. 58, p. 623–636.

Blair, T.C., 1989, Upper Jurassic–Lower Cretaceous facies and their plate tectonic and paleogeographic significance, southeastern Mexico and Guatemala: Geological Society of America Abstracts with Programs, v. 21, p. 4.

Blair, T.C., and W.L. Bilodeau, 1986, Tectono-sedimentological analysis of Jurassic–lowermost Cretaceous strata, southeastern Mexico and its significance to the initial rifting of the Gulf of Mexico basin: Geological Society of America Abstracts with Programs, v. 18, p. 542.

Blickwede, J.F., 1981, Petrology and stratigraphy of Triassic (?) "Nazas Formation," Sierra de San Julian, Zacatecas, Mexico: AAPG Bulletin, v. 65, p. 1012.

Blickwede, J.F., and T. Queffelec, 1988, Perdido Fold Belt; a new deep water frontier in the western Gulf of Mexico (abs): AAPG Bulletin, v. 72, p. 16.

Blickwede, J.F., and T. Queffelec, 1992, El sistema plegado de Perdido; una gran provincia petrolera en el Golfo de México para el siglo XXI: XI Convención Geológica Nacional, Sociedad Geológica Mexicana, Libro de Resúmenes, p. 36.

Bochlke, J.E., and P.L. Abbott, 1986, Punta Baja Formation, a Campanian submarine canyon fill, Baja California, Mexico, *in* P.L. Abbott, ed., Cretaceous Stratigraphy Western North America: Los Angeles, Pacific Section, SEPM, p. 91–101.

Bolaños, L., and B.E. Buitrón, 1984, Contribución al conocimiento de los inocerámidos de México: Memoria—Congreso Latinoamericano de Paleontología, p. 406–414.

Boles, J.R., 1986, Mesozoic sedimentary rocks in the Vizcaino Peninsula-Isla de Cedros area, Baja California, Mexico, *in* P.L. Abbott, ed., Cretaceous Stratigraphy Western North America, Los Angeles Pacific Section, SEPM, p. 63–77.

Boles, J.R., and C.A. Landis, 1984, Jurassic sedimentary melange and associated facies, Baja California: Geological Society of America Bulletin, v. 95, p. 513–521.

Booth-Bartolini, C., M. Morales, and S. Finney, 1992, Chronostratigraphic correlation of Ordovician deep-marine massive quartzarenites, central Sonora, Mexico: Geological Society of America Abstracts with Programs, v. 24, p. A359.

Bottjer, D.J., and M.H. Link, 1984, A synthesis of Late Cretaceous southern California and northern Baja California paleogeography, *in* J.K. Crouch and S.B. Bachman, eds., Tectonics and Sedimentation along the California Margin, Pacific Section, SEPM, v. 38, p. 171–188.

Boyd, K.P., 1992, Sedimentation, salt tectonics, and hydrocarbon occurrence of the northern Gulf of Mexico (abs.): II Sympósio Asociación Mexicana de Geológos Petroleros, Instituto Mexicano del Petróleo, México, D.F.

Bracken, B., 1984, Environments of deposition and early diagenesis, La Joya Formation, Huizachal Group red beds, northeastern Mexico, *in* P.S. Ventress, D.G. Bebout, B.F. Perkins, and C.H. Moore, eds., The Jurassic of the Gulf Rim: Gulf Coast Section SEPM, Proceedings of the Third Annual Research Conference, p. 19–26.

Bridges, L.W., 1984, Hydrocarbon potential of Chihuahua, *in* E.C. Kettenbrink, Geology and Petroleum Potential of Chihuahua: West Texas Geological Society Publication 84–80, p. 124–125.

Brown, M.L., and R. Dyer, 1987, Mesozoic geology of northwestern Chihuahua, Mexico, *in* W.R. Dickinson and M.A. Klute, eds., Mesozoic Rocks of Southern Arizona and Adjacent Areas: Arizona Geological Society Digest, v. 18, p. 381–394.

Brunner, P., 1992, Ostrácodos y micromoluscos del pozo Salsipuedes #1, Tab. SE de Mexico. Implicaciones paleoecológicas y cronoestratigráficas: XI Convención Geológica Nacional, Sociedad Geológica Mexicana, Libro de Resúmenes, p. 38.

Buch, I.P., 1984, Upper Permian(?) and Lower Triassic metasedimentary rocks, northeastern Baja California, Mexico, *in* V.A. Frizzell, Jr., ed., Geology of the Baja California Peninsula: Pacific Section SEPM, v. 39, p. 31–36.

Buitrón-Sánchez, B.E., 1984, Late Jurassic bivalves and gastropods from northern Zacatecas, México, and their biogeographic significance, *in* G.E.G Westermann, ed., Jurassic–Cretaceous biochronology and paleogeography of North America, Geological Association of Canada Special Paper 27, p. 89–98.

Buitrón-Sánchez, B.E., 1990, La presencia de *Angulocrinus Polyclonus* (Félix) en el Oxfordiano del Sur de México: Revista del Instituto Mexicano del Petróleo, v. 22, n. 3, p. 19–25.

Buitrón-Sánchez, B.E., S. Morales-Soto, and R. Vidal-Serratos, 1992, Nerineidos (gastropoda Mollusca) Cretácicos de la Formacion Morelos en el area Mezcaltepec-Campuzano, Estado de Guerrero: XI Convención Geológica Nacional, Sociedad Geológica Mexicana, Libro de Resúmenes, p. 40.

Burkart, B., T.C. Blair, and D. Moravec, 1989, Late Cretaceous back-arc deposits of Chiapas, Mexico and their relationships to the orogen: Geological Society of America Abstracts with Programs, v. 21, p. 5.

Busby-Spera, C.J., 1988a, Sedimentologic evolution of a submarine canyon in a forearc basin, Upper Cretaceous Rosario Formation, San Carlos, Mexico: AAPG Bulletin, v. 72, p. 717–737.

Busby-Spera, C.J., 1988b, Evolution of a Middle Jurassic back-arc basin, Cedros Island, Baja California: Evidence from a marine volcaniclastic apron: Geological Society of America Bulletin, v. 100, p. 218–233.

Busby-Spera, C.J., and J.R. Boles, 1986, Sedimentation and subsidence styles in a Cretaceous forearc basin, southern Vizcaino Peninsula, Baja California (Mexico), *in* P.L. Abbott, ed., Cretaceous Stratigraphy Western North America: Los Angeles, Pacific Section SEPM, p. 71–90.

Busby-Spera, C. J., J.M. Mattinson, and N.R. Riggs, 1987, Lower Mesozoic extensional continental arc, Arizona and California: a depocenter for craton derived quartz arenites: Geological Society of America Abstracts with Programs, v. 19, p. 607.

Busby-Spera, C.J., W.R. Morris, and D. Smith, 1988, Syndepositional normal faults across the width of the Cretaceous forearc region of Baja California (Mexico): Geological Society of America Abstracts with Programs, v. 20, p. 147.

Busch, D.A., 1989, Giant Chicontepec Field of east-central Mexico, *in* History of Petroleum, Symposium, Abstracts: AAPG Bulletin, v. 73, p. 1142.

Busch, D.A., and S.A. Govela, 1978, Stratigraphy and structure of Chicontepec turbidites, southeastern Tampico-Misantla basin, Mexico: AAPG Bulletin, v. 62, p. 235–246.

Butterlin, J., 1981, Claves para la determinación de macroforaminíferos de México y del Caribe, del Cretácico Superior al Mioceno medio: Instituto Mexicano del Petróleo, México, D.F., 218 p.

Cabrera-Castro, R., and J.E. Lugo-Rivera, 1984, Estratigrafía-sedimentología de las cuencas Tertiarias: Boletín de la Asociación Mexicana de Geólogos Petroleros, v. 36, p. 3–55.

Callaway, J.M., and J.A. Massare, 1989, *Shastasauras altispinus* (Ichthyosauria, Shastasauridae) from the Upper Triassic of the El Antimonio district, northwestern Sonora, Mexico: Journal of Paleontology, v. 63, p. 930–939.

Cantú-Chapa, A., 1982, The Jurassic–Cretaceous boundary in the subsurface of eastern Mexico: Journal of Petroleum Geology, v. 4, p. 311–318.

Cantú-Chapa, A., 1984, El Jurásico Superior de Tamán, San Luis Potosí, este de México, *in* C.M. Perrilliat, ed.: Memoria, III Congreso Latino americano de Paleontología, p. 207–215.

Cantú-Chapa, A., 1985, Is there a Chicontepec paleocanyon in the Paleogene of eastern Mexico? Journal of Petroleum Geology, v. 8, p. 423–434.

Cantú-Chapa, A., 1987, The Bejuco Paleocanyon (Cretaceous to Paleocene) in the Tampico district, Mexico: Journal of Petroleum Geology, v. 10, p. 207–218.

Cantú-Chapa, A., 1989, La Peña Formation (Aptian); a condensed limestone-shale sequence from the subsurface of NE Mexico: Journal of Petroleum Geology, v. 12, p. 69–83.

Cantú-Chapa, A., 1992, The Jurassic Huasteca series in the subsurface of Poza Rica: Journal of Petroleum Geology, v. 15, p. 259–283.

Cantú-Chapa, C.M., 1993, Sedimentation and tectonic subsidence during the Albian-Cenomanian in the Chihuahua basin, Mexico: *in* A.J.T. Simo, R.W.

Scott, and J.P. Masse, eds.: AAPG Memoir 56, p. 61–70.
Cantú-Chapa, C.M., and R. Arenas-Partida, 1992, Relaciones paleogeográficas y paleotectónicas durante de Triásico–Jurásico en el sector centro-norte de México: XI Convención Geológica Nacional, Sociedad Geológica Mexicana, Libro de Resúmenes, p. 44.
Cantú-Chapa, C.M., R. Sandoval-Silva, and R. Arenas-Partida, 1985, Evolución sedimentaria del Cretácico Inferior en el norte de México: Revista del Instituto Mexicano del Petróleo, v. 17, p. 14–37.
Cantwell, W.A., 1993, Platform-to-basin lithofacies and paleoenvironments of the Aurora and Upper Tamaulipas formations (Albian) southeastern Coahuila, Mexico: Master's Thesis, University of New Orleans.
Cantwell, W.A., and W.C. Ward, 1990, Shelf-to-basin lithofacies transition between the Aurora and upper Tamaulipas Formations and its relationship to postulated early Mesozoic transcurrent faults: Geological Society of America Abstracts with Programs, v. 22, p. A185.
Carán, S., and B.M. Winsborough, 1986, Depositional environments of the Cuatro Ciénegas basin northeastern Mexico—inland "sabkhas" and stromatolitic lakes: Geological Society of America Abstracts with Programs, v. 18, p. 557.
Cárdenas-López, J.G., 1993, Aspectos diagenéticos de la región Prados-Gaucho: Asociación Mexicana Geólogos Petroleros, I Simposio de Geología de Subsuelo, Ciudad del Carmen, Resúmenes, p. 92–93.
Carlsen, T.W., 1987, Stratigraphy and structural traverse of Santa Rosa Canyon, Nuevo León, Mexico: Actas de la Facultad de Ciéncias, Universidad Autónoma de Nuevo León, Linares, v. 2, p. 205–212.
Carranza-Castañeda, O., 1989, Rinocerontes de la fauna del Rancho El Ocote, Mioceno tardío (Henfiliano tardío) del Estado de Guanajuato: Universidad Nacional Autónoma de México, Instituto de Geología, Revista, v. 8, p. 88–99.
Carranza-Casteñeda, O., 1992, Los sedimentos continentales del Mioceno Tardío de la Mesa Central y sus implicaciones estratigráficas: XI Convención Geológica Nacional, Sociedad Geológica Mexicana, Libro de Resúmenes, p. 46.
Carranza-Castañeda, O., and J.M. Castillo-Cerón, 1992, Vertebrados fósiles del Mioceno Tardio, de Zietla, Estado de Hidalgo: XI Convención Geológica Nacional, Sociedad Geológica Mexicana, Libro de Resúmenes, p. 45.
Carranza-Castañeda, O., and W.E. Miller, 1988, Roedores caviomorfos de la Mesa Central de México, Blancano temprano (Pliocene tardío) de la fauna local Rancho Viejo, Estado de Guanajuato: Universidad Nacional Autónoma de México, Instituto De Geología, Revista, v. 7, p. 182–199.
Carranza-Castañeda, O., and A.H. Walton, 1992, Cricetid rodents from the Rancho El Ocote fauna, late Hemphillian (Pliocene), Guanajuato: Universidad Nacional Autónoma de México, Instituto de Geología, Revista, v. 10, p. 71–93.
Carreño, A.L., 1981 (1983), Ostrácodos y foraminíferos planctónicos de la Loma del Tirabuzón, Santa Rosalía, Baja California Sur, e implicaciones bioestratigráficas y paleoecológicas: Universidad Nacional Autónoma de México, Instituto de Geología, Revista, v. 5, p. 55-64.
Carreño, A.L., 1984, Los ostrácodos de edad Oligo-Miocenica de la Sub-Provincia Tampico Misantla, área centro-oriental de México, Part III: Memoria, Congreso Latinoamericano de Paleontología 3, p. 480–487.
Carreño, A.L., 1986, Los ostrácodos de edad Oligo-Miocenica de la sub-provincia Tampico-Misantla, area centro-oriental de México; Parte II, Mioceno; formaciones Meson y Tuxpan: Revista del Instituto de Geología, Universidad Nacional Autónoma México, v. 6, p. 178–192.
Carrillo, M., and Martínez, E., 1983, Evidencias de facies continentales en la Formación Matzitzi, Estado de Puebla: Revista del Instituto de Geología, Universidad Nacional Autónoma México, v. 5, p. 117–118.
Carrillo-Bravo, A., 1982, Exploración petrolera de la Cuenca Mesozóica del centro de México: Asociación Mexicana Geólogos Petroleros, v. 34, p. 21–46.
Carrillo-Bravo, J., 1980, Paleocañones Terciarios de la planicie costera del Golfo de México: Boletín Asociación Mexicana de Geólogos Petroleros, v. 32, p. 27–56.
Carrillo-Martínez, M., 1981 (1983), Contribución al estudio geológico del macizo calcáreo El Doctor, Querétaro: Universidad Nacional Autónoma de México, Instituto de Geología, Revista, v. 5, p. 25–29.
Carrillo-Martínez, M., 1989a, Estratigrafía tectónica de la parte centro-oriental del Estado de Querétaro: Universidad Nacional Autónoma de México, Instituto de Geología, Revista, v. 8, p. 188–193.
Carrillo-Martínez, M., 1989b, Structural analysis of two juxtaposed Jurassic lithostratigraphic assemblages in the Sierra Madre Oriental fold and thrust belt of central Mexico: Geofísica Internacional, v. 28, p. 1007–1028.
Carrillo-Martínez, M., G. Velázquez-G., and L. Cepeda, 1986, Contribución del estudio petrográfico y quimico de areniscas del Jurásico Superior, estados de Querétaro e Hidalgo: Universidad Nacional Autónoma de México, Instituto de Geología, Revista, v. 6, p. 269–271.
Castro-Mora, J., 1985, Análisis tectónico-estratigráfico de cuencas pre-Cretácicas: Petrólero Internacional, v. 43, p. 24, 26–27, 30, 32, 35.
Celestino-U., J.L., 1993, Objectivos petroleros en la porción norte del prospecto Tuxpan: Asociación Mexicana Geólogos Petroleros, I Simposio de Geologia de Subsuelo, Ciudad del Carmen, Resúmenes, p. 32–48.
Celis-Gutiérrez, S., 1981 (1983), Estudio paleoecológico preliminar de los foraminíferos bentónicos de

probable edad Pleistocénica de Salina Grande, Sonora: Universidad Nacional Autónoma de México, Instituto de Geología, Revista, v. 5, p. 121.

Celis-Gutiérrez, S., 1986, Interpretación paleoambiental de los depósitos marinos litorales Pleistocenicos de la localidad de Punta Chueca, Sonora: Universidad Nacional Autónoma de México, Instituto de Geología, Revista, v. 6, p. 259–268.

Cevallos-Ferriz, S.R.S., and R. Weber, 1992, Dicotyledonous wood from the Upper Cretaceous (Maastrichtian) of Coahuilla: Universidad Nacional Autónoma de México, Instituto de Geología, Revista, v. 10, p. 65–70.

Cevallos-Ferriz, S., A. Salcico-Reyna, and A. Pelayo-Ledesma, 1981 (1983), Una nueva sección del Precámbrico de Sonora; los estromatolitos y su importancia en estos estudios: Universidad Nacional Autónoma de México, Instituto de Geología, Revista, v. 5, p. 1–16.

Cevallos-Ferriz, S., A. Pelayo-Ledesma, and A. Salacido-Reyna, 1988, Presencia del estromatolito Colonnella Komar 1964, y su contribución al esquema paleoecológico de la Formación Gamuza (Rífico) de Caborca, Sonora: Universidad Nacional Autónoma de México, Instituto de Geología, Revista, v. 7, p. 206–216.

Chancellor, G.R., 1983, Cenomanian–Turonian ammonites from Coahuila, Mexico: Bulletin of the Geological Institutions of the University of Uppsala, New Series, 9, p. 77–129.

Charleston, S., 1981, Regional stratigraphic relationships and interpretation of the environments of deposition of the Lower Cretaceous, *in* C.I. Smith and S.B. Katz, eds., Lower Cretaceous Stratigraphy and Structure, northern Mexico: Field Trip Guidebook, West Texas Geological Society Publication 81–74, p. 85–88.

Charvat, W.A., and R.C. Grayson, Jr., 1981, Anoxic sedimentation in the Eagle Ford Group (Upper Cretaceous) of central Texas (abs.), *in* C.M Maggio et al., eds.: Gulf Coast Association of Geological Societies 31st annual meeting, Transactions, p. 256.

Chavarria, G.J., 1980, Geology of the Campeche Sound: 11th World Energy Conference, Proceedings, v. 1A, Munich, p. 362–371.

Chávez-Quirarte, R., 1982, El Cretácico Superior en el área del proyecto hidroeléctrico de San Juan Tetelcingo, Guerrero: Libro guía de la excursión geológica a la Cuenca del Alto Rio Balsas, Sociedad Geológica Mexicana, p. 55–58.

Chiodi, M., et al., 1988, Une discordance ante albienne datee par une faune d'Ammonites et de Brachiopodes de type Tethysien au Mexique central: Geobios, v. 21, p. 125–135.

Clark, J.M., and A. Hopson, 1985, Distinctive mammal-like reptile from Mexico and its bearing on the phylogeny of the *Tritylodontidae*: Nature, v. 315, p. 398–400.

Cocheme, J.S., and A. Demant, 1989, Geology of the Yecora area, northern Sierra Madre Occidental, Mexico, *in* Geology of Sonora Mexico, Decade of North American Geology: Boulder, Colorado, Geological Society of America Guidebook.

Collins, L.L., 1985, Rudist paleoecology of the Cretaceous El Abra limestone reef core, central Mexico: Master's Thesis, George Washinton University, St. Louis, Missouri, 129 p.

Collins, L.S., 1988, The faunal structure of a Mid-Cretaceous rudist reef core: Lethaia, 21, p. 271–280.

Contreras-Barrera, A.D., and R. Gio-Argáez, 1985, Consideraciones paleobiológicas de los icnofósiles de la Formación Chicontepec en el Estado de Puebla: Revista del Instituto de Geología, Universidad Nacional Autónoma México, v. 6, p. 73–85.

Contreras-Montero, B., and M. Nuñez, 1984, Estúdio bioestratitgráfico basado en amonitas de las rocas Liásicas de Honey-Pahuatlán, Puebla: Memoria III, Congreso Latinoamericano de Paleontología 3, p. 156–164.

Contreras-Montero, B., C.A. Martínez, and M.E. Gómez-Luna, 1988, Bioestratigrafía y sedimentología del Jurásico Superior en San Pedro del Gallo, Durango, México: Revista del Instituto Mexicano del Petróleo, v. 20, p. 5–49.

Corona-Esquivel, R.J., 1981 (1984), Estratigrafía de la región de Olinalá-Tecocoyunca, noreste del Estado de Guerrero: Revista del Instituto de Geologia, Universidad Nacional Autónoma de México, v. 5, p. 1–16.

Corpus Christi Geological Society, 1981, Geology of Peregrina & Novillo canyons, Ciudad Victoria, Mexico: Corpus Christi Geological Society, Corpus Christi, Texas, 23 p.

Cotter, E., J.C. Wilson, and E.F. McBride, 1982, Upper Cretaceous coals of Sabinas Basin, northeastern Mexico; products of back-barrier deposition and changing rates of progradation (abs.): International Congress on Sedimentology, v. 11, p. 57.

Cruz-Helu, P., V.R. Verdugo, and P.R. Barcenas, 1977, Origin and distribution of Tertiary conglomerates, Veracruz Basin, Mexico: AAPG Bulletin, v. 61, p. 207–226.

Cuevas-Leree, J.A., 1984, Análisis de subsidencia e historia térmica en la Cuenca de Sabinas: Boletín de la Asociación Mexicana de Geólogos Petroleros, v. 36, p. 56–100.

Cuevas-Leree, J.A., 1985, Analysis of subsidence and thermal history in the Sabinas Basin, northeastern Mexico: Master's Thesis, University of Arizona, Tucson, Arizona, 81 p.

Cuevas-Pérez, A., 1987, Stratigraphic and tectonic evolution of the Sabinas basin of northeastern Mexico: Geological Society of America Abstracts with Programs, v. 19, p. 632.

Cuevas-Pérez, E., 1983, Evolución geológica Mesozóica del Estado de Zacatecas, México: Zentralblatt, Geologie Paläontologie, Teil I (3/4), p. 190–201.

Cuevas-Pérez, E., 1985, Geologie des Alteren Mesozoikums in Zacatecas und San Luis Potosí, Mexiko: Unpublished Ph.D. Dissertation, Universität Marburg, 189 p.

Cunningham, A.B., and P.L. Abbott, 1986, Sedimentology and provenance of the Upper Cretaceous Rosario Formation south of Ensenada, Baja California, Mexico, *in* P.L. Abbott, ed., Cretaceous Stratigraphy Western North America: Los Angeles, Pacific Section SEPM, p. 103–118.

Dávila-Alcocer, V.M., 1981 (1983), Radiolarios del Cretácico Inferior de la Formación Plateros del distrito minero de Fresnillo, Zacatecas: Universidad Nacional Autónoma de México, Instituto de Geología, Revista, v. 5, p. 119–120.

Dávila-Alcocer, V.M., and E.A. Pessagno, Jr., 1986, Bioestratigrafía basada en radiolarios del Triásico marino en el NW de la Peninsula de Vizcaíno, Baja California Sur: Universidad Nacional Autónoma de México, Instituto de Geología, Revista, v. 6, p. 136–144.

Dávila-Alcocer, V.M., and F.J. Vega, 1992, Diagénesis temprana, sus implicaciones paleoambientales en un ejemplo para el Maestrichtiano del Grupo Difunta, Estado de Nuevo León: XI Convención Geológica Nacional, Sociedad Geológica Mexicana, Libro de Resúmenes, p. 63.

De Cserna, Z., 1990, La evolución de la geología en México (1500–1929): Universidad Nacional Autónoma de México, Instituto de Geología, Revista, v. 9, p. 1–20.

Delannoy, J.J., 1985, Le Karst; un témoin des mutations socio-économiques dans la Sierra de Zongolica (Mexique), *in* Colloque international de karstologie appliquée. Annales de la Société Géologique de Belgique, 77–83, 108 p.

Delgado-Argote, L.A., 1988, Geología preliminar de la secuéncia volcanosedimentária y serpentinitas asociadas del Jurásico(?) del area de Cuicatlán-Concepción Papalo, Oaxaca: Revista del Instituto de Geologia, Universidad Nacional Autónoma México, v. 7, p. 127–135.

Delgado-Argote, L.A., 1989, Regional implications of the Jurassic–Cretaceous volcanosedimentary Cuicateco terrane, Oaxaca, Mexico: Geofísica International, v. 28, p. 939–973.

Dellatre, M.P., 1984, Permian miogeoclinal strata at El Volcan, Baja California, Mexico, *in* V.A. Frizzell Jr., ed., Geology of the Baja California Peninsula: Pacific Section SEPM, v. 39, p. 23–29.

Deloffre, R., E. Fourcade, and F. Michaud, 1985, *Acroporella chiapasis*, n. sp., Maestrichtian Dasycladaceae algae of Chiapas, southeastern Mexico: Bulletin des Centres de Recherches Exploration-Production Elf-Aquitaine, v. 9, p. 115–125.

Dickinson, W.R., M.A. Klute, and P.N. Swift, 1986, The Bisbee Basin and its bearing on Late Mesozoic paleogeographic and paleoetectonic relations between the Cordilleran and Caribbean regions, *in* P.L. Abbott, ed., Cretaceous Stratigraphy Western North America: Los Angeles, Pacific Section SEPM, p. 51–62.

Dooley, R., 1989a, The Tamaulipas Formation, *in* K. Greier and J.L. Kowalski, eds., Geology of the Sierra de Catorce Uplift, Real de Catorce, San Luis Potosí, Mexico, Guide Book: The University of Texas Pan American Geological Society for the Gulf Coast Association of Geological Societies 39th Annual Convention, p. 57–64.

Dooley, R., 1989b, The La Caja Formation, *in* K. Greier and J.L. Kowalski, eds., Geology of the Sierra de Catorce Uplift, Real de Catorce, San Luis Potosí, Mexico, Guide Book: The University of Texas Pan American Geological Society for the Gulf Coast Association of Geological Societies 39th Annual Convention, p. 47–52.

Dooley, R., and C. Geil, 1989, The Taraises Formation, *in* K. Greier and J.L. Kowalski, eds., Geology of the Sierra de Catorce uplift, Real de Catorce, San Luis Potosí, Mexico: Guide Book, The University of Texas Pan American Geological Society for the Gulf Coast Association of Geological Societies 39th Annual Convention, p. 53–56.

Dueñas-García, J.C., 1992, Evaluación geológico-económica de los yacimientos caoliniferos de Huayacocotla, Veracruz: XI Convención Geológica Nacional, Sociedad Geológica Mexicana, Libro de Resúmenes, p. 68.

Dueñas Marco, A., 1992, Palinomorfos marinos y continentales de tres localidades de lechos rojos (Fm. Todos Santos) del sureste de México; su implicación cronoestratigráfica: XI Convención Geológica Nacional, Sociedad Geológica Mexicana, Libro de Resúmenes, p. 69.

Dyer, R., 1986, Precambrian and Paleozoic rocks of Sierra Carrizalillo, Chihauhua, Mexico, a preliminary report: Geological Society of America Abstracts with Programs, Rocky Mountain Section, v. 18, p. 353.

Dyer, R., ed., 1987, Paleozoico de Chihuahua: Sociedad Geológica Mexicana A.C., Universidad Autónoma de Chihuahua and University of Texas at El Paso, Excursion Geológica No. 2, Libreto Guia de Caminos, 104 p.

Dyer, R., and I.A. Reyes, 1987, The geology of Cerro El Carrizalillo, Chihuahua, Mexico, preliminary results: Universidad Autónoma de Chihuahua Gaceta Geológica, v. 1, p. 108–128.

Echanove-Echanove, O., 1986, Geología petrolera de la Cuenca de Burgos: Boletín de la Asociación Mexicana de Geólogos Petroleros, v. 38, p. 3–74.

Eguiluz de Antuñano, S., 1991, Discordancia Cenomaniana sobre la plataforma de Coahuila: Asociación Mexicana de Geólogos Petroleros Boletín, v. 41, p. 1–17.

Eguiluz de Antuñano, S., and M. Aranda-García, 1983, Posibilidades económico-petroleras en rocas clásticas del Neocomiano en la margen sur de la Paleoisla de Coahuila: XXI Congreso Nacional de la Asociación Ingenieros Petróleos Mexicanos, p. 5–13.

Eguiluz de Antuñano, S., and M. Aranda-García, 1984, Economic oil possibilities in clastic rocks of the Neocomian along the southern margin of the Coahuila Island, *in* J.L. Wilson, W.C. Ward, and J.M. Finneran, eds., A field guide to Upper Jurassic and Lower Cretaceous Carbonate Platform and Basin Systems, Monterrey-Saltillo area, northeast

Mexico: Gulf Coast Section, SEPM Foundation, p. 43–51.

Elías-Herrera, M., and J. L. Sánchez-Zavala, 1992, Relaciones tectonoestratigráficas del Terreno Guerrero en la porción suroccidental del Estado de México y su interpretación geodinámica: XI Convención Geológica Nacional, Sociedad Geológica Mexicana, Libro de Resúmenes, p. 70.

El Paso Geological Society, 1983, Geology and Mineral Resources of North Central Chihuahua; Guidebook for 1983 Field Conference (includes regional studies, areal studies on igneous rocks, areal studies on sedimentary rocks, general economic studies, economic studies of Sierra Pena Blanca and Sta. Eulalia).

El Paso Geological Society, 1988, Stratigraphy, tectonics, and resources of parts of Sierra Madre Occidental province, Mexico, *in* K. E. Clark, P.C. Goodell, and J.M. Hoffer, eds., Field Trip Guidebook. Twenty-nine papers on various subjects, 385 p.

El Paso Geological Society, 1992, Geology and mineral resources of northern Sierra Madre Occidental, *in* K.F. Clark, J. Roldan-Quintana, and R.H. Schmidt, eds., Field Trip Guidebook. Includes 44 short papers on various geological subjects.

Enciso-de la Vega, S., 1988, Una nueva localidad Pérmica con fusilinidos en Puebla: Revista del Instituto de Geología, Universidad Nacional Autónoma México, v. 7, p. 28–31.

Enciso-de la Vega, S., 1992, Propuesta de nomenclatura estratigráfica para la cuenca de México: Universidad Nacional Autónoma de México, Instituto De Geología, Revista, v. 10, p. 26–36.

Enciso-de la Vega, S., 1992, Una sucesión tipo flysch del Misisípico-Pensilvanico de la Sierra del Catorce in San Luis Potosí, México: XI Convención Geológica Nacional, Sociedad Geológica Mexicana, Libro de Resúmenes, p. 73.

Enos, P., 1974, Reefs, platforms, and basins of Middle Cretaceous in northeast Mexico; AAPG Bulletin, v. 58, p. 800–809.

Enos, P., 1983a, Late Mesozoic paleogeography of Mexico, *in* M.W. Reynolds and E.D. Dolly, eds., Mesozoic Paleogeography of the West-Central United States: Rocky Mountain Section, SEPM Paleogeography Symposium 2, p. 133–157.

Enos, P., 1983b, Sedimentation and Diagenesis of Mid-Cretaceous Platform Margin East-Central Mexico; Field trip, Dallas Geological Society, Dallas, Texas, p. 1–16.

Enos, P., 1985a, Cretaceous debris reservoirs, Poza Rica field, Veracruz, Mexico, *in* P.O. Roehle and P.W. Choquette, eds., Carbonate Petroleum Reservoirs: Springer Verlag, New York, p. 457–469.

Enos, P., 1985b, Diagenetic evolution of Cretaceous reefs in Mexico: Proceedings International Coral Reef Symposium, 5, p. 301–305.

Enos, P., 1986, Diagenesis of mid-Cretaceous rudist reefs, Valles Platform, Mexico, *in* J.H. Schroeder and B.H. Purser, eds., Reef Diagnesis: Springer Verlag, Berlin, p. 159–185.

Enos, P., 1988, Evolution of pore space in the Poza Rica trend (Mid-Cretaceous), Mexico: Sedimentology, 35, p. 287–325.

Enos, P., and B.P. Stephens, 1991, Basin-margin carbonates, Mid-Cretaceous, El Lobo, Mexico, *in* P. Enos et al., eds., Sedimentation and Diagenesis of Middle Cretaceous Platform Margins, East-Central Mexico: Dallas Geological Society Field Trip, p. 2–22.

Enos, P., and B.P. Stephens, 1993, Mid-Cretaceous basin margin carbonates, east-central Mexico: Sedimentology, v. 40, p. 539–556.

Estavillo, G.C., and C.J. Aguayo, 1985, Ambientes sedimentários recientes en Laguna Madre: Sociedad Geológica Mexicana Boletín, v. 46, p. 29–64.

Farek, M.M., 1987, The Cretaceous succession of La Chona quadrangle, Nuevo León, Mexico: Actas de la Facultad de Ciéncias, Universidad Autónoma de Nuevo León, Linares, v. 2, p. 41–49.

Farek, M.M., 1989, Lithostratigraphic studies in the eastern front of the Sierra Madre Oriental, Mexico; La Chona quadrangle: Abstracts with Programs, Geological Society of America, v. 21, p. 10–11.

Fastovsky, D.E., J.M. Clark, and J.A. Hopson, 1987, Preliminary report of a vertebrate fauna from an unusual paleoenvironmental setting, Huizachal Group, Early or Mid-Jurassic, Tamaulipas, Mexico, *in* P.M. Currie and E.H. Koster, eds., Fourth Symposium on Mesozoic Terrestrial Ecosystems, Short Papers; Occasional Paper of the Tyrell Museum of Paleontology, n. 3, p. 82–87.

Fastovsky, D.E., O.D. Hermes, J.M. Clark, and J.A. Hopson, 1988, Volcanos, debris flows, and Mesozoic mammals: Huizachal Group (Early or Middle Jurassic) Tamaulipas, Mexico: Geological Society of America Abstracts with Programs, v. 20, n. 7, p. 317–318.

Ferrusquia-Villafranca, I., and O. Comas-Rodríguez, 1988, Reptiles marinos Mesozóicos en el sureste de México y su significación geológico-paleontológica entre Sierra de Gamaon y Laguna de Santaiguillo, Estado de Durango: Revista del Instituto de Geología, Universidad Nacional Autónoma México, v. 7, p. 136–147.

Finneran, J.B., 1984, Zuloaga Formation (Upper Jurassic) shoal complex, Sierra de Enfrente, Coahuila, northeast Mexico (abs.): AAPG Bulletin, v. 68, p. 476.

Finneran, J.B., 1986, Carbonate petrography and depositional environments of the Upper Jurassic Zuloaga Formation, Sierra de Enfrente, Coahuila, Mexico: Master's Thesis, Stephen F. Austin University, 270 p.

Fitzpatrick, S., 1988, The lithologies and depositional environments of the Rara Formation, Sierra del Cuervo, Chihuahua, *in* K.F. Clark, P.C. Goodell, and J.M. Hoffer, eds., Stratigraphy, Tectonics and Resources of Parts of Sierra Madre Occidental Province, Mexico: Guidebook for the 1988 Field Conference, El Paso Geological Society, p. 209–216.

Flores de Dios-Gonzáles, L.A., M. Guerrero-Suástegui, and B.E. Buitrón-Sánchez, 1992, Determinación de fósiles del Paleozóico Superior de la secuéncia sedimentária de la región de San Juan Ihualtepec Edo de Oaxaca: XI Convención Geológica Nacional, Sociedad Geológica Mexicana, Libro de Resúmenes, p. 74.

Flores-Espinoza, E., 1983, Depositional environments and coal beds of the Bigford Formation (Eocene), northeastern Mexico: Master's Thesis University of Texas at Austin, 120 p.

Flores-Espinoza, E., 1989, Stratigraphy and sedimentology of the Upper Cretaceous terrigenous rocks and coal of the Sabinas-Monclova area, northern Mexico: Ph.D. Dissertation, University of Texas at Austin, 389 p.

Flowers G.C., S.A. Nelson, and W.C. Isphording, 1986, Origin and significance of volcanic ash from Loltun Caves, Yucatan, Mexico: Geological Society of America Abstracts with Programs, v. 18, p. 603.

Flynn, J.J., R.M. Cipolletti, and M.J. Novacek, 1989, Chronology of early Eocene marine and terrestrial strata, Baja California, Mexico: Geological Society of America Bulletin, v. 101, p. 1182–1196.

Fortunato, K.S., and W.C. Ward, 1982, Upper Jurassic–Lower Cretaceous fan-delta complex: La Casita Formation of the Saltillo area, Coahuila, Mexico: Gulf Coast Association of Geological Societies Transactions, v. 32, p. 473–482.

Fourner, G.R., 1982, Neogene palynostratigraphy of southern Mexico margin: DSDP leg 66, AAPG Memoir 34, p. 611.

Frame, A., and W.C. Ward, 1987, Lowermost Cretaceous coral-rich limestone in Nuevo León and Coahuila Mexico: Actas de la Facultad de Ciéncias, Universidad Autónoma de Nuevo León, Linares, v. 2, p. 33–39.

Franzen, J.L., and P. Meiburg, 1992, Excavación, preparación e investigación paleontológica de hallazgos de mamiferos del Pleistoceno en el Edo. de Nuevo León, México: XI Convención Geológica Nacional, Sociedad Geológica Mexicana, Libro de Resúmenes, p. 75–76.

Galloway, W.E., et al., 1992, Cenozoic, *in* A. Salvador, ed., The Gulf of Mexico basin, Boulder, Colorado: Geological Society of America, v. J, p. 245–324.

García Esparza, J., 1987, Estúdio preliminar de la estratigrafía y estilo estructural de la Sierra Rica, en el noroeste de Chihuahua México: Universidad Autónoma de Chihuahua Gaceta Geológica, v. 1, p. 163–175.

García Esparza, J., 1988, Preliminary stratigraphic and structural study of the Sierra Rica, northwest Chihuahua, *in* K.F. Clark, P.C. Goodell, J.M. Hoffer, eds., Stratigraphy, tectonics and resources of parts of Sierra Madre Occidental Province, Mexico, Guidebook for the 1988 Field Conference: El Paso Geological Society, p. 135–146.

García-Hernández, J., 1993, Roca almacenadora Jurásico Superior-Oxfordiano: Asociación Mexicana Geólogos Petroleros, I Simposio de Geología de Subsuelo, Ciudad del Carmen, Resúmenes, p. 49–61.

García-Luna, L.M., F. Verdugo-Diaz, and A. Uribe-Carvajal, 1990, Sedimentos del Grupo Claiborne presentes en el noreste de la Republica Mexicana: X Convención Geológica Nacional, Sociedad Geológica Mexicana, Libro de Resúmenes, p. 73.

Garza-Hernández, A., 1984, Bioestratigrafía del Cenozóico (Oligoceno–Mioceno) del borde oriental de la planicie costera y Faja de Oro Marina: Memoria III, Congreso Latinoamericano de Paleontología 3, p. 471–479.

Gastil, R.G., R.H. Miller, and M.F. Campa-Uranga, 1986b, The Cretaceous paleogeography of peninsular California and adjacent Mexico, *in* P.L. Abbott, ed., Cretaceous Stratigraphy Western North America: Los Angeles, Pacific Section SEPM, p. 41–50.

Geological Society of America, 1991, The Gulf of Mexico Basin, *in* A. Salvador, ed., D.N.A.G. vol. J, 568 p.

Gerstenhauer, A., 1987, Kalkkrusten und Karstformenschatz auf Yucatan, Mexico: Erdkunde, v. 41, p. 30–37.

Goldhammer, R.K. and P.J. Lehmann, 1991, Part 4—Stratigraphic Framework, *in* R.K. Goldhammer et al., eds., Sequence Stratigraphy and Cyclostratigraphy of the Mesozoic of the Sierra Madre Oriental, Northeast Mexico: Field guidebook, Gulf Coast Section SEPM, p. 15–32.

Goldhammer, R.K., and J.L. Wilson, 1991, Part 3, Tectonic development, *in* R.K. Goldhammer et al., eds., Sequence Stratigraphy and Cyclostratigraphy of the Mesozoic of the Sierra Madre Oriental, Northeast Mexico: Field guidebook, Gulf Coast Section SEPM, p. 7–14.

Goldhammer, R.K., P.J. Lehman, R.G. Todd, J.L. Wilson, W.C. Ward, and C.R. Johnson, 1991, Sequence Stratigraphy and Cyclostratigraphy of the Mesozoic of the Sierra Madre Oriental, Northeast Mexico: Gulf Coast Section SEPM Field Guidebook, 85 p.

Gonzáles, M.L., 1989, Evaluación geológico-geoquímica de la provincia de Chihuahua: Boletín de la Asociación Mexicana de Geológos Petroleros, v. 38, n. 2 (dated July–December, 1986), p. 3–58.

González-Arreola, C., and M. Carillo-Martínez, 1986, Presencia de amonitas heteromorfos del Jurásico Superior y Cretácico Inferior (Hauteriviano-Barremiano) en el área de San Joaquín-Vizarrón, Estado de Querétaro: Universidad Nacional Autónoma de México, Instituto de Geología, Revista, v. 6, p. 171–177.

González-García, R., 1984, Petroleum Exploration in the "Gulf of Sabinas" a new gas province in Northern Mexico, *in* J.L. Wilson, W.C. Ward, and J.M. Finneran, eds., A field guide to Upper Jurassic and Lower Cretaceous carbonate platform and basin systems, Monterrey-Saltillo area, Northeast Mexico: Gulf Coast Section SEPM Foundation, p. 64–76.

González-García, R., and N. Holguín-Quiñones, 1991, Geology of the source rocks of Mexico: World Petroleum Congress, Buenos Aires, Argentina, Proceedings, v. 2, p. 1–8.

González-León, C., 1986, Estratigrafía del Paleozóico de la Sierra del Tule, noreste de Sonora: Revista del

Instituto de Geología, Universidad Nacional Autónoma Mexico, v. 6, p. 117–135.

González-León, C., 1988, Estratigrafía Geología Estructural de las rocas sedimentarlas Cretácias del área de Lampazos, Sonora: Universidad Nacional Autónoma de México, Instituto de Geología, Revista, v. 7, p. 148–162.

González-León, C., and C. Jacques-Ayala, 1990, Paleogeografía del Cretácico Temprano en Sonora: Actas de la Facultad de Ciéncias de la Universidad Autónoma de Nuevo León, Linares, v. 4, p. 125–152.

Gonzáles-León., C., C. Jacques-A., E. Almazán-V., J.L. Rodríquez-C., and J.C. García-B., 1987, Paleogeografía del Cretácico Inferior de Sonora: Actas de la Facultad de Ciéncias, Universidad Autónoma de Nuevo León, Linares, v. 2, p. 191–193.

Götte, M., 1988, Estudio geológico-estructural de Galeana, N.L. (México) y sus alrededores: Actas de la Facultad de Ciéncias de la Tierra, Universidad Autónoma de Nuevo León, Linares, v. 2, p. 61–87.

Götte, M., 1990, Halotektonische Deformationsprozesse in Sulfatgesteinen der Minas Viejas-Formation (Ober-Jura) in der Sierra Madre Oriental, nordost-Mexiko: Unpublished Ph. D. Dissertation, Technischen Hochschule Darmstadt, 270 p.

Götte, M., and D. Michalzik, 1992, Stratigraphic relations and facies sequences of an Upper Jurassic evaporitic ramp in the Sierra Madre Oriental (Mexico): Zentralblatt für Geologie und Paläeontologie, Teil 1, v. 6, p. 1445–1466.

Graham, A.K., 1987, Paleoecological implications of Tertiary palynofloras from Central America: Programs and Abstracts, American Association of Stratigraphic Palynologists, 20th annual meeting, p. 80.

Graham, A., 1989, Paleofloristic and paleoclimatic changes in the Tertiary of northern Latin America: Review of Palaeobotany and Palynology, v. 60, p. 283–293.

Grajales-N., J.M., D. Terrell, R. Torres-Vargas, C. Jacques-Ayala, 1990, Late Cretaceous synorogenic volcanic/sedimentary sequences in eastern Sonora, Mexico: Geological Society of America Abstracts with Programs, v. 22, n. 3, p. 26.

Granier, B., and F. Michaud, 1987a, Dasycladales from the Upper Jurassic of southeastern Mexico: Abstracts and Programs, International Symposium on Fossil Algae, v. 4, p. 36.

Granier, B., and F. Michaud, 1987b, *Deloffrella quercifolipora* , n. gen., n. sp., une algue dasycladacée nouvelle du Kimméridgien et du Portlandien du sud-est du Mexique: Bulletin de la Société Géologique de France, Huitiéme Série, v. 3 (6), p. 1089–1096.

Granier, B., F. Michaud, and E. Fourcade, 1986, *Apinella jaffrezoi* n. gen., n. sp., Algue dasycladacée du Kimméridgien du Chiapas, sud-est du Mexique: Geobios, v. 19 (6), p. 801–813.

Greier, K., 1989, The La Caja Formation, *in* K. Greier and J.L. Kowalski, eds., Geology of the Sierra de Catorce uplift, Real de Catorce, San Luis Potosí, Mexico, Guide Book: The University of Texas-Pan American Geological Society for the Gulf Coast Association of Geological Societies 39th Annual Convention, p. 65–66.

Grimm, K.A., M. Ledesma, R.E. Garrison, and C. Fonseca, 1991, The Oligo-Miocene San Gregorio Formation of Baja California Sur, Mexico: An early record of coastal upwelling along the eastern Pacific margin: First International Meeting on Geology of the Baja California Peninsula, p. 35.

Guerrero-Suástegui, M., and J. Ramírez-Espinoza, 1992, Depósitos de tormenta en el límite de la sedimentación de carbonatos y terrígenos (Cretácico Medio) en la región norte del Estado de Guerrero: XI Convención Geológica Nacional, Sociedad Geológica Mexicana, Libro de Resúmenes, p. 83–84.

Guerrero-Suástegui, M., J.M. Ramírez-E., O. Talavera-M., R.N. Hiscott, 1992, Análisis de facies, paleocorrientes y proveniencia de los conglomerados de las formaciones Angao y San Lucas en la región de Huetamo, Mich: XI Convención Geológica Nacional, Sociedad Geológica Mexicana, Libro de Resúmenes, p. 85

Gursky, H.-J., and D. Michalzik, 1989, Lower Permian turbidites in the northern Sierra Madre Oriental, Mexico: Zentralblatt für Geologie und Paläontologie, Teil 1, v. 5/6, p. 821–838.

Gursky, H.-J., and C. Ramírez-Ramírez, 1986, Notas preliminares sobre el descubrimiento de volcanitas ácidas en El Canõn de Caballeros (Nucleo del anticlinorio Huizachal-Peregrina, Tamaulipas, México): Actas de la Facultad de Ciéncias de la Tierra, Universidad Autónoma de Nuevo León, Linares, v. 1, p. 11–22.

Gutiérrez-Moreno, I., 1993, Estudio sedimentológico y evaluación del potencial almacenador de las rocas del Cretácico Médio en la porción norte de la Sonda de Campeche: Asociación Mexicana Geólogos Petroleros, I Simposio de Geologia de Subsuelo, Ciudad del Carmen, Resúmenes, p. 62–78.

Guzmán, A.E., 1984, Geología petrolera del Golfo de California y áreas aledañas: Asociación Mexicana de Geólogos Petroleros, v. 36.

Guzmán, A.E., and J.F. González-Piñeda, 1993, Sistemas petrolíferos de la plataforma continental frente a Tampico: Asociación Mexicana Geólogos Petroleros, I Simposio de Geología de Subsuelo, Ciudad del Carmen, Resúmenes, p. 137–138.

Guzmán, E.J., and A.E. Guzmán, 1981, Petroleum geology of Reforma area, southeast Mexico, and exploratory effort in Baja California, northwest Mexico, *in* M.T. Halbouty, ed., Energy Resources of the Pacific Region: AAPG Studies in Geology, v. 12, p. 1–11.

Guzmán-Vega, M.A., 1992a, Estudio geoquímico integral de la Cuenca de Tampico-Tuxpan y sus perspectivas económico petroleras (abs.): II Symposio de Asociación Mexicana de Geológos Petroleros, Instituto Mexicano del Petróleo, México, D.F.

Guzmán-Vega, M.A., 1992b, Evolución de la subsidencia en la Cuenca de Tampico-Tuxpan: XI

Convención Geológica Nacional, Sociedad Geológica Mexicana, Libro de Resúmenes, p. 87.

Halley, R.B., B.J. Pierson, and W. Schlager, 1984, Alternative diagenetic models for Cretaceous talus deposits, Deep Sea Drilling Project Site 536, Gulf of Mexico *in* R.T. Buffler, et al., Initial reports 77: Proceedings of the Ocean Drilling Program, Part A, p. 397–408.

Hamilton, W., 1987, Mesozoic geology and tectonics of the Big Maria Mountains region, southeastern Arizona, *in* W.R. Dickinson and M.A. Klute, eds., Mesozoic Rocks of Southern Arizona and Adjacent Areas: Arizona Geological Society Digest, v. 18, p. 33–47.

Handschy, J.W., 1984, A preliminary report on the geology of Sierra del Cuervo near Villa Aldama, Chihuahua, Mexico, *in* E.C.Kettenbrink, ed., Geology and Petroleum Potential of Chihuahua, Mexico: West Texas Geological Society Publication 84–80, p. 195–201.

Hansen, D.K.T., 1988, Petrology, physical stratigraphy and biostratigraphy of the lower Sierra Madre Formation, Cretaceous, west-central Chiapas: Unpublished Master's Thesis, University of Texas at Arlington, 137 p.

Hardin, N.S., 1989, Salt distribution and emplacement processes, northwest Gulf lower slope; a suture between two provinces, *in* Gulf of Mexico Salt Tectonics; Associated Processes and Exploration Potential; 10th Annual Reserch Conference Program and Extended Abstracts: Gulf Coast Section SEPM, p. 54–59.

Harkey, D.L., 1985, Structural geology and sedimentary analysis (Las Vigas Formation), Sierra San Ignacio, Chihuahua, Mexico: Master's Thesis, University of Texas at El Paso, 122 p.

Hernández, M.T., 1981, Bioestratigrafía de la Serie Coahuila en sus facies de plataforma, área de Monclova, Coahuila: Revista del Instituto Mexicano del Petróleo, v. 13, p. 5–23.

Hernández-García, S., 1992a, Estado actual y perspectivas de la exploración en la Región Marina (abs): II Symposio e Asociación Mexicana de Geólogos Petroleros, Instituto Mexicano del Petróleo, México, D.F.

Hernández-García, S., 1992b, Calcarenitas del Eoceno Medio; otra opción de la Región Marina: XI Convención Geológica Nacional, Sociedad Geológica Mexicana, Libro de Resúmenes, p. 88.

Hernández-García, S., 1993, Subsistema almacenador en los horizontes Cretácico Superior y, Eoceno Médio de la región marina: Asociación Mexicana Geólogos Petroleros, I Simposio de Geología de Subsuelo, Ciudad Carmen, Resúmenes, p. 79–83.

Hernández-Guzmán, M., 1992, Diagénesis en sedimentos terrigenos: las formaciones Huayacocotla y Rosario de la Cuenca Tampico-Misantla, este de México: XI Convención Geológica Nacional, Sociedad Geológica Mexicana, Libro de Resúmenes, p. 89.

Hernández-Jiménez, M.A, and E. Rosales-C., 1992, Modelo de depósito de las facies orgánicas de la Formación La Peña, en el área de Piedras Negras, Coahuila: XI Convención Geológica Nacional, Sociedad Geólgica Mexicana, Libro de Resúmenes, p. 90.

Hernández-Michaca, J.L., and C. Santana-Palomino, 1992, Geología del Valle de Puebla: XI Convención Geológica Nacional, Sociedad Geológica Mexicana, Libro de Resúmenes, p. 93.

Herrera-Soto, M.E., and C.F. Estavillo-González, 1991, Análisis estratigráfico y modelo de sedimentación de la Formación Todos Santos en el Area del Alto Uzpanapa-Matías Romero, Oaxaca: Instituto Mexicano del Petróleo, Revista, v. 23, n. 1, p. 5–42.

Herrera-Soto, M.E., and C.F. Estavillo-González, 1992, Análisis estratigráfico y ambientes de depósito de la Formación Todos Santos en el Estado de Chiapas: XI Convención Geológica Nacional, Sociedad Geológica Mexicana, Libro de Resúmenes, p. 94.

Hickey, J.J., 1984, Stratigraphy and composition of a Jura-Cretaceous volcanic arc apron, Punta Eugenia, B.C. Sur, Mexico, *in* V.A. Frizzell, ed., Geology of the Baja California Peninsula: Pacific Section SEPM, v. 39, p. 149–160.

Hildebrand, A.R., et al., 1991, Chicxulub crater: a possible Cretaceous/Tertiary boundary impact crater on the Yucatan Peninsula, Mexico: Geology, v. 19, p. 867–871.

Hoffer, R.L., and J.M. Hoffer, 1984, Cretaceous stratigraphy of south-central New Mexico and northern Chihuahua, Mexico, *in* E.C. Kettenbrink, ed., Geology and Petroleum Potential of Chihuahua, Mexico, West Texas Geological Society Publication 84–80, p. 112–115.

Holguín-Quiñones, N., 1988, Evaluacíon geoquímica del sureste de México: Asociacíon Mexicana de Geólogos Petroleros, Boletín, v. 37, n. 1, p. 3–48 (1985 issue).

Holguín-Quiñones, N., 1991, La geoquímica del petróleo en México: Boletín de la Asociación Mexicana de Geólogos Petroleros, v. 41, n. 1, p. 37–50.

Hurtado-Carador, M., 1981, Aplicación de registros geofíscos de pozos a la exploración por carbón: Resúmenes de la reunión, Geos, 1 (4,A-B), p. C6–C7.

Hurtado-González, B., 1984, Implicaciones paleoecológicas de los Moluscos Cretácicos (Bivalvia y Gastropoda) de la región de Ciudad del Maíz, San Luis Potosí, México: Memoria, III Congreso Latinoamericano de Paleontología 3, p. 415–424.

Ice, R.G., and C.L. McNulty, 1980, Foraminifers and calcispheres from the Cuesta del Cura and lower Agua Nueva Formations (Cretaceous) in east-central Mexico: Gulf Coast Association of Geological Societies Transactions, v. 30, p. 403–425.

Imoto, Nobuhiro, and E.F. McBride, 1986, Volcanism recorded in Carboniferous Tesnus Formation, Marathon Basin, Texas: Geological Society of America Abstracts with Programs, v. 18, p. 643.

Ingle, J.C., 1987, Paleo-oceanographic evolution of the Gulf of California: foraminiferal and lithofacies evidence: Geological Society of America Abstracts with Programs, v. 19, p. 721.

Isphording, W.C., 1984, The Clays of Yucatan, Mexico; a contrast in genesis, *in* A. Singer et al., eds., Palygorskite-Sepiolite Occurrences, Genesis and Uses, 37: Elsevier Science Publishing, Amsterdam, Netherlands, p. 59–73.

Iturralde-Vinent, M.A., and M. Norell, 1992, Los Saurios del Jurásico (Oxfordiano) de Cuba occidental: XI Convención Geológica Nacional, Sociedad Geológica Mexicana, Libro de Resúmenes, p. 96.

Jacobo-A., J., 1986, El basamento del Distrito Poza Rica y su implicación en la generación de hidrocarburos: Revista del Instituto Mexicano del Petróleo, v. 18, p. 5–24.

Jacobo-A., J., 1992, El basamento del noreste de México: geocronología, petrografía, y geoquímica: XI Convención Geológica Nacional, Sociedad Geológica Mexicana, Libro de Resúmenes, p. 96.

Jacques-Ayala, C., 1987, Petrology of the Cretaceous El Chanate Group of northwest Sonora, Mexico, and its tectonic implications: preliminary results: Geological Society of America Abstracts with Programs, v. 19, p. 714.

Jacques-Ayala, C., 1989, Arroyo Sásabe Formation (Aptian-Albian), northwestern Sonora, México—marginal marine sedimentation in the Sonora backarc basin: Revista del Instituto de Geología, Universidad Nacional Autónoma México, v. 8, p. 171–178.

Jacques-Ayala, C., 1992, The Lower Cretaceous Glance conglomerate and Morita Formation of the Sierra El Chanate, northwestern Sonora: Universidad Nacional Autónoma de México, Instituto de Geología Revista, v. 10, p. 37–46.

Jacques-Ayala, C., and P.E. Potter, 1987, Stratigraphy and paleogeography of Lower Cretaceous rocks, Sierra el Chanate, northwest Sonora, México, *in* W.R. Dickinson, and M.A. Klute, eds., Mesozoic Rocks of Southern Arizona and adjacent Areas, Arizona Geological Society Digest, v. 18, p. 203–214.

Jacques-Ayala, C., J.C. García y Barragán, and K.A. DeJong, 1990, Caborca-Altar geology: Cretaceous sedimentation and compression, Tertiary uplift and extension, *in* G.E. Gehrels, and J.E. Spencer, eds., Geologic Excursions Through the Sonoran Desert Region, Arizona and Sonora: Tucson, Arizona Geological Survey Special Paper 7, p. 165–182.

Jacques-Ayala, C., J.M. Grajales-Nishimura, and D. Terrel, 1990, Early Cretaceous marine sedimentation and volcanism in some localities of northern Sonora, Mexico: Geological Society of America Abstracts with Programs, v. 22, p. 31–32.

Jaffe, D.G., 1985, Petroleum potential and depositional environments of the Norphlet, Louann, Werner (Jurassic), and Eagle Mills (Triassic) formations along the north rim of the East Texas Basin: Master's Thesis, Baylor University, Waco, Texas, 168 p.

Johnson, C.C., 1984, Paleoecology, carbonate petrology and depositional environments of lagoonal facies, Cupido and El Abra formations, northeastern Mexico: Master's Thesis, University of Colorado, Boulder, 147 p.

Johnson, C.C., and E.G. Kauffman, 1986, Biofacies and ecological succession in the Barremian–Aptian Cupido rudist framework, northeastern Mexico (Abs): Proceedings, North American Paleontological Convention 4, p. A22.

Kennedy, W.J., and W.A. Cobbán, 1988, Mid-Turonian ammonite faunas from northern Mexico: Geological Magazine, v. 125, p. 593–612.

Ketner, K.B., 1986, Eureka Quartzite in Mexico? Tectonic implications: Geology, v. 14, p. 1027–1030.

Ketner, K.B., 1990, Stratigraphy, structure, and regional correlatives of tectonically juxtaposed Paleozoic miogeoclinal and eugeoclinal assemblages, Cerro Cobachi, Sonora, Mexico: Geological Society of America Abstracts with Programs, v. 22, n. 3, p. 34.

Kettenbrink, E.C., Jr., ed., 1984, Geology and petroleum potential of Chihuahua, Mexico: West Texas Geological Society Publication 84–80, 218 p.

Kimbrough, D.L., 1984, Paleogeographic significance of the Middle Jurassic Gran Cañon Formation, Cedros Island, Baja California Sur, *in* V.A. Frizzell Jr., ed., Geology of the Baja California Peninsula: Pacific Section SEPM, v. 39, p. 107–117.

Kindred, F.R., 1988, Origin and diagenesis of carbonate mudstone, shallow to deeper shelf, Aurora Formation, Coahuila, Mexico: Master's Thesis, University of New Orleans, 77 p.

Kitz, M.B., 1984, Petrology of a Lower Cretaceous (Neocomian) conglomerate sequence of the San Marcos Formation, Sierra Mojada, west-central Coahuila, Mexico: Master's Thesis, University of Texas at Arlington, 100 p.

Klein, R.T., S.R. Hannula, N.W. Jones, J.W. McKee, and T.H. Anderson, 1990, Cerro El Pedernal: Block-in-melange features in northern Zacatecas, Mexico: Geological Society of America Abstracts with Programs, v. 22, n. 3, p. 35.

Krutak, P.R., 1990, Mesozoic faults and fanglomerates of northeastern Mexico: Implications for tectonism and continental drift: Geological Society of America Abstracts with Programs, v. 22, p. 185.

Krutak, P.R., and Raul, G.A., 1984, Ostracods as indicators of low-energy versus high-energy marine carbonates, northeastern Yucatan shelf, Mexico (abs.): AAPG Bulletin, v. 68, p. 497.

Lachkar, G., F. Michaud, and E. Fourcade, 1989, Nouvelles données palynologiques sur le Jurassique superieur et le Cretace inferieur du sud-est du Mexique (Formation Jerico, Chiapas): Review of Palaeobotany and Palynology, v. 59, p. 93–1007.

Laudon, R.C., 1984, Evaporite diapirs in the La Popa basin, Nuevo León, Mexico: Geological Society of America Bulletin, v. 95, p. 1219–1225.

Ledesma-Vázquez, J., 1992, Identificación de paleobatimetría por medio de estructuras sedimentárias primárias: XI Convención Geológica Nacional, Sociedad Geológica Mexicana, Libro de Resúmenes, p. 101.

Ledesma-Vázquez, J., and M.E. Johnson, 1992, Determinación de límites de cuencas sedimentárias: XI Convención Geológica Nacional, Sociedad Geológica Mexicana, Libro de Resúmenes, p. 100.

Leveille, G., and E.G. Frost, 1984, Deformed upper Paleozoic-lower Mesozoic cratonic strata, El Capitan, Sonora, Mexico: Geological Society of America Abstracts with Programs, v. 16, p. 575.

Lock, B.E., 1984, Channels in resedimented chalks, Cretaceous Gulf Coastal province of Texas and Mexico: Gulf Coast Association of Geological Societies Transactions, v. 34, p. 373–382.

Longman, M.W., et al., 1983, Description of a paraconformity between carbonate grainstones, Isla Cancun, Mexico: Journal of Sedimentary Petrology, v. 53, p. 533–542.

Longoria, J.F., 1982a, Jurassic microfacies analysis of the Cerro De La Silla Sequence, NE Mexico, its paleogeographic and tectonic implications (abs.), *in* The Jurassic of the Gulf Rim: Third Annual Research Conference Gulf Coast Section, SEPM Foundation, p. 49–50.

Longoria, J.F., 1982b, Anatomy of the Huayacocotla segment of the Sierra Madre Oriental; its stratigraphic, paleogeographic, and tectonic implications on the origin of the Gulf of Mexico: Geological Society of America, Abstracts with Programs, v. 14, p. 549.

Longoria, J.F., 1984a, Cretaceous biochronology from the Gulf of Mexico region based on planktonic microfossils: Micropaleontology, v. 30, p. 225–242.

Longoria, J.F., 1984b, Stratigraphic studies in the Jurassic of northeastern Mexico: Evidence for the origin of the Sabinas basin: Society of Economic Paleontologists and Mineralogists Gulf Coast Section, Third Research Conference Proceedings, p. 171–193.

Longoria, J.F., 1984c, Mesozoic tectonostratigraphic domains in east-central Mexico, *in* G.E.G Westermann, ed., Jurassic-Cretaceous Biochronology and Paleogeography of North America: Geological Association of Canada Special Paper 27, p. 65–76.

Longoria, J.F., 1987a, Oblique subduction and kinematics of the American plate: evidence from the Mesozoic record of Mexico (abs.), *in* T.W.C. Hilde and R.L. Carlson, conveners, Texas A&M University Geodynamics Symposium, p. 90–91.

Longoria, J.F., 1987b, Paleogeographic development of Mexico during the Cretaceous: Actas de la Facultad de Ciéncias, Universidad Autónoma de Neuvo León, Linares, México, v. 2, p. 177–190.

Longoria, J.F., 1987c, Regional unconformities in the Cretaceous succession of Mexico: Actas de la Facultad de Ciéncias, Universidad Autónoma de Neuvo León, Linares, México, v. 2, p. 217–220.

Longoria, J.F., 1988, Late Triassic–Jurassic paleogeography and origin of Gulf of Mexico basin: Discussion: AAPG Bulletin, v. 72, p. 1411–1418.

Longoria, J.F., 1990, Stratigraphic events in the Mesozoic of northeast Mexico: Their relation to the dynamics of the Gulf of Mexico: Geological Society of America Abstracts with Programs, v. 22, p. A185.

Longoria, J.F., and R. Monreal, 1991, Lithostratigraphy, microfacies, and depositional environments of the Mesozoic of Sierra La Nieve, Coahuila, northeast Mexico: Revista de la Sociedad Geológica Espana, v. 4, p. 7–31.

Longoria, J.F., J.F. Maytorena, and J.L. Enríquez, 1986, Evolución tectoestratigráfica de las rocas carbonatadas de la región de Córdoba–Orizaba (abs.): Sociedad Geológica Mexicana, VIII Convención Nacional Resúmenes, p. 81–82.

Longoria, J.F. et al., 1989, Litho and biostratigraphic correlations between the Mesozoic of Mexico and Cuba; tectonostratigraphic implications: Geological Society of America Abstracts with Programs, v. 21, p. 33.

López-Gómez, O., 1987, Importancia económica petrolera del atolón de la Faja de Oro (abs.): Boletín de la Sociedad Venezolana de Geológos, v. 31, p. 62–63.

López-Ramos, E., 1979, Geología de México, 2nd edition, v. III, 446 p. (privately printed).

López-Ramos, E., 1980, Geología de México, 2nd edition, v. II, 454 p. (privately printed).

López-Ramos, E., 1981, Paleogeografía y tectónica del Mesozóico de México: Revista de Instituto de Geología, Universidad Nacional Autónoma México, v. 5, p. 158–177.

López-Ramos, E., 1984, Oil possibilities of Mesozoic in Mexican High Plateau (abs.), *in* W.E. Pratt Memorial Conference, In Future Petroleum Provinces of the World: AAPG Bulletin, v. 68, p. 1204.

López-Ticha, D., 1988 (1985), Revisión de la estratigrafía y potencial petrolero de la Cuenca de Tlaxiaco: Boletín de la Asociación Mexicana de Geólogos Petroleros, v. 37, n. 1, p. 49–92.

Lothringer, C.J., 1984, Geology of a Lower Ordovician allochthon, Rancho San Marcos, Baja California, Mexico, *in* V. Frizzell, Jr., ed., Geology of the Baja California Peninsula: Pacific Section SEPM, v. 39, p. 17–22.

Lugo, H.J., 1985, Morfoestructuras del fondo oceanico Mexicano: Investigaciones Geográficas, Boletín del Instituto de Geografía, 15, p. 9–39.

Lugo-Rivera, J.E., 1985, Biostratigraphy and foraminifera of the Neogene Coatzacoalcos Formation, in the Isthmian Salt Basin, Southeastern Mexico: Unpublished M.A. Thesis, University of Texas at Austin.

Machain-Castillo, M.L., 1985, Pliocene Ostracoda of the Saline Basin, Veracruz, Mexico: Ph.D. Dissertation, Louisiana State University, Baton Rouge, 271 p.

Machain-Castillo, M.L., 1986, Ostracod biostratigraphy and paleoecology of the Pliocene of the Isthmian salt basin, Veracruz, Mexico: Tulane Studies in Geology and Paleontology, v. 19, p. 123–140.

Macias-Ortíz, J.A., and A. Segura-Cuervo, 1984, Estúdio bioestratigráfico del Cretácico y Jurásico

Superior en los pozos Faja de Oro Nos. 1, 2, y 3 en la Planicie Costera del Golfo de Mexico: Memoria III, Congreso Latinoamericano de Paleontología, p. 237–241.

Malpica-Cruz, V.M., 1986, Terrazas marinas Pleistocénicas en la costa central del Estado de Veracruz: 8a Convención Geológica Nacional Resúmenes, p. 164.

Malpica-Cruz, V.M., 1992, Definición de dos provincias sedimentológicas en el litoral del Golfo de México con base a los depósitos eólicos: XI Convención Geológica Nacional, Sociedad Geológica Mexicana, Libro de Resúmenes, p. 101.

Mandujano-Velásquez, J., 1992, Estudio estratigráfico-sedimentológico e implicaciones económicas de las rocas del Cretácico en el área Luna-Sen, Estado de Tabasco: XI Convención Geológica Nacional, Sociedad Geológica Mexicana, Libro de Resúmenes, p. 102.

Marshall, R.H., 1984, Petrology of the subsurface Mesozoic rocks of the Yucatan Platform: Unpublished Master's Thesis, University of New Orleans, 96 p.

Martill, D.M., 1989, A new "Solenhofen" in Mexico: Geology Today, v. 5, p. 25–28.

Martínez, M., J. Barceló-Duarte, V. González-Pacheco, and G. Hernández-Reyes, 1992, Ambientes y facies sedimentárias de la secuéncia Cretácica en la porción central de la Cuenca Morelos-Guerrero: XI Convención Geológica Nacional, Sociedad Geológica Mexicana, Libro de Resúmenes, p. 111.

Martínez-Hernández, E., 1992, Caracterización ambiental del Terciário de la región, Estado de Chiapas—un enfoque palinoestratigráfico: Universidad Nacional Autónoma de México, Instituto de Geología, Revista, v. 10, p. 54–64.

Martínez-Hernández, E., and A. C. Tomasini-Ortiz, 1989, Esporas, hifas y otros restos de hongos fósiles de la cuenca carbonífera de Fuentes-Rio Escondido (Campaniano-Maastrichtiano), Estado de Coahuila: Universidad Nacional Autónoma de México, Instituto de Geología, Revista, v. 8, p. 235–242.

Martínez-Palomares (1985), 1988, Distribución actual e isopacas de las rocas Paleozóicas en el norte de los Estados de Sonora y Chihuahua: Boletín de Asociación Mexicana de Geólogos Petroleros, v. 37, no 2., p. 3–30.

Massingill, J.V., R.N. Bergantino, H.S. Fleming, and R.H. Feden, 1973, Geology and genesis of the Mexican Ridges: Journal of Geophysical Research, v. 78, p. 2498–2507.

Masters, C.D., and J.A. Peterson, 1981a, Assessment of conventionally recoverable petroleum resources of southeastern Mexico, northern Guatemala, and Belize: U.S. Geological Survey Open-File Report 81-1144, p. 1–7.

Masters, C.D., and J.A. Peterson, 1981b, Assessment of conventionally recoverable petroleum resources of northeastern Mexico: U.S. Geological Survey Open-File Report 81-1143, p. 1–7.

Maurrasse, F.J.M.R., 1990, Stratigraphic correlation for the circum-Caribbean region, *in* G. Dengo and J.E. Case, eds., The Caribbean Region: Boulder, Colorado, Geological Society of America, The Geology of North America, v. H, Plates 4, 5A, and 5B.

Maytorena-Silva, J.F., and F.A. Esparza-Yañez, 1990, The Cambrian succession of central Sonora: tectonic implications: Geological Society of America Abstracts with Programs, v. 22, n. 3, p. 65.

McBride, E.F., 1983, Influence of basin history on reservoir quality of sandstones; Upper Cretaceous of northern Mexico (abs.): AAPG Bulletin, v. 67, p. 510.

McCloy, C., 1984, Stratigraphy and depositional history of the San Jose del Cabo Trough, Baja California Sur, Mexico, *in* V.A. Frizzell Jr., ed., Geology of the Baja California Peninsula: Pacific Section SEPM, v. 39, p. 267–273.

McCloy, C., 1987, Neogene biostratigraphy and depositional history of the southern Gulf of California: Geological Society of America Abstracts with Programs, v. 19, p. 764.

McFarlan, E., and L.S. Menes, 1991, Lower Cretaceous, *in* A. Salvador, ed., The Gulf of Mexico Basin: Boulder Colorado, Geological Society of America, The Geology of North America, v. J, Chapter 9.

McKee, J.W., and N.W. Jones, 1984, Pennsylvanian wildflysch at Los Piloncillos and Pico El Fraile, Las Delicias area, Coahuila, Mexico. Geological Society of America Abstracts with Programs, v. 16, p. 589.

McKee, J.W., N.W. Jones, and T.H. Anderson, 1986, The Late Paleozoic and Early Mesozoic geological history of northeastern Mexico: the sedimentary record: Geological Society of America Abstracts with Programs, v. 18, p. 689.

McKee, J.W., N.W. Jones, and T.H. Anderson, 1988, Las Delicias Basin: A record of Late Paleozoic arc volcanism in northeastern Mexico: Geology, v. 16, p. 37–40.

McKee, J.W., N.W. Jones, and L.E. Long, 1990, Stratigraphy and provenance of strata along the San Marcos fault, central Coahuila, Mexico: Geological Society of America Bulletin, v. 102, p. 593–614.

McKee, J.W., M.B. McKee, and T.H. Anderson, 1991, An Aptian–Albian history of northern Sonora and southern Arizona: Geological Society of America Abstracts with Programs, v. 23, p. A127.

McLean, H., B.P. Hausback, and J.H. Knapp, 1987, The geology of west-central Baja California Sur, Mexico: United States Geological Survey Bulletin 1579, p. 11–16.

McNulty, C.L., 1983, Micropaleontological stratigraphic framework for the Cretaceous black lime mudstone lithofacies of the Gulf of Mexico, *in* G.B. Martin and B.F. Perkins, eds., Habitat of Oil and Gas in the Gulf Coast: Program and Abstracts, Gulf Coast Section SEPM, p. 313–328.

Medrano-Morales, L.M., and M.A. Romero-Ibarra, 1993, Caracterización geoquímica de las rocas generadoras de los sistemas petrolíferos del Jurásico Superior en el área marina de Campeche:

Asociación Mexicana Geólogos Petroleros, I Simposio de Geología de Subsuelo, Ciudad del Carmen, Resúmenes, p. 5–16.

Meiburg, Peter, 1987, Paleogeografía y desarrollo estructural del Cretácico de la Sierra Madre Oriental septentrional, México: Actas de la Facultad de Ciéncias, Universidad Autónoma de Nuevo León, Linares, v. 2, p. 197–199.

Meiburg, P., 1992, Desarrollo estructural durante el Mesozóico y Cenozóico del Pre-y Post-Salinar de la Sierra Madre Oriental septentrional: XI Convención Geológica Nacional, Sociedad Geológica Mexicana, Libro de Resúmenes, p. 117.

Meiburg, Peter, et al., 1987, El basamento pre-Cretácico de Aramberri—estructura clave para comprender el decollement de la cubierta Jurásica/Cretácica de la Sierra Madre Oriental, México?: Actas de la Facultad de Ciéncias de la Tierra, Universidad Autónoma de Neuvo León, Linares, v. 2, p. 15–22.

Mejia-Dautt, O., et al., 1980, Evaluación geológica petrolera, plataforma Valles-San Luis Potosí: XI Excursion geológica, Petroleros Mexicanos, Superintendencia General de Distritos de Exploración Petrolera, Zona Norte, 123 p.

Mellor, E.I., and J.A. Breyer, 1981, Petrology of Late Paleozoic basin-fill sandstones, north-central Mexico: Geological Society of America Bulletin, part I, v. 92, p. 367–373.

Mendoza-Rosales, C., I. Arellana-Gil, and G. Silva-Romo, 1992, Nuevas localidades del contacto transicional de las formaciones Huizachal y Huayacocotla: XI Convención Geológica Nacional, Sociedad Geológica Mexicana, Libro de Resúmenes, p. 121.

Meneses-Rocha, J.J., 1987 (1990), Marco tectónico y paleogeografía del Triásico Tardio Jurasico en el sureste de México: Boletín Asociación Mexicana de Geólogos Petroleros, v. 30, p. 3–69.

Meyer, M.G., and W.C. Ward, 1984, Outer ramp limestones of the Zuloaga Formation, Astillero Canyon, Zacatecas, Mexico: Jurassic of the Gulf Rim, 3rd. Research Conference, Gulf Coast Section SEPM, p. 275–282.

Michalzik, D., 1986a, Stratigraphy and paleogeography of the northeastern Sierra Madre Oriental, Mexico (Triassic–lowermost Cretaceous): Zentralblatt für Geologie und Paläontologie, Teil 1, v. 1985, n. 9/10, p. 1161–1170.

Michalzik, D., 1986b, Procedencia y parámetros ambientales de los lechos rojos Huizachal en el área de Galeana, Nuevo León, México: Actas de la Facultad de Ciéncias de la Tierra, Universidad Autónoma de Nuevo León, Linares, v. 1, p. 23–41.

Michalzik, D., 1987, Sedimentación y sucesión de facies en un margen continental pasivo del Triásico al Cretácico Temprano del noreste de la Sierra Madre Oriental, México: Actas de la Facultad de Ciéncias, Universidad Autónoma de Nuevo León, Linares, v. 2, p. 27–31.

Michalzik, D., 1988, Trias bis tiefste Unter-Kreide der nordostlichen Sierra Madre Oriental, Mexico—Facielle Entwicklung eines passiven Kontinental Randes: Ph.D. Thesis, Techn. Hochschule Darmstadt, 247 p.

Michalzik, D., 1991, Facies sequence of Triassic–Jurassic red beds in the Sierra Madre Oriental (NE Mexico) and its relation to the early opening of the Gulf of Mexico: Sedimentary Geology, Amsterdam, Elsevier Science Publishers, v. 71, p. 243–259.

Michalzik, D., and D. Schumann, 1994, Facies relations and paleoecology of Upper Jurassic–Lower Cretaceous fan delta to shelf depositional system in the Sierra Madre Oriental, NE Mexico: Sedimentology.

Michaud, F., 1984a, Foraminiferos y Dacycladaceas del Jurásico Superior y del Cretácico Tardío del Estado de Chiapas, México: Memoria, III Congreso Latinoamericano de Paleontología, v. 3, p. 255–268.

Michaud, F., 1984b, Algunos fósiles de la Formación Ocozocuautla, Cretácico Superior de Chiapas, México: III Memoria, Congreso Latinoamericano de Paleontología, v. 3, p. 425–431.

Michaud, F., 1988a, *Neogyroporella? servaisi* n. sp., nouvelle dasycladacée du Maastrichtien du Chiapas sud-est du Mexique: Cretaceous Research, v. 9, p. 369–378.

Michaud, F., 1988b, Apports de la micropaleontologie a la connaissance stratigraphique de la Formation San Ricardo (Callovien–Neocomien), etát du Chiapas: Revista del Instituto de Geología, Universidad Nacional Autonóma México, v. 7, p. 35–40.

Michaud, F., and E. Fourcade, 1989, Stratigraphie et Paléogéographie du Jurassique et du Crétacé du Chiapas (Sud-Est du Mexique): Bulletin de la Societé Géologique France, v. 8, p. 639–650.

Michaud, F., E. Fourcade, and R. Gutiérrez-Coutino, 1984, Nouvelles données sur le Mésozoïque de la bordure Nord-Est du Batholite du Chiapas, Sud-Est du Mexique: Comptes Rendus de l'Académie des Sciences, Série 2, Mécanique, Physique, Chimie, Sciences de l'Université Sciences de la Terre, v. 299, p. 645–650.

Michaud, F., G. Termier, H. Termier, and E.R. Carranco, 1986, Spongiaires du Jurassique supérieur et du Néocomien du sud-est Mexicain: Comptes Rendus de l'Académie des Sciences, Série 2, Mécanique, Physique, Chimie, Sciences de l'Université, Sciences de la Terre, v. 303, p. 645–650.

Michaud, F., J. Bourgois, E. Barrier, and E. Fourcade, 1989, La série cretacée de Tecoman (Etát de Colima): consequences sur les rapports structuraux entre zones internes et zones de l'édifice montagneux mexicain: Comptes Rendus Académie Science Paris, v. 309, p. 587–593.

Millán, S.M., 1985, Preliminary stratigraphic lexicon of north and central Guatemala: United Nations Development Program, 122 p.

Minero, C.J., 1983a, Backreef facies, El Abra Formation, *in* Sedimentation and Diagenesis of Mid-Cretaceous Platform Margin, East-Central Mexico: Field Guide, Dallas Geological Society, p. 32–46.

Minero, C.J., 1983b, Discontinuity surfaces and correlation in the Sierra de El Abra, *in* Sedimentation and Diagenesis of Mid Cretaceous Platform Margin, East Central Mexico, Field Guide, Dallas Geological Society, p. 63–78.

Minero, C.J., 1983c, Sedimentary Environments and Diagenesis of the El Abra Formation (Cretaceous), Mexico: Ph.D. Dissertation, State University New York at Binghamton, 388 p.

Minero, C.J., 1988, Sedimentation and diagenesis along an island-sheltered platform margin, El Abra Formation, Cretaceous of Mexico, *in* N.P. James et al., eds., Paleokarst: New York, NY, Springer-Verlag, p. 385–405.

Minero, C.J., 1991, Sedimentation and diagenesis along open and island-protected windward carbonate platform margins of the Cretaceous El Abra Formation, Mexico: Sedimentary Geology, v. 71, p. 261–288.

Minero, C.J., P. Enos, and E.J. Aguayo, 1983, Sedimentation and diagenesis of mid-Cretaceous platform margin, east-central Mexico: Field Trip Guidebook, Dallas Geological Society, 168 p.

Minjarez-Sosa, I., J.A. Ochoa-G., P. Sosa-León, 1992, La secuéncia Paleozóica de plataforma de la Sierra Agua Verde: XI Convención Geológica Nacional, Sociedad Geológica Mexicana, Libro de Resúmenes, p. 124.

Mitre-Salazar, L.M., 1989, Secuencias estratigráficas invertidas en el área de la Presa del Junco, Estado de Zacatecas: Universidad Nacional Autónoma de México, Instituto de Geología, Revista, v. 9, p. 52–57.

Moldovanyi, E.P., 1982, Isotopic recognition of successive cementation events within the phreatic environment, Lower Cretaceous Sligo and Cupido formations: Master's Thesis, University of Michigan, Ann Arbor, Michigan.

Moldovanyi, E.P., and K.C. Lohmann, 1984a, Isotopic recognition of successive cementation events within the phreatic environment, Lower Cretaceous Sligo and Cupido Formations, *in* J.L. Wilson, W.C. Ward, and J.M. Finneran, eds., A Field Guide to Upper Jurassic and Lower Cretaceous Carbonate Platform and Basin Systems, Monterrey-Saltillo Area, Northeast Mexico: Gulf Coast Section, SEPM Foundation, p. 52–63.

Moldovanyi, E.P., and K.C. Lohmann, 1984b, Isotopic and petrographic record of phreatic diagenesis: Lower Cretaceous Sligo and Cupido formations: Journal of Sedimentary Petrology, v. 54, p. 972–985.

Monod, O., J. Ramírez, M. Faure, H. Sabaneros, and M.F. Campa, 1992, Preliminary assessment of stratigraphy and structure, San Lucas region, Michoacán and Guerrero States, SW Mexico: Discussion: Mountain Geologist, v. 29, p. 1–2.

Monreal, R., 1987, Regional stratigraphic studies in the Lower Cretaceous of eastern Chihuahua, Mexico: Actas de la Facultad de Ciéncias, Universidad Autónoma de Nuevo León, Linares, v. 2, p. 51–62.

Monreal, R., 1990a, Paleogeographic and tectonic significance of the Lower Cretaceous succession of north-central Mexico: Geological Society of America Abstracts with Programs, v. 22, p. 69.

Monreal, R., 1990b, The Chihuahua trough, its lower Cretaceous stratigraphy, paleogeography, and tectonics: Geological Society of America Abstracts with Programs, v. 22, p. A184.

Montellano-Ballesteros, M., 1990 (1992), Una edad del Irvingtoniano al Rancholabreano para la fauna Cedazo del Estado de Aguascalientes: Universidad Nacional Autónoma de México, Instituto de Geología, Revista, v. 9, p. 195–203.

Montgomery, H.A., 1987a, Microfacies and paleogeographic significance of the Permian patch reefs at Sierra Plomosa, Chihuahua: Universidad Autónoma de Chihuahua Gaceta Geológica, v. 1, n. 1, p. 70–81.

Montgomery, H.A., 1987b, Late Paleozoic paleogeographic development of northern Mexico: Actas de la Facultad de Ciéncias de la Tierra, Universidad Autónoma de Nuevo León, Linares, v. 2, p. 9–14.

Montgomery, H., and J.F. Longoria, 1987, Allochthonous carbonates of the Permian volcanic arc at Los Delicias, Coahuila, Mexico: Geological Society of America Abstracts with Programs, v. 19, p. 776.

Morales-Montaño, M., and J. Cota-Reyna, 1990, Nuevas localidades Cámbricas en Sonora: Geological Society of America Abstracts with Programs, v. 22, n. 3, p. 70.

Morales-Montaño, M., J. Cota-Reyna, and R. López-Soto, 1990, Cambrian-Jurassic relations in Sonora, Mexico: Geological Society of America Abstracts with Programs, v. 22, p. A114.

Morán-Zenteno, D.J., 1984, Geología de la República Mexicana: Universidad Nacional Autónoma México y INEGI, Mexico City, Mexico, 87 p.

Mora-Oropeza, G., and S. Chavez-Reguera, 1992, Estratigrafia de secuéncias sismicas en el Tertiario del sureste de Mexico y sus implicaciones petroleras: XI Convención Geológica Nacional, Sociedad Geológica Mexicana, Libro de Resúmenes, p. 128.

Mora-Oropeza, G., and M.E. Monroy-Audelo, 1992, Paleosedimentación niveles de desacoplamiento tectónico y potencial petrolero en el "Golfo de Huamuxtitlan": XI Convención Geológica Nacional, Sociedad Geológica Mexicana, Libro de Resúmenes, p. 129.

Morón-Rios, A., and M. del C. Perrilliat, 1987, Una especie nueva del género Griffithides Portlock (Arthropoda: Trilobita) del Paleozoico Superior de Oaxaca: Universidad Nacional Autónoma de Mexico, Instituto de Gología, Revista, v. 7, p. 67–70.

Muela, P., Jr., 1985, The geology of the northern portion of Sierra Santa Rita, Chihuahua, Mexico: Master's Thesis, University of Texas at El Paso, 101 p.

Mulsow, M.H., 1983, Petrography of the Lower Cretaceous Cupido Formation—San Lorenzo Canyon, Saltillo, Coahuila, Mexico: Master's Thesis, University of Southwestern Louisiana, Lafayette, Louisiana, 69 p.

Murchey, B.L., 1990, Radiolarian biostratigraphy of Paleozoic siliceous sedimentary rocks in central

Sonora, Mexico: Geological Society of America Abstracts with Programs, v. 22, n. 3, p. 71.

Murray, H.H., 1989, Clay resources and potential in Mexico (abs.): Industrial Minerals, v. 258, p. 61.

Navarro-Baca, J.F., and J.M. Brandi-Purata, 1990, Interpretación geológico-geofísica de la región Sal Somera, Edo de Veracruz: Boletín de la Asociación Mexicana de Geofísicos de Exploración, v. 30, n. 4, p. 11–52.

Neurauter, W.T., et al., 1980, Estratigrafía y geología superficial en la Bahía de Campeche, *in* Symposio Internacional de Mecánica de Suelos Marinos, v. 2, Sociedad Mexicana de Mecánica de Suelos, México, D.F., p. 27–40.

Nishimura, Jose M.G., 1990, Fechamiento K-Ar de arcillas: su aplicación potencial en la exploración petrolera: Boletín de la Asociación Mexicana de Geológos Petroleros, v. 40, n. 2, p. 36–41.

Oki, M., 1976, The biostratigraphy and paleomagnetic stratigraphy of the San Miguel and Olmos formations, Piedras Negras, Coahuila, Mexico: Master's Thesis, University of Southern California at Los Angeles.

Olóriz, F., 1987, El significado biogeográfico de las plataformas mexicanas en el Jurásico superior. Consideraciones sobre un modelo eco-evolutivo: Revista de la Sociedad Mexicana de Paleontología, p. 219–247.

Olóriz, F., et al., 1988a, Análisis isotópicos y consideraciones paleoecológicas en el Jurásico superior de México (Fm. La Casita, Cuencamé, Durango). Datos preliminares: II Congreso Geológico de España, Granada, p. 144–148.

Olóriz, F., et al., 1988b, Las plataformas mexicanas durante el Jurásico superior. Un ejemplo de áreas de recepción en biogeografía: IV Jornadas de Paleontología (Salamanca, Oct. 12-15, 1988), p. 277–287.

Omaña, P.L., 1991, Bioestratigrafía del Paleoceno–Eoceno Inferior basada en foraminíferos planctónicos en la Cuenca de Chicontepec (Ver., Hgo., y S.L.P.): Revista del Instituto Mexicano del Petróleo, v. 23, p. 14–21.

Ornelas de Hernández, M., 1984, El genero *Bonetocardiella* en México y su importancia bioestratigráfica: Memoria, III Congreso Latino americano de Paleontología, v. 3, p. 361–370.

Ornelas-Sánchez N., H. Alzaga-Ruiz, and J. Bustamante-García, 1992, Variaciones del nivel del mar y su influencia sobre la microflora y microfauna del Jurásico Superior-Cretácico Inferior en la plataforma de la Sierra de Chiapas: XI Convención Geológica Nacional, Sociedad Geológica Mexicana, Libro de Resúmenes, p. 137.

Ortega-Gutiérrez, F., 1981 (1984), La evolución tectónica premisisípica del sur de México, *in* J.C. Guerrero, ed., V Symposio sobre evolución tectónica de México: Revista del Instituto de Geología, Universidad Nacional Autónoma México, v. 5, n. 2, p. 140–157.

Ortega-Gutiérrez, F., 1984a, Evidence of Precambrian evaporites in the Oaxacan granulite complex of southern Mexico: Precambrian Research, v. 23, p. 377–393.

Ortega-Gutiérrez, F., 1984b, Relaciones estratigráficas del basamento pre-Oxfordiano de la región Caopas-Rodeo, Zacatecas, y su significado tectónico (abs.): Sociedad Geológica Mexicana, VII Convención Nacional, Resúmenes, p. 56.

Ortega-Gutiérrez, F., and C. González-Arreola, 1985, Una edad Cretácica de las rocas sedimentarias deformadas de la Sierra Juárez, Oaxaca: Revista del Instituto de Geología, Universidad Nacional Autónoma México, v. 5, p. 100–101.

Ortiz-Gómez, P. de L., M.A. Sánchez-Rios, R. Morales-Hernández, and A. Sosa-Patrón, 1992, El nanoplankton calcáreo (familia *Nannoconus*) del Cretácico Inferior del Pozo Tamaulipas 21D, Estado de Tamaulipas: Informe preliminar, Instituto Mexicano del Petróleo, p. 138.

Ortiz-Hernández, L.E., H. Lapierre, and A. Sánchez, 1992, Presencia de un vulcanismo calcoalcalino del Aptiano-Albiano en el borde suroccidental de la plataforma carbonatada de Valles-San Luis Potosí: XI Convención Geológica Nacional, Sociedad Geológica Mexicana, Libro de Resúmenes, p. 140.

Ortuño-Arzate, F., and T. Adatte, 1992, Contexto geodinámico de la Cuenca Mesozoíca de Chihuahua, Mexico: XI Convención Geológica Nacional, Sociedad Geológica Mexicana, Libro de Resúmenes, p. 144.

Ortuño-Arzate, F., and J. Delfaud, 1988, Estudios geodinámicos de evolución de la cuenca mesozoica de Chihuahua en el Norte de México. II Congreso Geológico de España, Granada, p. 309–326.

Oviedo-Pérez, A., and H.L. Martínez-Kemp, 1992, Posibilidades económico-petroleras de la Cuenca de Oaxaca: XI Convención Geológica Nacional, Sociedad Geológica Mexicana, Libro de Resúmenes, p. 147.

Oviedo-Pérez, A.E., J.V. Santamaría, and H.L. Martínez-Kemp, 1992, Estudio geológico económico petrolero integral de la Cuenca de Tlaxiaco (abs.): II Symposio de Asociación Mexicana de Geólogos Petroleros, Instituto Mexicano del Petróleo, México, D.F.

Padilla-Avila, P.E., 1992, Posibles nuevas especies de ostrácodos del Neogeno de la región en Achotal, sureste de México: XI Convención Geológica Nacional, Sociedad Geológica Mexicana, Libro de Resúmenes, p. 148.

Padilla y Sánchez, R.J., 1982, Geologic evolution of the Sierra Madre Oriental between Linares, Concepción del Oro, Saltillo, and Monterrey, Mexico: Ph.D. Dissertation, University of Texas at Austin, 217 p.

Palmer, A.R., W.D. DeMis, W.R. Muehlberger, and R.A. Robison, 1984, Geological implications of Middle Cambrian boulders from the Haymond Formation (Pennsylvanian) in the Marathon basin, West Texas: Geology, v. 12, p. 91–94.

Pantoja-Alor, J., 1992a, La Formación Mal Paso y su importancia en la estratigrafía del sur de México: XI Convención Geológica Nacional, Sociedad

Geológica Mexicana, Libro de Resúmenes, p. 149–151.

Pantoja-Alor, J., 1992b, Geología y paleoambiente de la Cantera Tlayua, Tepexi de Rodríguez, Estado de Puebla: Revista del Instituto de Geologia, Universidad Nacional Autónoma México, v. 9, n. 2, p. 156–169.

Pantoja-Alor, J., and S. Estrada-Barraza, 1986, Estratigrafía de los alrededores de la mina de fierro de El Encino, Jalisco: Sociedad Geológica Mexicana Boletín, v. 47, p. 1–16.

Pate, D.L., 1984, Reconnaissance in the Sierra la Viga, Nuevo León, Mexico: The Texas Caver, v. 29, p. 9–11.

Paterson, D.S., 1983, Sandstone diagenesis and its variation with deltaic depositional environments, Upper Cretaceous, southern Rio Escondido basin, Coahuila, Mexico (abs.): AAPG Bulletin, v. 67, p. 529–530.

Pedrín-Avilés, S., et al., 1990 (1992), Estratigrafía del Pleistoceno superior-Holoceno en el área de la laguna costera de Balandra, Estado de Baja California Sur: Universidad Nacional Autónoma de México, Instituto de Geología, Revista, v. 9, p. 170–176.

Pérez, C., S. Charleston, and R. Malpica, 1984, Una nueva localidad del Paleozóico en México: Sociedad Geológica Mexicana Memoria, 80a Aniversario, VII Convención Nacional, p. 11–22.

Pérez-Cruz, G.A., 1982, Algunas resultados de la investigación geológico-geofísica de la porción noroccidental del Golfo de California: Asociación Mexicana de Geólogos Petroleros, v. 34.

Pérez-Cruz, G.A., 1992, Estilos estructurales de la Cuenca de Burgos y sus perspectivas económico-petroleras (abs.): II Symposio Asociación Mexicana de Geológos Petroleros, Instituto Mexicano del Petróleo, México, D.F.

Pérez-Matus, J.D., 1984, Importancia del conocimiento geológico de los yacimientos petrolíferos para su mejor explotación: Revista del Instituto Mexicano del Petróleo, v. 16, n. 3, p. 18–50.

Pérez-Matus, J.D., and A. Espino-Moreno, 1984, Información geológica necesaria para el desarrollo de algunas estructuras recién perforadas en la Sonda de Campeche: Ingeniería Petrolera, v. 24, n. 5, p. 24.

Pérez-Ortiz, J.A., and C. Tejeda-Galicia, 1993, Estúdio optico de la maduración térmica de los materiales organicos fósiles en los pozos Costero-1, Troje-1 y Saraguatos-1: Asociación Mexicana Geólogos Petroleros, I Simposio de Geologia de Subsuelo, Ciudad del Carmen, Resúmenes, p. 3–4.

Pérez-Segura, E., and C. Jaques-Ayala, 1991, Studies of Sonoran Geology: Geological Society of America Special Paper 254, 136 p.

Perrilliat-Montoya, M. del C., 1984, Monografía de los moluscos del Mioceno medio de Santa Rosa, Veracruz, Parte VII, Pelecípodos; *Deissenidae* a *Verticordiidae*: Paleontología Mexicana, v. 48, 88 p.

Perrilliat-Montoya, M. del C., 1986 (1960), Moluscos del Mioceno de la Cuenca Salina del Istmo de Tehuantepec México: Instituto Geología y Paleontológia Mexicana, Unversidad Nacional Autónoma México, n. 8, p. 12–15. (English translation in U.S.G.S. files, Reston, Virginia.)

Perrilliat-Montoya, M. del C., 1986 (1963), Moluscos de la Formación Agueguexquite (Mioceno Medio) del Istmo de Tehuantepec, México: Instituto de Geológia y Paleontológia, Universidad Nacional Autonoma México, n. 14, p. 7. (English translation in U.S.G.S. files, Reston, Virginia.)

Perrilliat-Montoya, M. del C., 1986 (1977), Monografía de los moluscos de Mioceno medio de Santa Rosa, Veracruz, México, Parte IV: Instituto de Geología y Paleontología, Universidad Nacional Autónoma México, n. 43, p. 13–15, 29–30. (English translation in U.S.G.S. file, Reston, Virginia.)

Perrilliat-Montoya, M. del C., 1987, Gasterópodos y un cefalópodo de la Formación Ferrotepec (Mioceno medio) de Michoacán: Paleontología Mexicana v. 52, 58 p.

Perry, E., et al., 1987, Carbonate platform accretion by cementation of ribbons of dune ridge sediment, Yucatan, Mexico (abs.): AAPG Bulletin, v. 71, p. 601.

Perry, E., et al., 1988, Modern dolomite occurrence, Noc Ac Cenote, Yucatan, Mexico (abs.): EOS American Geophysical Union Transactions, v. 69, p. 355.

Perry, E., et al., 1989, Geologic and environmental aspects of surface cementation, north coast, Yucatan, Mexico: Geological Society of America, v. 17, p. 818–821.

Pessagno, E. A., Jr., and D. M. Hull, 1992, Stratigraphic and chronostratigraphic studies of Jurassic and Lower Cretaceous radiolaria from San Pedro del Gallo, Durango, Mexico: Geological Society of America Abstracts with Programs, v. 24, n. 7, p. A361.

Pessagno, E.A., Jr., et al., 1987, Studies of North American Jurassic Radiolaria; Part I, Upper Jurassic (Kimmeridgian–upper Tithonian) *Pantanelliidae* from the Taman Formation, east-central Mexico; tectonostratigraphic, chronostratigraphic, and phylogenetic implications: Special Publications—Cushman Foundation for Foraminiferal Research, v. 23, p. 1–51.

Phillips, J.D., 1986, Geology of the southern portion of Sierra de Palomas, northwestern Chihuahua, Mexico: Master's Thesis, University of Texas at El Paso, 95 p.

Phillips, J.R., 1984, "Middle" Cretaceous metasedimentary rocks of La Olvidada, northeastern Baja California, Mexico, *in* V.A. Frizzel, ed., Geology of the Baja California Peninsula: Pacific Section SEPM, v. 39, p. 37–41.

Pingatore, N.E., et al., 1984, Limestone stratigraphy, central Sierra Peña Blanca, Chihuahua, Mexico, *in* E.C. Ketternbrink, ed., West Texas Geological Society Publication 84–80, p. 116–123.

Poignant, A.F., and F. Michaud, 1985, *Lithophyllum-berriozabalense et Lithothamnium subguabairense,* deux nouvelles espéces de *Mélobésiées* du Crétacé

Supérieur Mexicain: Bulletin des Centres de Recherches Exploration-Production Elf-Aquitaine, v. 9, (1), p. 127–135.

Poole, F.G., and R.J. Madrid, 1986, Paleozoic rocks in Sonora (Mexico) and the relation to the southwestern continental margin of North America: Geological Society of America Abstracts with Programs, v. 18, p. 720–721.

Poole, F.G., and R.J. Madrid, 1988, Comparison of allochthonous Paleozoic eugeoclinal rocks in the Sonoran, Marathon, and Antler orogens: Geological Society of America Abstracts with Programs, v. 20, p. A267.

Pouyet, S., and Herrera, 1986, Systematics and paleogeographical studies of some species of Bryozoa (Cheilostomata) from the Gulf of Mexico: Universidad Nacional Autónoma de México, Instituto de Geología, Revista, v. 6, p. 204–221.

Prieto-Mendoza, J.J., 1992, Fosfogénesis actual en Baja California Sur: XI Convención Geológica Nacional, Sociedad Geológica Mexicana, Libro de Resúmenes, p. 157.

Puy-Alquiza, M.I. 1992, Implicación de las secuéncias litológicas de las islas San Jose, Espíritu Santo y la región de Punta Coyotes, Baja California Sur, México: XI Convención Geológica Nacional, Sociedad Geológica Mexicana, Libro de Resúmenes, p. 160.

Quezada-Muñetón, J.M., 1983, Las formaciones San Ricardo y Jericó del Jurásico Medio-Cretácico Inferior en el sureste de México: Boletín de Asociación Mexicana de Geólogos Petroleros, v. 35, n. 1, p. 37–64.

Quezada-Muñetón, J.M., 1984, El Grupo Zacatera del Jurásico Medio-Cretácico Inferior de la Depresion Istmica, 20 km al Norte de Matias Romero, Oax: Memorias, VII Convención Nacional, Sociedad Geológica Mexicana, p. 40–59.

Quezada-Muñetón, J.M., 1990 (1987), El Cretácico Medio-Superior, y el Límite Cretácico Superior-Terciario Inferior en la Sierra de Chiapas: Asociación Mexicana de Geólogos Petroleros, v. 39.

Quintero-L., O., 1984, El basamento Precámbrico de Chihuahua en el Rancho el Carrizalillo, Municipio de Aldama, Chihuahua: VII Convención Nacional, Sociedad Geológica Mexicana Resúmenes, Mexico D.F., p. 55.

Quintero-L., O., and J.C. Guerrero, 1985, Una nueva localidad del basamento Precámbrico de Chihuahua, en el área de Carrizalillo: Instituto de Geología, Revista, v. 6, p. 98–99.

Ramírez-Ramírez, C., 1986, Pre-Mesozoic tectonic evolution of Huizachal-Peregrina anticlinorium, and adjacent parts of eastern Mexico: Geological Society of America Abstracts with Programs, v. 18, p. 725.

Ramírez-Ramírez, C., 1991, Pre-Mesozoic geology of Huizachal-Peregrina anticlinorium, Ciudad Victoria Tamaulipas, and adjacent parts of eastern Mexico: Ph.D. Thesis, University of Texas at Austin, 313 p.

Reaser, D.F., R.W. Bacon, and S.E. de la Vega, 1989, Geology of Sierra de Catorce San Luis Potosí, Mexico (abs.): 28th International Geological Congress Abstracts, v. 2, p. 679–680.

Reyeros de Castillo, M.M., 1983, Corales de algunas formaciones Cretácicas del Estado de Oaxaca: Paleontología Mexicana, v. 47, 67 p.

Reyes-Cortés, I.A., 1986, Cerro Carrizalillo: localidad tipo Precámbrico-Paleozóico de Chihuahua, *in* I.A. Reyes-C., ed., Excursión Geológica-Precámbrico de Chihuahua: Sociedad Geológica Mexicana A. C., Delegación Chihuahua, p. 31–35.

Reyes-Cortés, I.A., and P.E. Potter, 1987, Interpretación sedimentológica de la Formación Falomir, en el Cerro de Carrizalillo, Chihuahua, México: Universidad Autónoma de Chihuahua Gaceta Geológica, v. 1, n. 1, p. 31–35.

Richards, Trenton H., 1986, Mineralogy and morphology of the Gruta Cuatro Palmas, Coahuila, Mexico: Master's Thesis, West Texas State University, Canyon, Texas, 101 p.

Ricoy, J.U., 1989, Tertiary terrigenous depositional systems of the Mexican Isthmus Basins: Unpublished Ph.D. Dissertation, University of Texas at Austin, 145 p.

Riggs, N.R., and C.M. Busby-Spera, 1987, Tectonic setting of Lower Jurassic rocks in southern Arizona: Geological Society of America Abstracts with Programs, v. 19, p. 820.

Ritchie, A.W., and R.C. Finch, 1985, Widespread Jurassic strata on the Chortis block of the Caribbean plate: Geological Society of America Abstracts with Programs, v. 17, p. 700–701.

Rivera, Jorge E.L., 1985, Biostratigraphy and foraminifera of the Neogene Coatzacoalcos Formation, in the Isthmian Salt Basin, southeastern Mexico: Master's Thesis, University of Texas at Austin, 157 p.

Rivera-Carranco, E., 1987, Condiciones paleoambientales de depósito de las formaciones cámbricas del área de Caborca, Sonora: Universidad Nacional Autónoma de Mexico, Instituto de Geologá, Revista, v. 7, p. 22-27.

Rivera-Carranco, E., 1988, Génesis de la Formación Proveedora (Cámbrico Inferior) del área de Caborca, Sonora noroccidental: Universidad Nacional Autónoma de México, Instituto de Geología, Revista, v. 7, p. 163-167.

Rivera-Carranco, E., E.A. Hernández, and B.E. Buitrón, 1984, *Septaliphoria potosina* n. sp. (Brachiopoda-Rhynchonellida) del Jurásico Tardío de la Sierra de Catorce, San Luis Potosí, México: Memoria, III Congreso Latinoamericano de Paleontología, p. 216–224.

Roberts, D.C., and R. Dyer, 1988, A preliminary report on the geology of the Cerro Panales area, east-central Chihuahua, Mexico, *in* K.F. Clark, P.C. Goodell, and J.M. Hoffer, eds., Stratigraphy, Tectonics and Resources of Parts of Sierra Madre Occidental Province, Mexico, Guidebook for the 1988 Field Conference: El Paso Geological Society, p. 159–172.

Rocha, V.R., C.C. Villegas, and T.C. Velásquez, 1990, Estudio sismoestratigráfico de la Franja Tamabra, en el Estado de Veracruz, México: Instituto Mexicano del Petróleo, Revista, v. 22, n. 3, p. 6–18.

Rodgers, R.W., 1989a, The La Caja Formation, *in* K. Greier and J.L. Kowalski, eds., Geology of the Sierra de Catorce Uplift, Real de Catorce, San Luis Potosí, Mexico, Guide Book: The University of Texas-Pan American Geological Society for the Gulf Coast Association of Geological Societies 39th Annual Convention, p. 67–72.

Rodgers, R.W., 1989b, Paleozoic metasediments, *in* K. Greier and J.L. Kowalski, eds., Geology of the Sierra de Catorce Uplift, Real de Catorce, San Luis Potosí, Mexico, Guide Book: The University of Texas-Pan American Geological Society for the Gulf Coast Association of Geological Societies 39th Annual Convention, p. 29–32.

Rodgers, R.W., 1990, Tectonics and stratigraphy of the Sierra Catorce uplift-north central Mexico: Geological Society of America Abstracts with Programs, v. 22, p. A113.

Rodríguez-Castañeda, J.L., 1988, Estratigrafía de la región de Tuape, Sonora: Revista del Instituto de Geología, Universidad Nacional Autónoma México, v. 7, p. 52–66.

Rodríguez-Martínez, J.M., 1983, Condiciones geotérmicas en el desarrollo de los ciclos generadores de hidrocarburos en la Plataforma El Burro-Picachos, Coah: Geomimet, v. 3, p. 39–46.

Roldán-González, G., 1984, Interpretación paleoambiental del Eocene Superior, área Tecolutla: Memoria, III Congreso Latinoamericano de Paleontología, v. 3, p. 461–470;

Román-Ramos, J.R., and B. Márquez-Domínguez, 1993, Evaluación geológico-geoquímico de los sistemas petroleros de la Cuenca de Sabinas: Asociación Mexicana Geólogos Petroleros, I Simposio de Geología de Subsuelo, Ciudad del Carmen, Resúmenes, p. 17–119.

Romero-Morales, P.R., 1987, Algunas observaciones sobre la geología en la Sierra Samalayuca: Universidad Autónoma de Chihuahua Gaceta Geológica, v. 1, n. 1, p. 160–162.

Rosales, E., and R. Bello-Montoya, 1992, Eventos anóxicos Mesozóicos registrados en secuéncias adyacentes a la costa del Golfo de México: XI Convención Geológica Nacional, Sociedad Geológica Mexicana, Libro de Resúmenes, p. 169.

Rosales-Domínguez, M. del C., 1984, Bioestratigrafía del Aptiano Superior-Albiano (Cretácico) en la porción oriental del Estado de Durango: Memoria, III Congreso Latinoamericana de Paleontología, v. 3, p. 296–307.

Rosales-Lomelí. J., M.L. Ayala-Nieto, E. Martínez-Hernández, and E. Rosales-Contreras, 1992, Palinoestratigrafía de la Formación San Ricardo en la sección Rio Negro, Municipio de Cintralpa, Chiapas: XI Convención Geológica Nacional, Sociedad Geológica Mexicana, Libro de Resúmenes, p. 170.

Ross, C.A., and J.R.P. Ross, 1985a, Carboniferous and Early Permian biogeography: Geology, v. 13, p. 27–30.

Ross, C.A., and J.R.P. Ross, 1985b, Paleozoic tectonics and sedimentation in west Texas, southern New Mexico, and southern Arizona, *in* P.W. Dickerson and W.R. Muehlberger, eds., Structure and Tectonics of Trans-Pecos Texas: Field Conference: West Texas Geological Society Publication 85-81, p. 221–230.

Ross, M.A., 1981, Stratigraphy of the Tamaulipas Limestone, Lower Cretaceous, Mexico, *in* E.I. Smith, ed., Lower Cretaceous Stratigraphy and Structure, Northern Mexico: West Texas Geological Society Pub. 81-74, p. 43–57.

Rozo-Vera, G.A., 1990 (1992), Implicaciones paleoceanográficas de los foraminíferos bentónicos cuaternários de la boca del Golfo de California: Universidad Nacional Autónoma de México, Instituto de Geología, Revista, v. 9, p. 177–194.

Rozo-Vera, G., and A.L. Carreño, 1988, Distribución de foraminíferos plantónicos en sedimentos superficiales del Golfo de California: Universidad Nacional Autónoma de México, Instituto de Geología, Revista, v. 7, p. 217–225.

Russell, J.L., J.A. Baskin, and J.A. Peterson, 1984, Structure and Mesozoic stratigraphy of northeast Mexico: Corpus Christi Geological Society, 113 p.

Saldívar, R., and V.H. Garduño, 1984, Estúdio estratigráfico y estructural de las rocas del Paleozóico Superior de Santa María del Oro, Durango y sus implicaciones tectónicas: Sociedad Geológico de México, VII Convención Nacional, Libro de Resúmenes, p. 37–38.

Salmerón-U., P., 1985, El límite Eoceno-Oligoceno en México: Revista del Instituto Mexicano del Petróleo, v. 17, n. 3, p. 5–16.

Salmerón-U., P., 1986, The Eocene-Oligocene boundary in Mexico, *in* Developments in Paleontology and Stratigraphy: Elsevier Science Publishing Co., Amsterdam, Netherlands, v. 9, p. 189–192.

Salvador, A., 1987, Late Triassic–Jurassic paleogeography and origin of the Gulf of Mexico Basin: AAPG Bulletin, v. 71, p. 419–451.

Salvador, A., 1988, Reply to "Late Triassic–Jurassic paleogeography and origin of Gulf of Mexico Basin: Comments by J. Longoria": AAPG Bulletin, v. 72, p. 1419–1422.

Salvador, A., 1991a, Triassic–Jurassic, *in* A. Salvador, ed., The Gulf of Mexico Basin: Geological Society of America, The Geology of North America, v. J, p. 131–180.

Salvador, A., 1991b, Origin and development of the Gulf of Mexico basin, *in* A. Salvador, ed., The Gulf of Mexico Basin: Geological Society of America, The Geology of North America, p. 389–444.

Sánchez-Barreda, L.A., 1981, Geologic evolution of the continental margin of the Gulf of Tehuantepec in southwestern Mexico: Unpublished Ph.D. Dissertation, University of Texas at Austin, 191 p.

Sánchez-Barreda, L.A., 1990, Why wells have failed in southern Belize area: Oil and Gas Journal, August 20, p. 97–103.

Sánchez-Martin, V., and J. Barceló-Duarte, 1990, Estudio sedimentológico de los Lechos Rojos (Fm

Huizachal y La Boca) en el área Huizachal-Peregrina, Tamps.: X Convención Geológica Nacional, Sociedad Geológica Mexicana, Libro de Resúmenes, p. 73.

Sánchez-Mejorada, P., 1986, Evaporite deposit of Laguna del Rey: Transactions of the Society of Petroleum Engineers, v. 280, p. 1923–1927.

Sánchez-R., M.A., J.C. González, J.A. Gómez, S. Sánchez-M., 1992, Nannoplancton calcareo, foraminíferos planctónicos y bentónicos del Mioceno Medio-Plioceno Inferior de Achotal (SE de Mexico): XI Convención Geológica Nacional, Sociedad Geológica Mexicana, Libro de Resúmenes, p. 174.

Sandoval, J., and G.E.G. Westermann, 1986, The Bajocian (Jurassic) ammonite fauna of Oaxaca, Mexico: Journal of Paleontology, v. 60 (6), p. 1220–1271.

Sandstrom, M.A., 1982, Stratigraphy and environments of deposition of the Zuloaga Group, Victoria, Tamaulipas, Mexico, *in* The Jurassic of the Gulf Rim: Third Annual Research Conference Gulf Coast Section SEPM Foundation, p. 94–97.

Sandstrom, M.A., 1986, Stratigraphy and sedimentation of limestones referred to the Zuloaga Group, Victoria, Tamaulipas, Mexico: Geological Society of America Abstracts with Programs, v. 18, p. 738.

Santiago-Acevedo, J.S., 1980, Giant fields of the southern zone; Mexico, *in* M.T. Halbouty, ed., Giant Oil and Gas Fields of the Decade 1968–1978: AAPG Memoir 30, p. 339–385.

Santiago-Acevedo, J.S., and A. Baro, 1992, Mexico's Giant Fields, 1978–1988 Decade, *in* M.T. Halbouty, ed., Giant Oil and Gas Fields of the Decade 1978–1988: AAPG Memoir 54, p. 73–79.

Santiago-Acevedo, J.S., and O. Mejia-Dautt, 1980, Giant fields in the southeast of Mexico: Gulf Coast Association of Geological Societies Transactions, v. 30, p. 1–31.

Santiago-Acevedo, J.S., J. Carrillo-Bravo, and A.B. Martell, 1984, Geología petrolera de México, *in* Evaluatión de formaciones en México: Schlumberger, p. 1–36.

Sassen, R., 1990, Geochemistry of carbonate source rocks and crude oils in Jurassic salt basins of the Gulf Coast, *in* J. Brooks, ed., Classic Petroleum Provinces: London, The Geological Society Special Publication n. 50, p. 265–277.

Schlatter, R., and R. Schmidt-Effing, 1984, Bioestratigrafía y fauna de amonites del Jurásico Inferior (Sinemuriana) del área de Tenango de Doris, Estado de Hidalgo, México: Memoria, III Congreso Latinoamericano de Paleontología, v. 3, p. 154–155.

Schmidt-Effing, R., 1980, The Huayacocotla aulacogen in Mexico (Lower Jurassic) and the origin of the Gulf of Mexico, *in* R.H. Pilger Jr., ed., The origin of the Gulf of Mexico and the Early Opening of the Central North Atlantic Ocean: Baton Rouge, Louisiana, Louisiana State University, p. 79–86.

Schmitt, R., 1986a, Faziesmodell einer Unterkretazischen Karbonatplattform in Nordost, Mexiko (abs.): Geologie und Palaeontologie, Sonderband, p. 202–203.

Schmitt, R., 1986b, Desarrollo de una plataforma carbonatada durante el Cretácico Inferior en el noreste de México: Actas de la Facultad de Ciéncias de la Tierra, Universidad Autónoma de Nuevo León, Linares, v. 1, p. 42–48.

Schmitz, K.R., and E.C. Perry, Jr., 1984, Textural, chemical, and isotopic variations in Plio-Pleistocene dolomitic limestones; the northeastern Yucatan Peninsula (abs.): SEPM Meeting 1, p. 72.

Schoenherr, P., 1988, Litho- und Mikrobiostratigraphie der Mittel und Oberkreide Nordmexikos (Alb bis Campan): Ph.D. Dissertation, University Hannover, Hannover, Germany, 179 p.

Schumann, D., 1985, Diagénesis de los cefalópodos y reconstrucción del paleo-ambiente de la Formación La Casita del Jurásico de Nuevo León (nordeste de Mexico) (abs.): Memorias Congreso Latinoamericano de Geológia, v. 6, p. 182–183.

Schumann, D., 1986, Fazies und Sedimentation im Oberjura (Kimmeridge/Tithon) von Nord-Ost-Mexiko (abs): Geologie und Paleontologie, Sonderband, p. 75–76.

Schumann, D., 1987, Sedimentologische Analyse eines Oberjurassischen Meeresraumes in der Provinz Nuevo León (NE-Mexiko), *in* Treffen 2 Deutschsprachiger Sedimentologen, Heidelberger Geowissenschaft Arbeiten, 8, p. 217.

Schumann, D., 1988, Environment and postmortem history of Upper Jurassic ammonites in Nuevo León, NE Mexico: *in* J. Wiedmann et al., eds, Second International Schindewolf Symposium, v. 2, p. 731–736.

Scott, R.W., 1984a, Significant fossils of the Knowles Limestone, Lower Cretaceous, Texas, *in* W.P.S. Ventress, D.G. Bebout, B.F. Perkins, and C.H. Moore, eds., Jurassic of the Gulf Rim: Gulf Coast Section SEPM, p. 333–346.

Scott, R.W., 1984b, Mesozoic biota and depositional systems of the Gulf of Mexico-Caribbean region, *in* G.E.G. Westerman, ed., Jurassic–Cretaceous Biochronology and Paleogeography of North America: Geological Association of Canada Special Paper 27, p. 49–64.

Scott, R.W., 1990, Models and stratigraphy of Mid-Cretaceous reef communities, Gulf of Mexico, *in* Concepts in Sedimentology and Paleontology, SEPM Publication, v. 2, 102 p.

Scott, R.W., and C. González-León, 1991, Paleontology and biostratigraphy of Cretaceous rocks, Lampazos area, Sonora, Mexico, *in* E. Pérez-Segura and C. Jacques-Ayala, eds., Studies of Sonoran Geology: Geological Society of America Special Paper 254, p. 51–67.

Sedlock, R.L., and Y. Isozaki, 1990, Lithology and biostratigraphy of Franciscan-like chert and associated rocks, west-central Baja, California, Mexico. Concepts in Sedimentology and Paleontology: Geological Society of America Bulletin, v. 102, p. 852–864.

Seibertz, E., 1986a, Strömungs-Richtungen und

Sediment-Transport in der Mittelkreide Nord-Mexikos (abs.) Berliner Geowissenschaftliches Abhandlungen, Reihe A: Geologie und Palaeontologie, Sonderband, p. 72–73.

Seibertz, E., 1986b, Paleogeography of the San Felipe Formation (Mid-Cretaceous, NE Mexico) and facial effects upon the *inoceramids* of the Turonian/Coniacian transition: Zentralblatt für Geologie und Paläontologie, Teil I: Allgemeine, Angewandte, Regionale und Historische Geologie, 1985 (9–10), p. 1171–1181.

Seibertz, E., 1986c, Die Belemniten aus der Mittelkreide von Tepexi de Rodríguez bei Puebla (Mexiko) *in* 10, Geowissenschaftliches Latein amerika-Kolloquium; Kurzfassungen der Beiträge. Berliner Geowissenschaftliche Abhandlungen, Reihe A: Geologie und Palaeontologie, Sonderband, p. 187.

Seibertz, E., 1987, Interpretación genética de un dique de basalto en el Turoniano Inferior de la Sierra de Tamaulipas y su datación bioestratigráfica con *Inoceramidos* (Cretácico Medio, NE de México): Actas de la Facultad de Ciéncias, Universidad Autónoma de Neuvo León, Linares, v. 2, p. 147–150.

Seibertz, E., 1990, El desarrollo Cretácico del Archipiélago de Tamaulipas. II Génesis y datación de un dique de basalto y su efecto al ambiente deposicional Medio Cretácico de la Sierra de Tamaulipas (Cenomaniano/Turoniano, NE México): Actas de la Facultad de Ciéncias de la Universidad Autonoma de Nueva León, Linares, v. 4, p. 99–124.

Seibertz, E., and B.E. Buitrón, 1988, La localidad tipo de la Formación Xilitla, San Luis Potosí (Cretácico Superior basal): Revista del Instituto de Geología, Universidad Nacional Autónoma Mexico, v. 7, n. 1, p. 116–118.

Selvius, D.B., and J.L. Wilson, 1985, Lithostratigraphy and algal-foraminiferal biostratigraphy of the Cupido Formation, Lower Cretaceous, northeast Mexico: Gulf Coast Section, SEPM 4th Annual Research Conference Proceedings, p. 285–311.

Silva-Piñeda, A., 1981 (1983), *Asterotheca* plantas asociadas de la Formación Huizachal (Triásico Superior) del Estado de Hidalgo: Revista del Instituto de Geología, Universidad Nacional Autónoma de México, v. 5, p. 47–54.

Silva-Piñeda, A., 1984, Frutos del Cretácico Superior del Estado de Coahuila, México: Memoria, III Congreso Latinoamericano de Paleontología, v. 3, p. 432–437.

Silva-Piñeda, A., 1992, Presencia de Otozamites (Cycadophyta) y Podozamites (Coniferophyta) en el Jurásico Superior (Kimeridgiano-Titoniano) del sur del Estado de Veracruz: Universidad Nacional Autónoma de México, Instituto de Geología, Revista, v. 10, p. 94.

Silva-Piñeda, A., and R.H. Alzaga, 1991, Una nueva localidad con plantas del Jurásico en el Estado de Puebla, Mexico: Revista del Instituto Mexicano del Petróleo, v. 23, n. 2, p. 13–16.

Silva-Piñeda, A., and S. González-Gallardo, 1988, Algunas bennettitales (Cycadophyta) y coniferales (Coniferophyta) del Jurásico medio del área de Cualac, Guerrero: Universidad Nacional Autónoma de México, Instituto de Geología, Revista, v. 7, p. 244.

Silva-Piñeda, A., J. Pantoja-Alor, and B.E. Buitrón-Sánchez, 1992, El Paleozóico Tardio de México de acuerdo a su paleobiota: XI Convención Geológica Nacional, Sociedad Geológica Mexicana, Libro de Resúmenes, p. 182–184.

Sivils, D.J., 1987, Stratigraphy and structure of Sierra de Palomas: Chihuahua, Mexico: Universidad Autónoma de Chihuahua, Gaceta Geológica, v. 1, n. 1, p. 176–202.

Sivils, D.J., and J.D. Phillips, 1986, Geology of Sierra de Palomas, Chihuahua, Mexico, *in* Geology of south central New Mexico, El Paso Geological Society Field Trip: El Paso Geological Society, El Paso, Texas, p. 60–66.

Smith, B.A., 1986, Upper Cretaceous stratigraphy and the Mid-Cenomanian unconformity of east central Mexico: Ph.D. Dissertation, University of Texas at Austin, 191 p.

Smith, C.I., 1981, Review of geologic setting, stratigraphy, and facies distribution of the Lower Cretaceous in northern Mexico, *in* Lower Cretaceous Stratigraphy and Structure, Northern Mexico: West Texas Geological Society Field Trip Guidebook, Publication 81-74, p. 1–27.

Smith, D.L., 1987, Fault-controlled sedimentation in a mid-Cretaceous forearc basin—Valle Formation, Cedros Island, Baja California, Mexico: Geological Society of America Abstracts with Programs, v. 19, p. 452.

Smith, J.T., 1984, Miocene and Pliocene marine mollusks and preliminary correlations, Vizcaino Peninsula to Arroyo La Purisima, northwestern Baja California Sur, Mexico, *in* V.A. Frizzell Jr., ed., Geology of the Baja California Peninsula: Pacific Section SEPM, v. 39, p. 197–217.

Smith, S.A., R.F. Sloan, T.L. Pavlis, and L.F. Serpa, 1989, Reconnaissance study of Cretaceous rocks in the State of Colima, Mexico: A volcanic arc deformed by right-lateral transpression: EOS, American Geophysical Union Transactions, v. 70, p. 1313.

Socki, R., S. Gaona-Vizcayno, E. Perry, and M. Villasuso-Pino, 1984, A chemical drill; sulfur isotope evidence for the mechanism of formation of deep sinkholes in tropical karst, Yucatan, Mexico (abs.): Abstracts with Programs, Geological Society of America, v. 16, p. 662.

Sohl, N.F., R.E. Martínez, P. Salmerón-Ureña, and F. Soto-Jaramillo, 1991, Upper Cretaceous, *in* A. Salvador, ed., The Gulf of Mexico Basin: Boulder, Colorado, Geological Society of America, The Geology of North America, v. J., p. 205–244.

Soto-Jaramillo, F., 1980, Zonificación microfaunística de parte de los estratos Cretácicos del Cañón de la Borrega, Tamaulipas: V. Convención Geológica Nacional, Sociedad Geológica Mexicana, Resúmenes, p. 42–43.

Soto-Navarro, P.R., S. Ortuño-Arzate, J. Zaldivar-Ruiz, and Ma. de los Angeles Hernández-J., 1992, Evolución geodinámica de la Cuenca Mesozóica de Tlaxiaco: XI Convención Geológica Nacional, Sociedad Geológica Mexicana, Libro de Resúmenes, p. 187.

Southworth, C.S., 1984, Structural and hydrogeologic applications of remote sensing data, eastern Yucatan Peninsula, Mexico, *in* B.F. Beck, ed., Sinkholes: Their Geology, Engineering, and Environmental Impact: Rotterdam, Netherlands, A.A. Balkema, p. 59–64.

Stanley, G.D., M.R. Sandy, and C. Gonzáles-León, 1991, Upper Triassic fossils from the Antimonio Formation, northern Sonora, support the Mojave-Sonora megashear: Geological Society of America Abstracts with Programs, v. 23, p. A127–128.

Stass, K. 1988, Fazielle Entwicklung der Formation Papantón (Alb und Cenoman) am Westrande der zentralamerikanische Hochebene: Ph.D. Dissertation, Universitaet Hannover, Hannover, Germany, 176 p.

Steele, D.R., 1982, Physical stratigraphy and petrology of the Cretaceous Sierra Madre limestone, west-central Chiapas, Mexico: Unpublished Master's Thesis, University of Texas at Arlington, 174 p.

Steele, D.R., 1986, Physical stratigraphy and petrology of the Cretaceous Sierra Madre Limestone, west-central Chiapas, *in*, D.R. Steele et al., eds., Contributions to the Stratigraphy of the Sierra Madre Limestone (Cretaceous) of Chiapas: Boletín Instituto de Geología, Universidad Naciónal Autónoma México, v. 102, p. 1–101.

Stevens, C.H., 1982, The Early Permian *Thysanophyllum* coral belt: another clue to Permian plate tectonic reconstructions: Geological Society of America Bulletin, v. 93, p. 798–803.

Stewart, J.H., 1990, Latest Proterozoic and Paleozoic history of the southern margin of North America in Mexico and the United States: Geological Society of America Abstracts with Programs, v. 22, p. A113.

Stewart, J.H., and J. Roldán-Quintana, 1986, Late Triassic rift basins in northern Mexico: New information from the Barranca Group: Geological Society of America Abstracts with Programs, v. 18, p. 764.

Stewart, J.H., and J. Roldán-Quintana, 1991, Upper Triassic Barranca Group: Nonmarine and shallow-marine rift-basin deposits of northwestern Mexico, *in* E. Pérez-Segura, and C. Jacques-Ayala, Studies in Sonoran Geology: Geological Society of America Special Paper 254, p. 19–36.

Stewart, J.H., M.A.S. McMenamin, and J.M. Morales-Ramírez, 1984, Upper Proterozoic and Cambrian rocks in the Caborca region, Sonora, Mexico: Physical stratigraphy, biostratigraphy, paleocurrent studies and regional relations: United States Geological Survey Professional Paper 1309, 36 p.

Stewart, J.H., T.H. Anderson, G.B. Haxel, L.T. Silver, and J.E. Wright, 1986, Late Triassic paleogeography of the southern Cordillera: The problem of a source for voluminous volcanic detritus in the Chinle Formation of the Colorado Plateau region: Geology, v. 14, p. 567–570.

Stewart, J.H., F.G. Poole, K.B. Ketner, R.J. Madrid, J. Roldán-Quintana, and R. Amaya-Martínez, 1990, Tectonics and stratigraphy of the Paleozoic and Triassic southern margin of North America, Sonora, Mexico, *in* G.E. Gehrels, and J.E. Spencer, eds., Geologic Excursions Through the Sonoran Desert Region, Arizona and Sonora: Arizona Geological Survey Special Paper 7, p. 183–202.

Stone, P., and C.H. Stevens, 1987, Permian continental margin of southwestern North America: Geological Society of America Abstracts with Programs, v. 19, p. 857.

Stone, P., and C.H. Stevens, 1988, Pennsylvanian and Early Permian paleogeography of east-central California: implications for the shape of the continental margin and the timing of continental truncation: Geology, v. 16, p. 330–333.

Stump, T.E., 1981, Marine Miocene faunas of Tiburón Island, Sonora, Mexico and their zoogeographic implications, *in* L. Ortlieb and J. Róldan Quintana, eds., The Geology of Northwestern Mexico and Southern Arizona: Instituto de Geología Universidad Nacional Autónoma México, Estación del Noreste, p. 105–124.

Suarez-Vidal, F., 1992, Formación de bancos orgánicos en ambiente de post-arco, durante el Aptiano-Albiano en Baja California: XI Convención Geológica Nacional, Sociedad Geológica Mexicana, Libro de Resúmenes, p. 188.

Suter, M., 1984, Cordilleran deformation along the eastern edge of the Valles-San Luis Potosí carbonate platform, Sierra Madre Oriental fold-thrust belt, east-central Mexico: Geological Society of America Bulletin, v. 95, p. 1387–1397.

Suter, M., 1984, Interpretación geológica de los datos adquiridos en el Pozo Cantarell 2239: Ingeniería Petrolera, v. 24, p. 25.

Szymoniak, R., 1982, The Mexican Coal Basins (in Russian): Przeglad Geologiczny, v. 30, (10), p. 554–556.

Tania-Jiménez, P., 1983, Nueva especie de género *Bishopella* en el Albiano medio del Estado de Coahuila, México: Asociación Mexicana de Geólogos Petroleros, v. 35.

Tavitas-Galván, J.E., and B.J. Solano-Maya, 1984, Estudio bioestratigráfico del subsuelo en el oriente de la plataforma de Valles-San Luis Potosí, Estados SE de Tamaulipas y Oeste de San Luis Potosí: Memoria, III Congreso Latinoamericano de Paleontología, v. 3, p. 225–236.

Taylor, D.G., J.H. Callomon, R. Hall, P.L. Smith, H.W. Tipper, and G.E.G. Westerman, 1984, Jurassic ammonite biogeography of western North America: The tectonic implications, *in* G.E.G. Westerman, ed., Jurassic–Cretaceous Biochronology and Paleogeography of North America: Geological Association of Canada Special Paper 27, p. 121–141.

Téllez-Duarte, M.A., and J.C. Navarro-Fuentes, 1990, Estratigrafía y paleoambientes del Grupo Rosario

(Cenomaniano—Maastrichtiano) en la Mesa de la Sepultura, Baja California, México: Actas de la Facultad de Ciéncias de la Universidad Autónoma de Nuevo León, Linares, v. 4, p. 47–60.

Téllez-Duarte, M.A., F.J. Aranda-Manteca, and G. Gascón-Romero, 1992, Procesos tafronómicos en un conglomerado del Plioceno en La Mesa, Baja California: XI Convención Geológica Nacional, Sociedad Geológica Mexicana, Libro de Resúmenes, p. 189.

Téllez-Duarte, M.A., M. G. Rendón-Márquez, and A. Martin-Barajas, 1992, Ambientes de depósito de la secuencia marina de Puertecitos, B.C., y su correlación con otras localidades del Neogeno en el noroeste de la depresíon de Golfo de California: XI Convención Geológica Nacional, Sociedad Geológica Mexicana, Libro de Resúmenes, p. 189.

Téllez-Girón, C., 1983, Microfacies y zonificación del Pérmico de las Delicias, Coahuila, México: Revista del Instituto Mexicano del Petróleo, v. 15, n. 3, p. 6–45.

Tinker, S.W., 1985, Lithostratigraphy and biostratigraphy of the Aptian La Peña Formation, Northeast Mexico and South Texas: Master's Thesis, University of Michigan, Ann Arbor, Michigan, 63 p.

Torres-E., J.A., C.R. Salinas-H., and A. López-F., 1987, Estratigrafía preliminar del Paleozóico en las áreas de La Vinata y Sierra Azarate, noroeste de Chihuahua, México, Universidad Autonóma de Chihuahua Gaceta Geológica, v. 1, n. 1, p. 203–217.

Torres-Roldán, V., and J.L. Wilson, 1986, Tectonics and facies in the late Paleozoic Plomosas Formation of the Pedregosa Basin of Chihuahua: Geological Society of America Abstracts with Programs, v. 18, p. 774.

Tosdal, R.M., G.B. Haxel, and J.E. Wright, 1989, Jurassic geology of the Sonoran Desert region, southern Arizona, southeastern California, and northernmost Sonora: Construction of a continental-margin magmatic arc, *in* J.P. Jenney, and S.J. Reynolds, eds., Geologic Evolution of Arizona: Arizona Geological Society Digest, v. 17, p. 397–434.

Trejo, H.M., 1980, Distribución estratigráfica de los Tintínidos Mesozóicos Mexicanos: Revista del Instituto Mexicano del Petróleo, v. 12, n. 4, p. 4–13.

Troughton, G.H., 1974, Stratigraphy of the Vizcaíno Peninsula near Asunción Bay, Territorio de Baja California, México: Unpublished Master's Thesis, San Diego State University, 83 p.

Tyler, N., and W.A. Ambrose, 1986, Depositional systems and oil and gas plays in the Cretaceous Olmos Formation, south Texas: Texas University, Bureau of Economic Geology, Report of Investigations No. 152, p. 1–42.

Vachard, D., and A. Oviedo-Pérez, 1992, Fusulinidos del Pérmico Superior de Olinalá, Guerrero: XI Convención Geológica Nacional, Sociedad Geológica Mexicana, Libro de Resúmenes, p. 201.

Vachard, D., and C. Téllez-Girón, 1986a, El alga *Nuia* en el Ordovícico de México: hipótesis diversas: Revista del Instituto Mexicano del Petróleo, v. 28, n. 2, p. 12–25.

Vachard, D., and C. Téllez-Girón, 1986b, El género *Polyderma* y nuevas soluciones a la sistomática de las calcísferas, microfósiles problemáticos del Paleozóico: Revista del Instituto Mexicano del Petróleo, v. 28, n. 3, p. 6–44.

Valdes, M.L.C., 1993, Integración e interpretación geológica-geoquímica el área Ocuapan: Asociación Mexicana Geólogos Petroleros, I Simposio de Geológia de Subsuelo, Ciudad del Carmen, Resúmenes, p. 1–2.

Valdes-Lourdes, C., and R.L Villanueva, 1992, Integración e interpretación geológico-geoquímica del prospecto Ocuapan: XI Convención Geológica Nacional, Sociedad Geológica Mexicana, Libro de Resúmenes, p. 57.

Valencia-Islas, J.J., and J.P. Fortune, 1992, La relación entre la diagénesis de la materia mineral y orgánica en el Jurásico Superior del centro de México: XI Convención Geológica Nacional, Sociedad Geológica Mexicana, Libro de Resúmenes, p. 202.

Valencia-Islas, J.J., and F.J. Pol, 1992, La relación entre la diagénesis de la materia mineral y orgánica en el Jurásico Superior del Centro de México: Revista del Instituto Mexicano del Petróleo, v. 24, n. 3 , p. 5–11.

VanBerg, J., 1989, The Zuloaga Formation, *in* K. Greier and J.L. Kowalski, eds., Geology of the Sierra de Catorce Uplift, Real de Catorce, San Luis Potosí, Mexico, Guide Book: The University of Texas-Pan American Geological Society for the Gulf Coast Association of Geological Societies 39th Annual Convention, p. 41–46.

Vega-Vera, F.J., 1987, Importancia geológico-estratigráfica de la transición Cretácico Superior-Terciário en la Cuenca de la Popa (Grupo Difunta), Nuevo León: Actas de la Facultad de Ciéncias, Universidad Autónoma de Nuevo León, Linares, v. 2, p. 107–110.

Vega-Vera, F.J., and M. del C. Perrilliat, 1989a, La presencia del Eoceno marino en la cuenca de La Popa (Grupo Difunta), Nuevo León: Universidad Nacional Autónoma de México, Instituto de Geología, Revista, v. 8, p. 67–70.

Vega-Vera, F.J., and M. del C. Perrilliat, 1989b, Una nueva especie del género *Costacopluma* (Arthropoda: Decapoda) del Maastrichtianao de Nuevo León: Universidad Nacional Autónoma de México, Instituto de Geología, Revista, v. 8, p. 84–87.

Vega-Vera, F.J., L.M. Mitre-Salazar, and E. Martínez-Hernández, 1989, Contribución al conocimiento de estratigrafía del Grupo Difunta (Cretácico Superior-Terciario) en la noreste de México: Revista del Instituto de Geología, Universidad Nacional Autónoma México, v. 8, p. 179–187.

Velázquez-C., E., 1992, Interés económico petrolero de los terrigenos de la Formación Chicontepec (area Huhutla, Hidalgo): XI Convención Geológica Nacional, Sociedad Geológica Mexicana, Libro de Resúmenes, p. 208.

Vera-Morán, A., 1993, Sincronización de eventos del sistema petrolífero Tithoniano-Plio-Pleistoceno para una porción de la Cuenca de Comalcalco: Asociación Mexicana Geólogos Petroleros, I Simposio de Geología de Subsuelo, Ciudad del Carmen, Resúmenes, p. 22–31.

Vidal, F.S., 1984, Jurassic stratigraphy, depositional environment and paleogeography on the east flank of the Tamaulipas Paleopeninsula: Master's Thesis, San Diego State University, 138 p.

Villarino, R.V., 1992, Estado actual y perspectivas de la exploración de la región norte de Petroleos Mexicanos (abs.): II Symposium de Asociación Mexicanos de Geólogos Petroleros, Instituto Mexicano del Petróleo, Mexico D.F.

Villaseñor-Martínez, A.B., and C. González-Arreola, 1987, Fauna de amonitas y presencia de *Lamellaptychus murocostatus* Trauth del Jurásico Superior de la Sierra de Palotes, Durango: Universidad Nacional Autónoma de Mexico, Instituto de Geología, Revista, v. 7, p. 71–77.

Villaseñor-Martínez, A.B., F. Olóriz Sáez, and C. González Arreola, 1991, Las plataformas marinas del Jurásico Superior en el área de Sierra de Catorce, Estado de San Luis Potosí; una aproximación a la interpretación ecoestratigráfica: Memoria de la Convención sobre la Evolución Geológica de México, Primer Congreso Mexicano de Mineralogía, p. 238–240.

Villaseñor-Martínez, A.B., F. Olóriz-Sáez, C. González-Arreola, and L. Lara-Morales, 1992, Ammonites del Jurásico Superior de Mazapil (Zacatecas, México): XI Convención Geológica Nacional, Sociedad Geológica Mexicana, Libro de Resúmenes, p. 212.

Vinas-Gomez, F., 1992, Estudio bioestratigráfico basado en nanoplancton calcáreo del Pozo Kukulkan no. 1: XI Convención Geológica Nacional, Sociedad Geológica Mexicana, Libro de Resúmenes, p. 214.

Vinet, M.J., 1975, Geology of Sierra Baluartes and Sierra de Pajaros Azules, Coahuila, Mexico: Master's Thesis, University of New Orleans.

Viniegra, O.F., 1981, Great carbonate bank of Yucatan, southern Mexico: Journal Petroleum Geology, v. 3, p. 247–278.

Vokes, E.H., 1984a, A new species of *Turbinella* (Mollusca; Gastropoda) from the Pliocene of Mexico, with a revision of the geologic history of the line: Tulane Studies in Geology and Paleontology, v. 18, p. 47–52.

Vokes, E.H., 1984b, The genus *Harpa* (Mollusca; Gastropoda) in the New World: Tulane Studies in Geology and Paleontology, v. 18, p. 53–60.

Waite, L.E., 1983, Biostratigraphy and paleoenvironmental analysis of the Sierra Madre Limestone (Cretaceous), Chiapas, southern Mexico: Master's Thesis, University of Texas at Arlington, 192 p.

Waite, L.E., 1986, Biostratigraphy and paleoenvironmental analysis of the Sierra Madre Limestone (Cretaceous), Chiapas, *in* D.R. Steele et al., eds., Contributions to the stratigraphy of the Sierra Madre Limestone (Cretaceous) of Chiapas: Boletín, Instituto de Geologia, Universidad Nacional Autonóma de México, v. 102, p. 103–245.

Waite, L.E., and B.F. Perkins, 1983, Biofacies and biostratigraphy of the Cretaceous Sierra Madre Limestone, Chiapas, southern Mexico (abs.): 32nd Annual Meeting Southeastern Section Geological Society of America, Abstracts with Programs, v. 15, p. 48.

Walker, J.D., 1988, Permian and Triassic rocks of the Mojave Desert and their implications for timing and mechanisms of continental truncation: Tectonics, v. 7, p. 685–709.

Ward, W.C., and R.B. Halley, 1985, Dolomitization in a mixing zone of near-seawater composition, late Pleistocene, northeastern Yucatan Peninsula: Journal of Sedimentary Petrology, v. 55, p. 407–420.

Wardlaw, B., W.M. Furnish, and M.K. Nestell, 1979, Geology and paleontology of the Permian beds near Las Delicias, Coahuila, Mexico: Geological Society of America Bulletin, v. 90, part 1, p. 111–116.

Weidie, A.E., and W.C. Ward, 1987, Laramide tectonics and Upper Cretaceous–Lower Tertiary centers of deposition, NE Mexico: Actas de la Facultad de Ciéncias Universidad Autónoma de Nuevo León, Linares, v. 2, p. 195–196.

Weidie, A.E., W.C. Ward, and R. Smith, 1987, Upper Jurassic–Lower Cretaceous depositional systems, paleogeography and depositional environments, NE Mexico: Actas de la Facultad de Ciéncias, Universidad Autónoma de Nuevo León, Linares, v. 2, p. 25–26.

Westermann, G.E.G., 1981, The upper Bajocian and lower Bathonian (Jurassic) ammonite faunas of Oaxaca, Mexico and West-Tethyan affinities: Paleontologia Mexicana, v. 46, 63 p.

Westermann, G.E.G., 1984a, Summary of symposium papers on the Jurassic–Cretaceous biochronology and paleogeography of North America, *in* G.E.G. Westermann, ed., Jurassic–Cretaceous Biochronology and Paleogeography of North America: Geological Association of Canada Special Paper 27, p. 307–315.

Westermann, G.E.G., 1984b, The late Bajocian *Duashnoceras* association (Jurassic Ammonitina) of Mixtepec in Oaxaca, Mexico: Memoria, III Congreso Latinoamericano de Paleontología, p. 192–199.

Westermann, G.E.G., R. Corona, and R. Carrasco, 1984, The Andean Mid-Jurassic *Neuqueniceras* ammonite assemblage of Cualac, Mexico, *in* G.E.G. Westermann, ed., Jurassic–Cretaceous biochronology and paleogeography of North America: Geological Association of Canada Special Paper 27, p. 99–112.

West Texas Geological Society, 1984, Geology and Petroleum Potential of Chihuahua, Mexico, Field Trip Guidebook, West Texas Geology Society Publication 84–80, 218 p.

Whitaker, T.M., 1988, The caves of Chiapas, southern Mexico: Cave Science (1982), v. 15, p. 51–81.

Wilbert, W.P., 1985a, Early Cretaceous carbonate ramp, Sierra de la Paila, Coahuila, Mexico: Bulletin South Texas Geological Society, v. 25, p. 41–49.

Wilbert, W.P., 1985b, Emil Böse and the discovery of the Coahuila Peninsula: Bulletin South Texas Geological Society, v. 25, p. 31–35.

Wilson, H.H., 1987, The structural evolution of the Golden Lane, Tampico Embayment, Mexico: Journal of Petroleum Geology, v. 10, p. 5–40.

Wilson, J.L., 1981, Lower Cretaceous stratigraphy in the Monterrey-Saltillo area, *in* C.I. Smith, ed., Lower Cretaceous Stratigraphy and Structure, Northern Mexico: West Texas Geological Society Publication 81-74, p. 78–84.

Wilson, J.L., 1984, Late Jurassic and Cretaceous facies around the Gulf of Mexico, *in* W.P.S. Ventress, D.G. Bebout, B.F. Perkins, and C.H. Moore, eds., The Jurassic of the Gulf Rim: Gulf Coast Section SEPM, Proceedings of the Third Annual Research Conference, p. 54–55.

Wilson, J.L., 1987a, The late Paleozoic geologic history of southern New Mexico and Chihuahua: Universidad Autónoma de Chihuahua Gaceta Geológica, v. 1, n. 1, p. 36–53.

Wilson, J.L., 1987b, Controls on carbonate platform-basin systems in northeast Mexico: Actas de la Facultad de Ciéncias de la Tierra, Universidad Autónoma de Nuevo León, Linares, v. 2, p. 23–24.

Wilson, J.L., 1989, Lower and Middle Pennsylvanian strata in the Oro Grande and Pedrogosa Basins, New Mexico: New Mexico Bureau of Mines and Mineral Resources, Bulletin 124, 16 p.

Wilson, J.L., 1990a, Basement structural controls on Mesozoic carbonate facies in northeastern Mexico—a review: Special Publications of the International Association of Sedimentologists, v. 9, p. 235–255.

Wilson, J.L., 1990b, Basement structural controls on Mesozoic carbonate facies in northeastern Mexico—a review: Actas de la Facultad de Ciéncias de la Tierra de la Universidad Autónoma de Nuevo León, Linares, v. 4, p. 5–45.

Wilson, J.L., and C.F. Jordan, 1988, Late Paleozoic–Early Mesozoic rifting in southern New Mexico and northern Mexico—Controls on subsequent platform development, *in* S.R. Robichaud and C.M. Gallick, eds., Basin to Shelf Facies Transition of the Wolfcampian Stratigraphy of the Oro Grande Basin: Permian Basin Section-SEPM Annual Field Seminar, Publication 88-28, p. 79–88.

Wilson, J.L., and G. Pialli, 1977, A Lower Cretaceous shelf margin in northern Mexico, *in* D.G. Bebout and R.G. Loucks, eds., Cretaceous Carbonates of Texas and Mexico: Bureau of Economic Geology, University of Texas, Report of Investigations 89, p. 286–294.

Wilson, J.L., and D.B. Selvius, 1984, Early Cretaceous in the Monterrey area of Northern Mexico, *in* J.L. Wilson, W.C. Ward, and J.M. Finneran, eds., A Field Guide to Upper Jurassic and Lower Cretaceous Carbonate Platform and Basin Systems, Monterrey-Saltillo Area, Northeast Mexico: Gulf Coast Section SEPM, p. 28–42.

Wilson, J.L., and W.C. Ward, 1993, Early Cretaceous carbonate platforms of northeastern and east-central Mexico: *in* A. Simo et al., eds., Atlas of Cretaceous Platforms, AAPG Memoir 56, p. 35-49.

Wilson, J.L., W.C. Ward, and J.M. Finneran, eds., 1984, A Field Guide to Upper Jurassic and Lower Cretaceous Carbonate Platform and Basin Systems, Monterrey-Saltillo Area, Northeast Mexico: Gulf Coast Section SEPM Foundation, 76 p

Witebsky, S., 1986, Paleontology and stratigraphy of the new Haymond Boulder Bed locality, southeastern Marathon Basin, West Texas: Geological Society of America Abstracts with Programs, v. 18, p. 792.

Woo, K.S., 1986, Coordinated textural-isotopic-chemical investigation of Mid-Cretaceous rudist limestones, Texas and Mexico; implications for diagenetic histories (abs.): Journal of the Geological Society of Korea, v. 22, p. 381.

Woods, R.D., A. Salvador, and A.E. Miles, 1991, Pre-Triassic, *in* A. Salvador, ed., The Gulf of Mexico Basin: Geological Society of America, p. 109–130.

Yang, Q., 1988, Upper Jurassic (Upper Tithonian) Radiolaria from the Taman Formation, east-central Mexico: Ph.D. Dissertation, University of Texas at Dallas, Richardson, Texas, 300 p.

Yang, Q., and E.A. Pessagno, Jr., 1989, Upper Tithonian *Vallupinae* (Radiolaria) from the Taman Formation, east-central Mexico: Micropaleontology, v. 35, p. 114–134.

Yin, D.D., 1988, Microfacies analysis and depositional systems of Potrero Chico (Northeast Mexico): Master's Thesis, University of Texas at Dallas, Richardson, Texas, 70 p.

Young, K., 1983, Mexico, *in* M. Mouyllade and A.E. Nairn, eds., The Phanerozoic Geology of the World II.: The Mesozoic B: Amsterdam, Elsevier, p. 61–88.

Young, K., 1984, Biogeography and stratigraphy of selected Middle Cretaceous rudists of southwestern North America: Memoria; III Congreso Latinoamericano de Paleontologia, v. 3, p. 341–360.

Young, K., 1986a, History of geology of northern Mexico and Texas—the German connection: Geological Society of America Abstracts with Programs, v. 18, p. 798.

Young, K., 1986b, The Albian–Cenomanian (Lower Cretaceous–Upper Cretaceous) boundary in Texas and northern Mexico: Journal of Paleontology, v. 60, p. 1212–1219.

Young, K., 1991, Interpretation of adjacent La Caja and La Casita outcrops, Mina Plomosas, Chihuahua: Geological Society of America Abstracts with Programs, v. 23, p. A127.

Young, K., 1992, Late Albian (Cretaceous) ammonites from Sierra Mojada, western Coahuila, Mexico: Texas Journal of Science, v. 44, p. 413–420.

Young, K., 1993, Middle Albian ammonites from El Madero, west-central Chihuahua: Texas Journal of Science, v. 45, p. 165–176.

Zaldivar, R.J., and V.H. Garduno-M., 1984, Estudio estratigráfico y estructural de la rocas del Paleozóico Superior de Santa María del Oro, Durango y sus implicaciones tectónicas: Sociedad Geológica Mexicana, Memorias VII Convención Nacional, p. 28–37.

Zamudio-Ángeles, D.J., 1992, Bioestratigrafía de las formaciones del Aptiano Tardio-Cenomaniano Temprano en el area Delícias, Chihuahua: XI Convención Geológica Nacional, Sociedad Geológica Mexicana, Libro de Resúmenes, p. 218.

Zenteno-B., M.A., A. Juárez, and J. Meneses, 1984, Exploration and development of the Campeche Sound and Chiapas-Tabasco areas: Congres Mondial du Petrole, v. 11, n. 2, p. 101–109.

Zozaya-Saynes, M., 1992, Estado actual y perspectivas de la exploración en la región sur de Petróleos Mexicanos (abs.): II Symposio Asociación Mexicana de Geólogos Petroleros, Instituto Mexicano del Petróleo, Mexico, D.F.

2. Plate Tectonics, Paleomagnetic Studies, and Regional Structural Terrane Studies

Aguirre-Díaz, G.J., and F.W. McDowell, 1993, Nature and timing of faulting and synextensional magmatism in the southern Basin and Range, central-eastern Durango, Mexico: Geological Society of America Bulletin, v. 105, p. 1435–1444.

Ahuja, C.M., and C.L.V. Aiken, 1984, Regional gravity and magnetic study of the tectonics of SE Chihuahua and northern Coahuila, Mexico, and West Texas: Geological Society of America Abstracts with Programs, v. 14, p. 106.

Aiken, C.L.V., R.W. Schellhorn, and M.F. de la Fuente, 1986, In search of the elusive Sonora-Mojave megashear in northern Mexico by gravity and magnetics: Geological Society of America Abstracts with Programs, v. 18, p. 523.

Aiken, C.L.V., R.W Schellhorn, and M.F. de la Fuente, 1988, Gravity of northern Mexico, *in* K.F. Clark, P.C. Goodell, and J.M. Hoffer, eds., Stratigraphy, Tectonics and Resources of Parts of Sierra Madre Occidental Province Mexico; Guidebook for the 1988 Field Conference: El Paso Geological Society, p. 119–134.

Alaniz-Alvarez, S., and F. Ortega-Gutiérrez, 1988, Constituye el Complejo Xolapa realmente las raices de un arco?: Union Geofisica Mexicana, Colima, GEOS, Numero extraordinario, Epoca 2, FQIT 15/57.

American Geophysical Union, 1993, Fall Meeting EOS, p. 574–577 and 590–591. Twenty nine abstracts of papers on tectonics of Mexico.

Anderson, T.H., 1988, The Mojave-Sonora megashear in northeastern Mexico: additional constraints on position and displacement: Geological Society of America Abstracts with Programs, v. 20, n. 7, p. A59.

Anderson, T.H., and P. Campbell, 1992, Mylonite at the Mojave-Sonora megashear, northwestern Mexico: Geological Society of America Abstracts with Programs, v. 24, p. A147.

Anderson, T.H., and V.A. Schmidt, 1983, The evolution of Middle America and the Gulf of Mexico-Caribbean Sea region during Mesozoic time: Geological Society of America Bulletin, v. 94, p. 941–966.

Anderson, T.H., and L.T. Silver, 1981, An overview of Precambrian rocks in Sonora: Revista del Instituto de Geología, Universidad Nacional Autónoma Mexico, v. 5, p. 131–139.

Anderson, T.H., et al., 1973, Geology of the western Altos Cuchumatanes, northwestern Guatemala: Geological Society of America Bulletin, v. 84, p. 805–826.

Anderson, T.H., R.J. Erdlac, and M.A. Sandstrom, 1985, Late Cretaceous allochthons and post-Cretaceous strike-slip displacement along the Cuilco-Chixoy-Polochic fault, Guatemala: Tectonics, v. 4, p. 453–475.

Anderson, T.H., R.J. Erdlac, and M.A. Sandstrom, 1986, Reply to comment on "Late Cretaceous allochthons and post-Cretaceous strike-slip displacement along the Cuilco-Chixoy-Polochic fault, Guatemala": Tectonics, v. 5, p. 473–475.

Anderson, T.H., J.W. McKee, and N.W. Jones, 1987, A northwesterly trending Jurassic fold nappe, northernmost Zacatecas, Mexico: Geological Society of America Abstracts with Programs, v. 19, p. 573.

Anderson, T.H., J.W. McKee, and N.W. Jones, 1991, A northwest trending, Jurassic fold nappe, northernmost Zacatecas, Mexico: Tectonics, v. 10, p. 383–401.

Anderson, T.H., J.W. McKee, N.W. Jones, and J.L. Rodriguez-Castañeda, 1991, Transpressional structures along the Late Jurassic, Mojave-Sonora megashear: Geological Society of America Abstracts with Programs, v. 23, p. A251.

Aranda-García, M., 1990, Reactivación Cenozóica del lineamiento Shafter, en la porción norte del Estado de Chihuahua: X Convención Geológica Nacional, Sociedad Geológica Mexicana, Libro de Resúmenes, p. 58.

Aranda-García, M., 1991, El segmento San Felipe del cinturón cabalgado, Sierra Madre Oriental, Estado de Durango, México: Boletín Asociación Mexicana de Geólogos Petroleros, v. 41, p. 18–36.

Aranda-Gómez, J.J., and J.A. Pérez-Venzor, 1987, Estudio geológico de Punta Coyotes, Baja California Sur: Universidad Nacional Autónoma de México, Instituto de Geología, Revista, v. 7, p. 1–21.

Aranda-Gómez, J.J., J.M. Aranda-Gómez, and A.F. Nieto-Samaniego, 1989, Consideraciones acerca de la evolución tectónica durante el Cenozóico de la

Sierra de Guanajuato y la parte meridional de la Mesa Central: Revista del Instituto de Geología, Universidad Nacional Autónoma México, v. 8, p. 33–46.

Ballard, M.M., R. Vander Voo, and J. Urrutia-Fucugauchi, 1989, Paleomagnetic results from the Grenvillian-aged rocks from Oaxaca, Mexico: evidence for a displaced terrane?: Precambrian Research, v. 42, p. 343–352.

Barros, J.A., C.A. Johnson, and C.G.A. Harrison, 1989, Tectonic evolution of south central Mexico: Geological Society of America Abstracts with Programs, v. 21, p. 3.

Bartok, P., 1993, Pre-breakup geology of the Gulf of Mexico-Caribbean: its relation to Triassic and Jurassic rift systems of the region: Tectonics, v. 12, p. 441–459.

Bartolini, C., C. González-León, and M. Morales, 1989, Laramide deformation in northwestern Mexico: Facts and uncertainties: Geological Society of America Abstracts with Programs, v. 21, p. A91.

Bazán-Perkins, S.D., and S. Bazán-Barron, 1992, Evolución geodinámica para el Golfo de Mexico y de Las Antillas: XI Convención Geológica Nacional, Sociedad Geológica Mexicana, Libro de Resúmenes, p. 30–32.

Bickford, M.E., 1988, The formation of continental crust: Part 1, A review of some principles; Part 2, An application to the Proterzoic evolution of southern North America: Geological Society of America Bulletin, v. 100, p. 1375–1391.

Bickford, M.E., W.R. Van Schmus, and I. Zietz, 1986, Proterozoic history of the midcontinent region of North America: Geology, v. 14, p. 492–496.

Blake, M.C., et al., 1984, Tectonostratigraphic terranes of Magdalena Island, Baja California Sur, *in* Geology of the Baja California Peninsula, V.A. Frizzell, Jr., ed.: Pacific Section SEPM, v. 39, p. 183–191.

Böhnel, E., 1982, Plattentektonische Bewegungen Mexikos seit dem Trias, *in* H.J. Behr et al., eds., Geowissenschaftliches Lateinamerika Kolloquium: Geol. Palöntol. Inst., Universitaet Göttingen, Germany, p. 12.

Böhnel, H., 1982, Paläomagnetische Untersuchungen zur plattentektonischen Entwicklung Mexikos seit der Trias (abs): Jahrestagung der Deutschen Geophysikalischen Gesellschaft, v. 42, p. 192.

Böhnel, H., 1985, Paläomagnetische Daten zur Tektonik Mexikos: Jahrestagung der Deutschen Geophysikalischen Gesellschaft, v. 45, p. 182.

Böhnel, H., et al., 1988, Paleomagnetism and ore petrology of three Cretaceous–Tertiary batholiths of southern Mexico: Neues Jahrbuch für Geologie und Paläontologie, Monatshefte, p. 97–127.

Böhnel, H.L., et al., 1989a, Paleomagnetic data and stability of tectonostratigraphic terranes in southern Mexico: 28th International Geological Congress Abstracts, v. 1, p. 167.

Böhnel, H.L., et al., 1989b, Paleomagnetic data and the accretion of the Guerrero terrane, southern Mexico continental margin, deep structure and past kinematics of accreted terranes: American Geophysical Union, Geophysical Monograph 50, p. 73–92.

Bourgois, J., et al., 1988, Fragmentation en cours de bord ouest du Continent Nord Américain: les frontièrs sous-marine du Bloc Jalisco (Mexique) de fracture de Riviera au large du Mexique: Comptes Rendus Académie Science Paris, v. 307, p. 1121–1130.

Brown, M.L., and R. Dyer, 1986a, Latest Paleozoic to mid-Tertiary tectonics of northwestern Chihuahua, Mexico: Geological Society of America Abstracts with Programs, v. 18, p. 344.

Brown, M.L., and R. Dyer, 1986b, Structural geology of Sierra de los Chinos and Cerro La Cueva, northwest Chihuahua, Mexico, *in* J. M. Hoffer, ed., Geology of South-Central New Mexico: El Paso Geological Society Guidebook, p. 141–151.

Brown, M.L., and J.W. Handschy, 1984, Tectonic framework of Chihuahua, Mexico, *in* E. C. Kettenbrink, ed., Geology and Petroleum Potential of Chihuahua, Mexico: West Texas Geological Society Publication 84-80, p. 161–173.

Buffler, R.T., 1983, Structure of the Mexican Ridges foldbelt, southwest Gulf of Mexico, *in* A.W. Bally, ed., Seismic Expression of Structural Styles; A Picture and Work Atlas: AAPG Studies in Geology, 15, p. 2.3.3 – 2.3.3-21.

Buffler, R.T., and D.S. Sawyer, 1985, Distribution of crust and early history, Gulf of Mexico basin: Gulf Coast Geological Society Transactions, v. 35, p. 333–344.

Buffler, R.T., F.J. Shaub, J.S. Watkins, and J.L. Worzel, 1979, Anatomy of the Mexican Ridges, southwestern Gulf of Mexico, *in* J.S. Watkins, L. Montadert, and P.W. Dickerson, eds., Geological and Geophysical Investigations of Continental Margins: AAPG Memoir 29, p. 319–327.

Buffler, R.T., et al., 1986, Continent-ocean transect F-1; Phanerozoic evolution of the Gulf of Mexico Basin: Geological Society of America Abstracts with Programs, v. 18, p. 552–553.

Buffler, R.T., et al., 1990, Crustal types and northwest trending structural features: Constraints on reconstructing the Gulf of Mexico basin: Geological Society of America Abstracts with Programs, v. 22, p. A186.

Burkart, B., 1983, Neogene North America-Caribbean plate boundary across northern Central America—offset along the Polochic fault: Tectonophysics, v. 99, p. 251–270.

Burkart, B., 1990, Contrast in effects of Late Cretaceous convergence between Chiapas, Mexico and Guatemala: Geological Society of America Abstracts with Programs, v. 22, A338.

Burkart, B., and S. Self, 1985, Extension and rotation of crustal blocks in northern Central America and effect on the volcanic arc: Geology, v. 13, p. 22–26.

Burkart, B., B.C. Deaton, C. Dengo, and G. Moreno, 1987, Tectonic wedges and offset Laramide structures along the Polochic fault of Guatemala and Chiapas, Mexico: Reaffirmation of large Neogene displacement: Tectonics, v. 6, p. 411–422.

Burke, K., C. Cooper, J.F. Dewey, J.P. Mann, and J. Pindell, 1984, Caribbean tectonics and relative plate motions, *in* W.E. Bonini, R.B. Hargraves, and R. Shagam, eds., The Caribbean-South American Plate Boundary and Regional Tectonics: Geological Society of America Memoir 162, p. 31–64.

Busby-Spera, C. J., 1988, Speculative tectonic model for the early Mesozoic arc of the southwest Cordilleran United States: Geology, v. 16, p. 1211–1225.

Busby-Spera, C.J., and B.P. Kokelaar, 1991, Controls of the Sawmill Canyon fault zone on Jurassic magmatism and extension/transtension in southern Arizona: Geological Society of America Abstracts with Programs, v. 23, p. A1250.

Busby-Spera, C. J., J.M. Mattinson, and B. Adams, 1989, The Jurassic transition from continental intra-arc extension to Gulf of Mexico—related rifting in the southwest Cordillera (abs.): EOS Transactions of the American Geophysical Union, v. 70, p. 1300.

Busby-Spera, C. J., J.M. Mattinson, and E.R. Schermer, 1990, Stratigraphic and tectonic evolution of the Jurassic arc: New field and U-Pb zircon geochronological data from the Mojave Desert: Geological Society of America Abstracts with Programs, v. 22, p. 11–12.

Caballero-Miranda, C., et al., 1990, Paleogeography of the northern portion of the Mixteca terrane, southern Mexico, during the Middle Jurassic: Journal of South American Earth Science, v. 3, p. 195–211.

Cameron, K.L., and S.H. Gunn, 1987, Cenozoic crustal growth and recycling in northwestern Mexico: Geological Society of America Abstracts with Programs, v. 19, p. 609.

Campa-Uranga, M.F., 1985, The Mexican thrust belt, *in* D.G. Howell, ed., Tectonostratigraphic Terranes of the Circum-Pacific Region: Houston, Texas, Circum-Pacific Council for Energy and Mineral Resources, Earth Science Series, No. 1, p. 299–313.

Carfantan, J.C., 1983, Les ensembles géologiques du Mexique meridional. Evolution geodynamique durante le Mesozoique et le Cenozoique: Geofísica Internacional, v. 22, p. 9–37.

Carfantan, J.C., 1984, Evolución estructural del sureste de Mexico: Paleogeografía e história tectónica de las zonas internas Mesozóicas: Revista del Instituto de Geologia, Universidad Nacional Autónoma México, v. 5, p. 207–216.

Carfantan, J.C., and M. Tardy, 1984, Heterochronology of structural belts in the western margin of the Gulf of Mexico, *in* The Geological Society of America, South-Central Section, 18th Annual Meeting, Geological Society of America Abstracts with Programs, v. 16, p. 80.

Carlsen, T.W., 1989, Regional geologic studies in a wrench tectonic regime; Iturbide Quadrangle, eastern front, Sierra Madre Oriental, Mexico: Geological Society of America Abstracts with Programs, v. 21, p. 6.

Carlsen, T.W., 1990, The eastern Front Ranges of the Sierra Madre Oriental, Victoria segment of the Mexican Cordillera: A new analogue for wrench tectonics: Geological Society of America Abstracts with Programs, v. 22, n. 3, p. 12.

Carrillo-Martínez, M., 1990, Geometría estructural de la Sierra Madre Oriental, entre Peñamiller y Jalpan, Estado de Querétaro: Universidad Nacional Autónoma de México, Instituto de Geología, Revista, v. 9, p. 62–70.

Castro-Mora, J., 1985, Análisis tectónico-estratigráfico de cuencas pre-Cretácicas: Petrolero Internacional, v. 43, p. 24, 26–27, 30–35.

Cebull, S.E., and D.H. Shurbet, 1987, Mexican volcanic belt: an intraplate transform?: Geofísica Internacional, v. 26, p. 1–13.

Cedillo-P., E., and M.N. Grajales, 1992, El Arco Permo-Triásico en Coahuila (Arco Delicias): XI Convención Geológica Nacional, Sociedad Geológica Mexicana, Libro de Resúmenes, p. 54.

Centeno-García, E., F. Ortega-Gutiérrez, and R. Corona-Esquivel, 1990, Oaxaca fault: Cenozoic reactivation of the suture between the Zapoteco and Cuicateco terranes, southern Mexico: Geological Society of America Abstracts with Programs, v. 22, p. 13.

Centeno-García, E., J. Ruiz, P.J. Coney, P.J. Patchett, and F. Ortega-Gutiérrez, 1992a, The pre-Cretaceous tectonics of western Guerrero terrane, Mexico: Geochemical and isotopic evidence of an accreted marginal oceanic basin: Geological Society of America Abstracts with Programs, v. 24, p. A64.

Centeno-García, E., J. Ruiz, P.J. Coney, J.P. Patchet, and F. Ortega-Gutiérrez, 1992b, El complejo metamórfico de Tumbiscato-Arteaga, Michoacán: Una secuéncia oceánica marginal acrecionada: XI Convención Geológica Nacional, Sociedad Geológica Mexicana, Libro de Resúmenes, p. 55–56.

Champion, D.E., D.G. Howell, and M. Marshall, 1986, Paleomagnetism of Cretaceous and Eocene strata, San Miguel Island, California, borderland and the northward translation of Baja California: Journal of Geophysical Research, v. 91, p. 11,557–11,570.

Charleston, S., 1981, A summary of the structural geology and tectonics of the State of Coahuila, Mexico, *in* C.I. Smith and S.B. Katz, eds., Lower Cretaceous Stratigraphy and Structure, Northern Mexico: West Texas Geological Society Publication 81-74, p. 28–36.

Chauve, P., E. Fourcade, and M. Carillo, 1985, Les rapports structuraux entre les domaines cordillérain et mésogéen dans la partie centrale du Mexique: Comptes Rendus de Académie des Sciences, Série 2, Mécanique, Physique, Chimie, Sciences de Univers, Sciences de la Terre, v. 301, p. 335–340.

Cohen, K. K., T.H. Anderson, and V.A. Schmidt, 1986, A paleomagnetic test of the proposed Mojave-Sonora megashear in northwestern Mexico: Tectonophysics, v. 131, p. 23–51.

Coney, P.J., 1983, Un modelo tectónico de México y sus relaciones con América del Norte, América del Sur y el Caribe: Revista del Instituto Mexicano del Petróleo, v. 15, p. 6–16.

Coney, P.J., 1987, The regional tectonic setting and possible causes of Cenozoic extension in the North

American Cordillera, *in* M.P. Coward, J.F. Dewey, and P.L. Hancock, eds., Continental Extensional Tectonics: London, The Geological Society Special Publication 28, p. 177–186.

Coney, P.J., 1989, Structural aspects of suspect terranes and accretionary tectonics in western North America: Journal of Structural Geology, v. 11, p. 107–125.

Coney, P.J., and M.F. Campa, 1987, Lithotectonic terrane map of Mexico (west of the 91st meridian): U.S. Geological Survey Miscellaneous Field Studies Map MF-1874 - D, Scale 1:250,000.

Connors, C. D., T.H. Anderson, and L.T. Silver, 1989, Expression and structural analysis of the Mojave-Sonora megashear in northwestern Sonora, Mexico: Geological Society of America Abstracts with Programs, v. 21, n. 6, p. A91.

Corbitt, L.L., 1984, Tectonics of fold and thrust belt of northwestern Chihuahua, *in* E.C. Kettenbrink, ed., Geology and Petroleum Potential of Chihuahua, México, Publication 84-80, West Texas Geological Society Field Trip Guidebook, p. 174–180.

Couch, R., and S. Woodcock, 1981, Gravity and structure of continental margins of southwestern Mexico and northwestern Guatemala: Journal of Geophysical Research, v. 86, p. 1829–1840.

Cuevas-Pérez, E., and W. Vortisch, 1986, La subcuenca de Zacatecas (abs.): GEOS, Boletín, epoca II, Unión Geofísica Mexicana, p. 12.

Cuevas-Pérez, E., et al., 1988 (1985), Una interpretación tectónica de Sinaloa a San Luis Potosí: Boletín de Asociación Mexicana de Geólogos Petroleros, v. 37.

Culotta, R., et al., 1992, Deep structure of the Texas Gulf passive margin and its Ouachita-Precambrian Basement: Results of the COCORP San Marcos Arch Survey: AAPG Bulletin, v. 76, p. 270–283.

Curray, J.R., and D.G. Moore, 1984, Geologic history of the Gulf of California, *in* J.K. Crouch and S.B. Bachman, eds., Tectonics and Sedimentation along the California Margin: Pacific Section SEPM, v. 38, p. 17–35.

Damon, P.E., and P.J. Coney, 1983, Rate of movement of nuclear Central America along the coast of Mexico during the last 90 M.A.: Geological Society of America Abstracts with Programs, v. 15, p. 553.

Damon, P.E., M. Shafiquillah, and J. Roldán-Quintana, 1984, The Cordilleran Jurassic arc from Chiapas (southern Mexico) to Arizona: Geological Society of American Abstracts with Programs, v. 16, p. 482.

Davis, G.H., 1984, Margen continental de colisión activo en la parte suroccidental del Golfo de Mexico: Revista del Instituto de Geología, Universidad Nacional Autónoma México, v. 5, p. 255–261.

Deaton, B.C., and B. Burkart, 1984, Time of sinistral slip along the Polochic fault of Guatemala: Tectonophysics, v. 102, p. 297–313.

De Cserna, Z., 1981 (1984), Margen continental de colisión activo en la parte suroccidental del Golfo de México: Revista del Instituto de Geología, Universidad Nacional Autónoma México, v. 5, p. 255–261.

De Cserna, Z., 1989, An outline of the geology of México, *in* A.W. Bally, and A.R. Palmer, eds., The Geology of North America—An Overview, v. A: Geological Society of America, p. 233–264.

De Cserna, Z., and F. Ortega-Gutiérrez, 1978, Reinterpretation of isotopic age data from the Granjeno Schist, Ciudad Victoria, Tamaulipas; y reinterpretación tectónica del Esquisto Granjeno, Ciudad Victoria, Tamaulipas; Contestación: Revista del Instituto de Geologia, Universidad Nacional Autónoma Mexico, v. 2, p. 212–215.

De Cserna, Z., J.L. Graf Jr., and F. Ortega-Gutiérrez, 1977, Alóctono del Paleozóico Inferior en la región de Ciudad Victoria, Estado de Tamaulipas: Revista del Instituto de Geologia, Universidad Nacional Autónoma de México, v. 1, p. 33–43.

De Cserna, Z., F. Ortega-Gutiérrez, and M. Palacios-Nieto, 1980, Reconocimiento geológico de la parte central de la cuenca del Alto Rio Balsas, estados de Guerrero y Puebla: Libro guía de la excursion geológica a la parte central de la cuenca del Alto Rio Balsas, estados de Guerrero y Puebla: Sociedad Geológica Mexicana, p. 1–33.

De Cserna, Z., et al., 1988, Estructura geólogica, gravimétria, sismicidad y relaciones neotectónicas regionales de la Cuenca de México: Boletín del Instituto de Geología, Universidad Nacional Autónoma México, v. 104, 71 p.

DeJong, K.A., 1986, The Mojave-Sonora Megashear near Caborca: A major fault but not necessarily a Jurassic strike-slip fault: Geological Society of America Abstracts with Programs, v. 18, p. 582.

DeJong, K. A., J.C. Garcia y Barragan, P.E. Damon, M. Miranda, C. Jacques-Ayala, and E. Almazán-Vázquez, 1990, Untangling the tectonic knot of Mesozoic Cordilleran orogeny in northern Sonora, NW Mexico: Geological Society of America Abstracts with Programs, v. 22, p. A327.

DeJong, K.A., et al., 1991, Mesozoic structural evolution of NW Mexico: Laramide thrust belt covers the Mojave-Sonora megashear: Geological Society of America Abstracts with Programs, v. 23, n. 5, p. A128.

Delgado-Argote, L.A., and E.A. Carballido-Sánchez, 1990, Análisis tectónico del sistema transpresivo neogénico entre Macuspana, Tabasco, y Puerto Ángel, Oaxaca: Universidad Nacional Autónoma de México, Instituto de Geología, Revista, v. 9, p. 21–32.

DeMets, C., and S. Stein, 1990, Present-day kinematics of the Rivera plate and implications for tectonics in southwestern Mexico: Journal of Geophysical Research, v. 95, p. 21,931–21,948.

DeMets, C., R.G. Gordon, D.F. Argus, and S. Stein, 1990, Current plate motions: Geophysical Journal International, v. 101, p. 425–478.

Dengo, C.A., 1985, Mid America: tectonic setting for the Pacific margin from southern Mexico to northwestern Colombia, *in* A.E.M. Nairn and F.G. Stehli, eds., The Ocean Basins and Margins, v. 7A: New York, Plenum Press, p. 123–180.

Dengo, C.A., 1986, Comment on "Late Cretaceous allochthons and post-Cretaceous strike-slip displacement along the Cuilco-Chixoy-Polochic fault, Guatemala" by T.H. Anderson, R.J. Erdlac, Jr., and M.A. Sandstrom: Tectonics, v. 5, p. 469–472.

Dengo, G., 1983, Informe preliminar de la geología regional de la cuenca media del Rio Usumacinta Guatemala y México: Instituto Nacional de Electrificación (INDE), 48 p.

Denison, R.E., 1986, Transverse foreland structures adjacent to the Ouachita foldbelt: Geological Society of America Abstracts with Programs, v. 18, p. 583.

Dickerson, P.W., 1985a, Evidence for Late Cretaceous Early Tertiary transpression, West Texas and north-central Mexico: 6th International Conference on Basement Tectonics Abstracts with Program, v. 6, p. 15.

Dickerson, P.W., 1985b, Evidence for Late Cretaceous early Tertiary transpression in Trans-Pecos Texas and adjacent Mexico: West Texas Geological Society, Field Trip Guidebook, Publication 85-81, p. 185–194.

Dickerson, P.W., 1986, Evidence for Laramide transpression in West Texas and adjacent Mexico: Proceedings, 6th International Conference on Basement Tectonics, p. 52–63.

Dickerson, P.W., 1987a, Structural and depositional setting of southwest U.S. and northern Mexico along a Paleozoic transform plate margin: Universidad Autónoma de Chihuahua Gaceta Geológica, v. 1, p. 129–159.

Dickerson, P.W., 1987b, Influence of antecedent structures upon Laramide deformation in southwestern United States and northern Mexico: Geological Society of America Abstracts with Programs, v. 19, p. 271.

Dickerson, P.W., and W.R. Muehlberger, eds., 1985, Structure and Tectonics of Trans-Pecos Texas: West Texas Geological Society, Field Trip Guidebook, Publication 85-81, 278 p.

Dickerson, P.W., D.H. Shurbet, and S.E. Cebull, 1988, Comment and Reply on "Tectonic interpretation of the westernmost part of the Ouachita-Marathon (Hercynian) orogenic belt, West Texas, Mexico": Geology, v. 16, p. 377.

Dickinson, W.R., and P.J. Coney, 1980, Plate tectonic constraints on the origin of the Gulf of Mexico, *in* R.H. Pilger, ed., The Origin of the Gulf of Mexico and the Early Opening of the Central North Atlantic: Proceedings of Symposium at Louisiana State University, Baton Rouge, p. 27–36.

Dilles, J.H., and J.E. Wright, 1988, The chronology of early Mesozoic arc magmatism in the Yerington district of western Nevada and its regional implications: Geological Society of America Bulletin, v. 100, p. 644–652.

Doert, Ulrich, 1987, Contribución al desarrollo estructural laramídico de la Sierra Madre Oriental Media (Nuevo León, Mexico): Actas de la Facultad de Ciéncias, Universidad Autónoma de Nuevo León, Linares, v. 2, p. 201–204.

Donnelly, T.W., 1989, Geologic history of the Caribbean and Central America, *in* A.W. Bally and A.R. Palmer, eds., The Geology of North America: An Overview; v. A, Geological Society of America, p. 299–321.

Donnelly, T.W., G.S. Horne, R.C. Finch, and E. López-Ramos, 1990, Northern Central America; the Maya and Chortis blocks, *in* G. Dengo and J.E. Case, eds., The Caribbean Region: Geological Society of America, The Geology of North America, v. H, p. 37–76.

Donnelly, T.W., et al., 1990, History and tectonic setting of Caribbean magmatism, *in* G. Dengo and J.E. Case, eds., The Caribbean Region: Geological Society of America, The Geology of North America, v. H, p. 339–374.

Dunbar, J.A., and D.S. Sawyer, 1987, Implications of continental crust extension for plate reconstruction: An example from the Gulf of Mexico: Tectonics, v. 6, p. 739–755.

Duncan, R.A., and R.B. Hargraves, 1984, Plate tectonic evolution of the Caribbean region in the mantle reference frame, *in* W.E. Bonini, R.B. Hargraves, and R. Shagan, eds., The Caribbean-South American Plate Boundary and Regional Tectonics: Geological Society of America Memoir 162, p. 81–84.

Dyer, R., R. Chavez-Quirarte, and R.S. Guthrie, 1987, Mesozoic evolution of Trans-Pecos Texas and Chihuahua tectonic belt: Geological Society of America Abstracts with Programs, v. 19, p. 650.

Dyer, R., R. Chavez-Quirarte, and R.S. Guthrie, 1988, Cordilleran orogenic belt of northern Chihuahua, Mexico: AAPG Bulletin, v. 72, p. 99.

Eguiluz de Antuñano, S., 1984a, Tectónica Cenozóica del norte de México: Boletín, Asociación Mexicana de Geólogos Petroleros, v. 36.

Eguiluz de Autuñano, S., 1984b, Fallas transcurrentes en el norte de México: Memorias de la VII Convención Nacional, Sociedad Geológica Mexicana, p. 60–74.

Eguiluz de Antuñano, S., 1991, Interpretación geológica y geofísica de la curvatura de Monterrey, en el noreste de México: Asociación de Ingenieros Petroleros de México. Ingeniería Petrolera, v. 31, p. 25–39.

Eguiluz de Antuñano, S., and M.F. Campa-Uranga, 1982, Problemas tectónicos del sector San Pedro del Gallo, en los estados de Chihuahua y Durango: Boletín Asociación Mexicana de Geológos Petroleros, v. 34, p. 5–42.

Erikson, J.P., 1990, The Montagua fault zone of Guatemala and Cenozoic displacement on the Caribbean-North America plate boundary: Geological Society of America Abstracts with Programs, v. 22, p. A220.

Ewing, T.E., 1985, Westward extension of the Devils River uplift—implications for the Paleozoic evolution of the southern margin of North America: Geology, v. 13, p. 433–436.

Ewing, T.E., 1991, Structural framework, *in* A. Salvador, ed., The Gulf of Mexico Basin: Geological

Society of America, The Geology of North America, v. J, p. 31–52.
Ewing, T.E., et al., 1990, Tectonic map of Texas: University of Texas at Austin, Bureau of Economic Geology, scale 1:750,000.
Farg, W., R. Vander Voo, R. Molina-Garza, D. Morán-Zenteno, and J. Urrutia-Fucugauchi, 1989, Paleomagnetism of the Acatlán terrane, southern Mexico: evidence for terrane rotation: Earth and Planetary Science Letters, v. 94, p. 131–142.
Faul, C.L., et al., 1986, A paleomagnetic study of the remagnetized Jurassic–Cretaceous boundary limestone section at Nuevo León, Mexico: Geological Society of America Abstracts with Programs, v. 18, p. 599.
Faust, M.J., 1984, Seismic stratigraphy of the Middle Cretaceous unconformity in the central Gulf of Mexico Basin: Master's Thesis, University of Texas at Austin, 164 p.
Faust, M.J., 1986, Seismic stratigraphy of Middle Cretaceous unconformity in central Gulf of Mexico basin (abs.): AAPG Bulletin, v. 70, p. 588.
Feldman, M.L., 1986, Paleozoic framework of the Gulf of Mexico: Gulf Coast Association of Geological Societies Transactions, v. 36, p. 97–107.
Fernández, S., et al., 1992, Estúdio de sismicidad en la Sierra Madre Oriental: XI Convención Geológica Nacional, Sociedad Geológica Mexicana, Libro de Resúmenes, p. 74.
Filmer, P.E., and J.L. Kirschvink, 1989, A paleomagnetic constraint on the Late Cretaceous paleoposition of northwestern Baja California, Mexico: Journal of Geophysical Research, v. 94, p . 7332–7342.
Fuente, M.F., C.L.V. Aiken, R.W. Schellhorn, and M. Mena, 1988, Gravity anomalies and structure of Mexico: Geological Society of America Abstracts with Programs, v. 20, p. A325.
Galicia-Barrios, L.G., 1992, Rasgos tectónicos de la interacción de los dominios Cordillerano y Tethysiano en el terreno Toliman: XI Convención Geológica Nacional, Sociedad Geológica Mexicana, Libro de Resúmenes, p. 77.
García-Abdeslem, J., and L.A. Delgado-Argote, 1992, Geología, magnetometría, profundidad a la isoterma de Curie y flexura de la litosfera en la Isla Guadalupe, México: XI Convención Geológica Nacional, Sociedad Geológica Mexicano, Libro de Resúmenes, p. 79.
Gastil, R.G., 1985, Terranes of peninsular California and adjacent Sonora, *in* D.G. Howell, ed., Tectonostratigraphic Terranes of the Circum-Pacific Region: Circum-Pacific Council for Energy and Mineral Resources, Earth Science Series, Number 1, p. 273–283.
Gastil, R.G., 1990, The Cordilleran geocline in peninsular California, USA and Mexico: Geological Society of America Abstracts with Programs, v. 22, p. 24.
Gastil, R.G., and R.H. Miller, 1983, Prebatholithic terranes of southern and peninsular California, U.S.A. and Mexico: status report, *in* C.H. Stevens, ed., Pre-Jurassic Rocks in Western North American Suspect Terranes: Pacific Section SEPM, p. 49–61.
Gastil, R.G., and R.H. Miller, 1984, Prebatholithic paleogeography of peninsular California and adjacent Mexico, *in* V.A. Frizzell, Jr., ed., Geology of the Baja California Peninsula: Pacific Section SEPM, v. 39, p. 9–16.
Gerstenhauer, A., U. Radtke, and A. Mangini, 1983, Neue Ergebnisse zur quartären Küstenentwicklung der Halbinsel Yucatan, Mexico: Essener Geographische Arbeiten (Paderborn) v. 6, p. 187–199.
Gleason, J.D., P.J. Patchett, W.R. Dickinson, and J. Ruiz, 1992, Paleozoic tectonics of the Ouachita orogen through Nd isotopes: Geological Society of America Abstracts with Programs, v. 24, p. A238.
Goetz, C.W., G.H. Girty, and R.G. Gastil, 1988, East over west ductile thrusting along a terrane boundary in the Peninsular Ranges: Rancho El Rosarito, Baja California, Mexico: Geological Society of America Abstracts with Programs, v. 20, p. 165.
Goetz, L.K., and P.W. Dickerson, 1985, A Paleozoic transform margin in Arizona, New Mexico, West Texas, and northern Mexico: West Texas Geological Society, Field Trip Guidebook, Publication 85-81, p. 173–184.
Goodell, P.C., 1990, An integrated geochemical, geophysical, and geological terrane analysis of Mexico: Geological Society of America Abstracts with Programs, v. 22, p. A326–A327.
Goodell, P.C., J.R. Dyer, and G.R. Keller, 1985, Initiation and reactivation of Proterozoic aulacogen, northern Mexico (abs): AAPG Bulletin, v. 69, p. 258.
Goodell, P.C., G.R. Keller, and J.R. Dyer, 1986, The Sierra del Nido tectonic block—a newly recognized cratonic feature in northern Mexico: Geological Society of America Abstracts with Programs, v. 18, p. 618.
Gordon, M.B., 1989, Mesozoic igneous rocks on the Chortis block: Implications for Caribbean reconstructions: EOS, American Geophysical Union Transactions, v. 70, p. 1342.
Göse, W.A., 1983, Tectonic evolution of Middle America: EOS, American Geophysical Union Transactions, v. 70, p. 317.
Göse, W.A., 1986, Paleomagnetism and tectonics in northern Mexico: Geological Society of America Abstracts with Programs, v. 18, p. 618.
Göse, W.A., and L.A. Sánchez-Barreda, 1981, Paleomagnetic results from southern Mexico: Geofísica Internacional, v. 20, p. 163–175.
Göse, W.A., R.C. Belcher, and G.R. Scott, 1982, Paleomagnetic results from northeastern Mexico: Evidence for large Mesozoic rotations: Geology, v. 10, p. 50–54.
Grajales-N., J.M., E. Cedillo-P., and D. Terrell, 1992, El arco Triásico-Jurásico en el norte de México (abs.): Sociedad Geológica Mexicana, A. C., XI Convención Geología Nacional, p. 81–82.
Grajales-N., J.M., E. Cedillo-P., I. Gallo-P., D.J. Terrell, 1992, Los arcos magmáticos del borde occidental del norte de México son la respuesta a un régimén de subduccion establecido desde el Triásico Tardío (abs): II Symposio de Asociación Mexicana de

Geólogos Petroleros, Instituto Mexicano del Petróleo, México D.F.

Griffith, R.C., 1987, Geology of the southern Sierra Calamjué area; structural and stratigraphic evidence for latest Albian compression along a terrane boundary, Baja California, Mexico: Unpublished Master's Thesis, San Diego State University, 115 p.

Griffith, R.C., and C.W. Goetz, 1987, Structural and geochronological evidence for mid-Cretaceous compressional tectonics along a terrane boundary in the Peninsular Ranges: Geological Society of America Abstracts with Programs, v. 19, p. 384.

Griggs, L.J., C. Aiken, M.F. de la Fuente, 1990, Delineation of major crustal boundaries using gravity and magnetic anomalies of Sonora, Mexico: Geological Society of America Abstracts with Programs, v. 22, p. A184.

Guerrero, J.C., E. Herrero-Bervera, and C.E. Helsley, 1990, Paleomagnetic evidence for post-Jurassic stability of southeastern Mexico: Maya Terrane: Journal of Geophysical Research, v. 95, p. 7091–7100.

Guerrero-García, J.C., 1989, Vertical tectonics in southern Mexico and its relation to trench migration: EOS, American Geophysical Union Transactions, v. 70, p. 1319.

Guerrero-García, J.C., L.T. Silver, and T.H. Anderson, 1978, Estudios geocronológicos en el complejo Xolapa: Boletín Sociedad Geológica Mexicana, v. 39, p. 22–23.

Guzmán-Speziale, M., W.D. Pennington, and T. Matumoto, 1985, Extensions of the North America-Caribbean plate boundary in southern Mexico (abs):., EOS, American Geophysical Union Transactions, v. 66, p. 1987.

Guzmán-Speziale, M., W. D. Pennington, and T. Matumoto, 1989, The triple junction of the North America, Cocos, and Caribbean plates: seismicity and tectonics: Tectonics, v. 8, p. 981–997.

Hagstrum, J.T., and R.L. Sedlock, 1990, Remagnetization and northward translation of Mesozoic red chert from Cedros Island and the San Benito Islands, Baja California, Mexico: Geological Society of America Bulletin, v. 102, p. 983–991.

Hagstrum, J.T., M. McWilliams, D.G. Howell, and C.S. Gromme, 1985, Mesozoic paleomagnetism and northward translation of the Baja California peninsula: Geological Society of America Bulletin, v. 96, p. 221–225.

Hagstrum, J.T., et al., 1987, Miocene paleomagnetism and tectonic setting of the Baja California peninsula, Mexico: Journal of Geophysical Research, v. 92, p. 2627–2640.

Hall, D.J., et al., 1982, The rotation origin of the Gulf of Mexico based on regional gravity data, *in* J.S. Watkins and C.L. Drake, eds., Studies in Continental Margin Geology: AAPG Memoir 34, p. 115–126.

Handschy, J.W., 1990, Late Paleozoic tectonism in north-central Mexico: Geological Society of America Abstracts with Programs, v. 22, p. 28.

Handschy, J.W., and J.R. Dyer, 1987, Polyphase deformation in Sierra del Cuervo, Chihuahua, Mexico: Evidence for Ancestral Rocky Mountain tectonics in the Ouachita foreland of northern Mexico: Geological Society of American Bulletin, v. 99, p. 618–632.

Handschy, J.W., J.R. Dyer, and M.L. Brown, 1985, Late Paleozoic deformation in north-central Mexico: a possible extension of the Ancestral Rockies: Geological Society of America Abstracts with Programs, v. 17, p. 222.

Handschy, J.W., G.R. Keller, and K.J. Smith, 1987, The Ouachita System in northern Mexico: Tectonics, v. 6, p. 323–330.

Handschy, J.W., P.C. Goodell, G.R. Keller, J.D. Hoover, and J.R. Dyer, 1985, Evidence for Late Proterozoic rifting in north-central Mexico and adjacent parts of West Texas and southern New Mexico: Geological Society of America Abstracts with Programs, v. 17, p. 223.

Hanus, V., and J. Vanek, 1984, Subduction induced fracture zones and distribution of hydrothermal activity in Mexico, *in* V. Cermak, ed., Terrestrial Heat Flow Studies and the Structure of the Lithosphere: Tectonophysics, v. 103, p. 297–305.

Harkey, D.A., and R. Dyer, 1985, Structural Geology of Sierra San Ignacio, Chihuahua, Mexico: West Texas Geological Society, Field Trip Guidebook, Publication 85-81, p. 195–204.

Harlan, S.S., J.W. Geissman, J.G. Price, C.D. Henry, T.C. Onstaff, and J. Oldow, 1987, Paleomagnetic documentation of the cessation of Chihuahua fold thrust belt deformation, S.W. Texas: Geological Society of America Abstracts with Programs, v. 19, p. 693.

Hausback, B.P., 1984, Cenozoic volcanic and tectonic evolution of Baja California Sur, Mexico, *in* V.A. Frizzell Jr., ed., Geology of the Baja California Peninsula: Pacific Section SEPM, v. 39, p. 219–236.

Hennings, P.H., and W.R. Muehlberger, 1988, Structural transect of the Chihuahua tectonic belt between Ojinaga and Aldama, Chihuahua, Mexico: Geological Society of America Abstracts with Programs, v. 20, n. 7, p. A385.

Hennings, P.H., and W.R. Muehlberger, 1990, Laramide fault inversion, basement buttressing, and evaporite tectonics in the Cordilleran thrust front along the Texas/Chihuahua border: Geological Society of America Abstracts with Programs, v. 22, p. 184.

Henry, C.D., 1987, Basin and Range faulting in mainland Mexico adjacent to the Gulf of California: Geological Society of America Abstracts with Programs, v. 19, p. 701.

Henry, C.D., and J.J. Aranda-Gómez, 1990, The real southern Basin and Range: Mid-Late Cenozoic extension in Mexico: Geological Society of America Abstracts with Programs, v. 22, p. A228.

Henry, C.D., and J.J. Aranda-Gómez, 1992, The real southern Basin and Range: Mid- to Late Cenozoic extension in Mexico: Geology, v. 20, p. 701–704.

Henry, C.D., and J.G. Price, 1985, Summary of the tectonic development of Trans-Pecos Texas: The

University of Texas at Austin, Bureau of Economic Geology, Miscellaneous Map No. 36, 8 p.

Henry, C.D., J.G. Price, and F.W. McDowell, 1983, Presence of the Rio Grande rift in West Texas and Chihuahua, *in* Geology and Mineral resources of north-central Chihuahua: Guidebook for the 1983 Field Conference, El Paso Geological Society, p. 108–119.

Henry, C.D., J.G. Price, and F.W. McDowell, 1986, Concurrent magmatic and tectonic variations in Trans-Pecos Texas and adjacant Chihuahua during the Tertiary: Geological Society of America Abstracts with Programs, v. 18, p. 635.

Henry, C.D., J.G. Price, G. Aguirre-Díaz, F.W. McDowell, and E.W. James, 1990, Mid-Cenozoic stress evolution and magmatism in the southern Cordillera, Texas and Mexico: Transition from continental arc to intraplate extension: Geological Society of America Abstracts with Programs, v. 22, p. 29.

Henry, C.D., J.G. Price, and E.W. James, 1991, Mid-Cenozoic stress evolution and magmatism in the southern Cordillera, Texas and Mexico: Transition from continental arc to interplate extension: Journal of Geophysical Research, v. 96, p. 13,545–13,560.

Herrero-Bervera, E., J. Urrutia-Fucugauchi, and M.A. Khan, 1990, A paleomagnetic study of remagnetized Upper Jurassic red beds from Chihuahua, northern México: Physics of Earth and Planetary Interiors, v. 62, p. 307–322.

Herrero-Bervera, E., J. Urrutia-Fucugauchi, and M.G. Bocanegra-Noriega, 1990, Paleomagnetism of Mesozoic units of the Maya Terrane: A paleotectonic interpretation: Geological Society of America Abstracts with Programs, v. 22, p. A186.

Herrmann, U., B.K. Nelson, and L. Rainsbacher, 1991, Structural, isotopic, and petrogenetic evidence for the origin of the Xolapa terrane, southern Mexico: Geological Society of America Abstracts with Programs, v. 23, p. A479.

Hickman, R.G., R.J. Varga, and R.M. Altany, 1985, Structural style and tectonic evolution of the Marathon thrust belt, West Texas: Geological Society of America Abstracts with Programs, v. 17, p. 609.

Horak, R.L., 1985, Trans-Pecos tectonism and its effect on the Permian basin, *in* P.W. Dickerson and W.R. Muehlberger, eds., Structure and Tectonics of Trans-Pecos Texas: West Texas Geological Society Field Conference, Publication 85-81, p. 81–85.

Howell, D.C., D.L. Jones, and E.R. Schermer, 1985, Tectonostratigraphic terranes of the circum-Pacific region, *in* D.G. Howell, ed., Tectonostratigraphic Terranes of the Circum-Pacific Region: Circum-Pacific Council for Energy and Mineral Resources, Earth Science Series, v. 1, p. 3–30.

Humphreys, E., and R. Weldon, 1990, Kinematic constraints on the rifting of Baja California: AAPG Memoir 47, p. 217–229.

Jacques-Ayala, C., J.C. García y Barragán, and K.A. DeJong, 1991, The interpreted trace of the Mojave-Sonora megashear in northwest Sonora—A Laramide thrust front and Middle Tertiary detachment zone: Memoria Convención Evolución Geológica de México, Primer Congreso Mexicano de Mineralogía, p. 78–80.

James, E.W., and C.D. Henry, 1990, Accreted late Paleozoic terrane in Trans-Pecos Texas: isotopic mapping of the buried Ouachita front: Geological Society of America Abstracts with Programs, v. 22, p. A329.

James, E.W., and C.D. Henry, 1993, Southeastern extent of the North American craton in Texas and northern Chihuahua as revealed by Pb isotopes: Geologic Society of America Bulletin, v. 105, p. 116–126.

Johnson, C.A., 1987, Regional tectonics in central Mexico: active rifting and transtension within the Mexican volcanic belt: EOS, American Geophysical Union Transactions, v. 68, p. 423.

Johnson, C.A., and C.G.A. Harrison, 1988, The Chapala-Oaxaca fault zone: a major trend-parallel fault in southwestern Mexico: EOS, American Geophysical Union Transactions, v. 69, p. 1450.

Johnson, C.A., H. Lang, E. Cabralcano, C. Harrison, and J.A. Barros, 1990, A new Cordilleran fold-thrust model for Laramide deformation in Tierra Caliente, Michoacan and Guerrero States, Mexico: Geological Society of American Abstracts with Programs, v. 22, p. A186–187.

Johnson, C.A., H. R. Lang, E. Cabralcano, C.G.A. Harrison, and J.A. Barros, 1991, Preliminary assessment of stratigraphy and structure, San Lucas region, Michoacan and Guerrero states, SW Mexico: Mountain Geologist, v. 28, p. 121–136.

Johnson, C.A., et al., 1992, Preliminary assessment of stratigraphy and structure, San Lucas region, Michoacan and Guerrero States, SW Mexico: Reply: Mountain Geologist, v. 29, p. 3–4.

Johnson, K.R., and P.D. Muller, 1986, Late Cretaceous collisional tectonics in the southern Yucatan (Maya) block of Guatemala: Geological Society of America Abstracts with Programs, v. 18, p. 647.

Jones, N.W., R. Dula, L.E. Long, and J.W. McKee, 1982, An exposure of a fundamental fault in Permian basement granitoids, Valle San Marcos, Coahuila, Mexico: Geological Society of America Abstracts with Programs, v. 14, p. 523–524.

Jones, N.W., J.W. McKee, B. Marquez, J. Tovar, L.E. Long, and T.S. Laudon, 1984, The Mesozoic La Mula Island, Coahuila, Mexico: Geological Society of America Bulletin, v. 95, p. 1226–1241.

Jones, N.W., J.W. McKee, T.H. Anderson, and L.T. Silver, 1990, Nazas Formation: A remnant of the Jurassic arc of western North America in north-central Mexico: Geological Society of America Abstracts with Programs, v. 22, p. A327.

Karl, J.H., T.S. Laudon, and D.G. Kilday, 1984, Gravity and magnetic evidence for a major NW-SE basement feature in Coahuila, Mexico: Geological Society of America Abstracts with Programs, v. 16, p. 554.

Kaygi, P.B., G.P. O'Donnell, and M.J. Welland, 1985, Stratigraphy and tectonic development of the southern Ouachita thrust belt—implications of new subsurface data, Arkansas: Geological Society of America Abstracts with Programs, v. 17, p. 624.

Keller, G.R., 1986, Geophysical constraints on the location of the Late Proterozoic to Permian southern margin of the North American Craton: Geological Society of America Abstracts with Programs, v. 18, p. 653.

Keller, G.R., and R. Dyer, 1989, The Paleozoic margin of North America in west Texas and northern Mexico: Geofísica Internacional, v. 28, p. 897–906.

Keller, G.R., and W.J. Peeples, 1984, Gravity anomalies in northern Chihuahua: an update, *in* E.C. Kettenbrink, ed., Geology and Petroleum Potential of Chihuahua: West Texas Geological Society Publication 84-80, p. 206–212.

Keller, G.R., and K.J. Smith, 1985, Structural relations between Marfa, Marathon, Val Verde and Delaware basins of West Texas (abs.): AAPG Bulletin, v. 69, p. 272.

Keller, G.R., K.L. Mickus, D.M. Jurick, and K.R. Libert, 1988, An overview of crustal structure across the Paleozoic continental margin of southern North America (abs.): EOS, American Geophysical Union Transactions, v. 69, n. 44, p. 1412.

Keller, G.R., K.J.M. Kruger, K.J. Smith, and W.M. Voight, 1989, The Ouachita system, a geophysical overview, *in* R.D. Hatcher Jr., W.A. Thomas, and G.W. Viele, eds., The Appalachian and Ouachita Orogen in the United States: Geological Society of America, The Geology of North America, v. F-2, p. 689–694.

Kesler, S.E., 1971, Nature of ancestral orogenic zone in nuclear Central America: AAPG Bulletin, v. 55, p. 2116–2129.

Kesler, S.E., 1973, Basement rock structural trends in southern Mexico: Geological Society of America Bulletin, v. 84, p. 1059–1064.

Kheaton, J.R., et al., 1989, The Amargosa Fault: A major late Quaternary intraplate structure in northern Chihuahua, Mexico: Geological Society of America Abstracts with Programs, v. 21, p. A148.

Kilmer, F.H., 1984, Geology of Cedros Island, Baja California, Mexico: Arcata, California: Master's Thesis, Humboldt State University, 69 p.

Kimbrough, D.L., 1982, Structure, petrology, and geochronology of Mesozoic paleo-oceanic terranes on Cedros Island and the Vizcaíno Peninsula, Baja California Sur, Mexico: Unpublished Ph.D. Dissertation, University of California at Santa Barbara, 395 p.

Kimbrough, D.L., 1985, Tectonostratigraphic terranes of the Vizcaino Peninsula and Cedros and San Benito Islands, Baja California, Mexico, *in* D.G. Howell, ed., Tectono-stratigraphic Terranes of the Circum-Pacific Region: Circum-Pacific Council for Energy and Mineral Resources, Earth Science Series, Number 1, p. 285–298.

Kleist, R., S.A. Hall, and I. Evans, 1984, A paleomagnetic study of the Lower Cretaceous Cupido Limestone northeast Mexico: Evidence for local rotation within the Sierra Madre Oriental: Geological Society of America Bulletin, v. 95, p. 55–60.

Klitgord, K.D., and H. Schouten, 1987a, Plate kinematics of the central Atlantic, *in* B.E. Tucholke and P.R. Vost, eds., The Geology of North America: Western Atlantic Region: Geological Society of America Decade of North American Geology (DNAG), Boulder, Colorado, v. M, p. 325–333.

Klitgord, K.D., and H. Schouten, 1987b, Constraints imposed by plate tectonics on the geometry and time of opening of the Gulf of Mexico: Geological Society of America Abstracts with Programs, v. 19, p. 729.

Klitgord, K., P. Popenoe, and H. Schauten, 1984, Florida: a Jurassic transform plate boundary: Journal of Geophysical Research, v. 89, p. 7753–7772.

Kowalski, J.L., 1989, The structural geology of the Sierra Catorce uplift, *in* K. Greier and J.L. Kowalski, eds., Geology of the Sierra de Catorce Uplift, Real de Catorce, San Luis Potosí, Mexico: Guide Book, The University of Texas, Pan American Geological Society for the Gulf Coast Association of Geological Societies 39th Annual Convention, p. 13–20.

Kruger, J.M., and G.R. Keller, 1986, Interpretation of crustal structure from regional gravity anomalies, Ouachita Mountains area and adjacent Gulf coastal plain: AAPG Bulletin, v. 70, p. 667–689.

Laubach, S.E., and M.L.W. Jackson, 1990, Origin of arches in the northwestern Gulf of Mexico basin: Geology, v. 18, p. 595–598.

Lewis, J.F., and G. Draper, 1990, Geology and tectonic evolution of the northern Caribbean margin, *in* G. Dengo and J.E. Case, eds., The Caribbean Region: Geological Society of America, The Geology of North America, v. H., p. 77–140.

Lillie, R.J., 1985, Tectonically buried continental ocean boundary, Ouachita Mountains, Arkansas: Geology, v. 13, p. 18–21.

Lipman, P.W., and J.T. Hagstrum, 1992, Jurassic ash-flow sheets, calderas, and related intrusions of the Cordilleran volcanic arc in southeastern Arizona: Implications for regional tectonics and ore deposits: Geological Society of America Bulletin, v. 104, p. 32–39.

Locker, S.D., and R.T. Buffler, 1983, Comparison of Lower Cretaceous carbonate shelf margins, northern Campeche Escarpment and northern Florida Escarpment, Gulf of Mexico, *in* A.W. Bally, ed., Seismic Expression of Structural Styles; a Picture and Work Atlas, Volume 2: AAPG Studies in Geology 15, p. 2.2.3-123–2.2.3-128.

Longoria, J.F., 1985a, Tectonic transpression in the Sierra Madre Oriental, northeastern Mexico: An alternative model: Geology, v. 13, p. 453–456.

Longoria, J.F., 1985b, Tectonic transpression in northeastern Mexico (abs.): Gulf Coast Association of Geological Society Transactions, v. 35, p. 199.

Longoria, J.F., 1985c, Tranpressional tectonic history of northeastern Mexico (abs.), *in* M.J. Aldrich, Jr., and

A.W. Laughlin, eds., Sixth International Conference on Basement Tectonics, Proceedings: International Basement Tectonics Association, p. 200–201.

Longoria, J.F., 1986, Mesozoic plate tectonic reconstruction of Mexico: evidence from the stratigraphic record (abs.), *in* W. Sager and C. Scotese, conveners, Mesozoic and Cenozoic Plate Reconstructions: Texas A&M University Geodynamics, Symposium, 3 p.

Longoria, J.F., 1990a, Mesozoic evolution of the Mexican Cordillera: Geological Society of America Abstracts with Programs, v. 22, n. 3, p. 38.

Longoria, J.F., 1990b, Structural traverse across the Sierra Madre Oriental fold-thrust belt in east-central Mexico: Alternative interpretation: Geological Society of America Bulletin, v. 102, p. 261–264.

Lonsdale, P., 1989, Geology and tectonic history of the Gulf of California, in E.L. Winterer, D.M. Hussong, and R.W. Decker, eds., Decade of North American Geology, The Eastern Pacific Region: Geological Society of America, v. N, p. 499–521.

Luhr, J.F., S.A. Nelson, J.F. Allan, and I.S.E. Carmichael, 1985, Active rifting in southwestern Mexico: Manifestations of an incipient eastward spreading-ridge jump: Geology, v. 13, p. 54–57.

Lund, S.P., D.J. Bottjer, K.J. Whidden, and J.E. Powers, 1991, Paleomagnetic evidence for the timing of accretion of the Peninsular Ranges terrane in southern California: Geological Society of America Abstracts with Programs, v. 23, p. 74.

Maher, D.J., N.W. Jones, J.W. McKee, and T.H. Anderson, 1991, Volcanic rocks at Sierra de Catorce, San Luis Potosí, México: A new piece for the Jurassic-arc puzzle: Geological Society of America Abstracts with Programs, v. 23, p. A133.

Mandujano-Velásquez, J., M. Vázquez-Menes, E. Rosales-Contreras, 1992, Geodinámica de las fosas de la Sierra de Chiapas: XI Convención Geologica Nacional, Sociedad Geológica Mexicana, Libro de Resúmenes, p. 103.

Martin del Pozzo, A.L., F.J. Urrutia, and H. Böhnel, 1985, Magnetostratigraphy of Tertiary and Quaternary volcanic rocks from central Mexico (abs.): EOS, American Geophysical Union Transactions, v. 66, p. 873.

Martínez-Reyes, J., and A.F. Nieto-Samaniego, 1990, Efectos geológicos de la tectónica reciente en la parte central de México: Universidad Nacional Autónoma de México, Instituto de Geología, Revista, v. 9, p. 33–50.

Mata-Jurado, M.E., and N. Pintado-Moscoso, 1992, Secciones estructurales regionales de la provincia geológica "Cuencas de Sureste": XI Convención Geológica Nacional, Sociedad Geológica Mexicana, Libro de Resúmenes, p. 114.

Mauger, R.L., 1986, Tertiary volcanism and Basin and Range faulting in the Sierra Tinaja Lisa, north-central Chihuahua, Mexico: Geological Society of America Abstracts with Programs, v. 18, p. 686.

McCabe, C., R. Vander Voo, and J. Urrutia-Fucugauchi, 1984, Paleomagnetism of the Tremadocian Tinú Limestone, State of Oaxaca, Mexico (abs.): EOS, American Geophysical Union Transactions, v. 65, p. 868.

McCabe, C., R. Vander Voo, and J. Urrutia-Fucugauchi, 1988, Late Paleozoic or early Mesozoic magnetizations in remagnetized Paleozoic rocks, State of Oaxaca, Mexico: Earth and Planetary Science Letters, v. 91, p. 205–213.

McKee, J.W., N.W. Jones, and L.E. Long, 1984, History of recurrent activity along a major fault in northeastern Mexico: Geology, v. 12, p. 103–107.

McKee, J.W., N.W. Jones, and T.H. Anderson, 1985, The Late Paleozoic volcanic arc in southern Coahuila Mexico: further definition: Geological Society of America Abstracts with Programs, v. 17, p. 659.

McKee, J.W., N.W. Jones, and T.H. Anderson, 1986, The Jurassic southern margin of North America: Geological Society of America Abstracts with Programs, v. 21, p. 34.

McKee, J.W., N.W. Jones, and T.H. Anderson, 1990, Paleozoic puzzle pieces, northeastern Mexico: Geological Society of America Abstracts with Programs, v. 22, p. 66–67.

Mendoza-Borunda, R., and L.A. Delgado-Argote, 1992, Análisis de esfuerzos a partir de patrones de fallamiento en el bloque de Santo Tomas, Baja California: XI Convención Geológica Nacional, Sociedad Geológica Mexicana, Libro de Resúmenes, p. 119.

Meneses-Rocha, J.J., 1985, Tectonic evolution of the strike-slip fault province of Chiapas, Mexico: Unpublished Master's Thesis, University of Texas at Austin, 315 p.

Meneses-Rocha, J.J., 1987 (1990), Marco tectónico y paleogeografía del Triásico Tardío-Jurásico en el sureste de México: Boletín Asociación Mexicana de Geólogos Petroleros, v. 39, p. 3–69.

Meneses-Rocha, J.J., 1992a, Evolución tectónica de la Fosa de Ixtapa, Chiapas, Mexico (abs.): II Symposium de Asociación Mexicana de Geológos Petroleros, Instituto Mexicano del Petróleo, México, D.F.

Meneses-Rocha, J.J., 1992b, La fosa de Ixtapa de Chiapas, Mexico: un nuevo modelo de cuenca asociada a fallas de transcurrencia: XI Convención Geológica Nacional, Sociedad Geológica Mexicana, Libro de Resúmenes, p. 121.

Meschede, M., L. Ratschbacher, W. Frisch, U. Hermann, and U. Riller, 1992, Movimientos transtensivos a lo largo del límite norte del Complejo Xolapa: la geodinámica en el Sur de Mexico: XI Convencion Geológica Nacional, Sociedad Geológica Mexicana, Libro de Resúmenes, p. 122.

Mickus, K.L., and G.R. Keller, 1992, Lithospheric structure of the south-central United States: Geology, v. 20, p. 335–338.

Miranda-Canseco, E., 1992, Estructuras transpresivas del sistema transcurrente Malpaso, extremo NW de la Sierra de Chiapas: XI Convención Geológica Nacional, Sociedad Geológica Mexicana, Libro de Resúmenes, p. 126.

Mitre-Salazar, L.M., 1981 (1983), Las imágenes LANDSAT—una herramienta útil en la interpretación geológico-estructural; un ejemplo en el noreste de México: Universidad Nacional Autónoma de México, Instituto de Geología, Revista, v. 5, p. 37–46.

Mitre-Salazar, L.M., 1989, La megafalla laramídica de San Tiburcio, Estado de Zacatecas: Revista del Instituto de Geología, Universidad National Autónoma México, v. 8, p. 47–51.

Mitre-Salazar, L.M., and A.R. Huizar, 1984, La Sierrita, Zacatecas: una estructura Laramídica anómala en el Altiplano Central (abs.): VII Convención Nacional de la Sociedad de Geólogos Mexicanos, p. 18–19.

Mitre-Salazar, L.M., and J. Roldán-Quintana, 1990, La Paz to Saltillo, northwestern and northern Mexico: H-1 Geological Society of America Centennial Continent/Ocean Transect #14.

Mitre-Salazar, L.M., et al., 1990, Southern Baja California to Coahuila: H-1: Geological Society of America Centennial Continent/Ocean Transect #13.

Molina-Garza, R.S., and J. Urrutia-Fucugauchi, 1985, Crustal structure beneath central and southern Mexico (abs.): EOS, American Geophysical Union Transactions, v. 66, p. 379.

Molina-Garza, R.S., R. Vander Voo, and J. Urrutia-Fucugauchi, 1990, Paleomagnetic data from Chiapas, southern Mexico: Implications for tectonic evolution of the Gulf of Mexico (abs.): EOS, American Geophysical Union Transaction, v. 71, p. 491.

Molina-Garza, R.S., R. Vander Voo, and J. Urrutia-Fucugauchi, 1992, Paleomagnetism of the Chiapas Massif, southern Mexico: Evidence for rotation of the Maya Block and implications for the opening of the Gulf of Mexico: Geological Society of America Bulletin, v. 104, p. 1156–1168.

Monreal, R., and J. Longoria, 1989, Morphotectonic analysis in a transpressional regime: The Chihuahua Tectonic Belt, Mexico: Geological Society of America Abstracts with Programs, v. 21, p. A91.

Montgomery, H. A., 1985, Tectostratigraphic evidence for late Paleozoic Pacific Plate collision and post-Upper Jurassic transpression in northeastern Chihuahua, Mexico: Geological Society of America Abstracts with Programs, v. 17, p. 667.

Montgomery, H., and J.F. Longoria, 1987, Paleozoic tectonics of Chihuahua—a paleogeographic perspective: Universidad Autónoma de Chihuahua Gaceta Geológica, v. 1, p. 54–69.

Montgomery, H., and J.F. Longoria, 1990, Permian paleogeographic reconstruction of the west-central (Mexico) margin of Pangea: Geological Society of America Abstracts with Programs, v. 22, n. 3, p. 69.

Montijo-González, A., and I. Minjárez-Sosa, 1990, Structural relationship between Paleozoic platform and basinal Paleozoic sequences of central Sonora, Mexico: Geological Society of America Abstracts with Programs, v. 22, n. 3, p. 69.

Moore, T.E., 1985, Stratigraphy and tectonic significance of the Mesozoic tectonostratigraphic terranes of the Vizcaino Peninsula, Baja California Sur, Mexico, *in* D.G. Howell, ed., Tectonostratigraphic terranes of the Circum-Pacific region: Houston, Texas, Circum-Pacific Council for Energy and Mineral Resources, Earth Science Series Number 1, p. 315–329.

Morán-Zenteno, D.J., 1986, Breve revisión sobre la evolución tectónica de México: Geofísica Internacional, v. 25, p. 9–38.

Morán-Zenteno, D.J., et. al., 1986, Magnetic fabrics and paleogeography of Oaxaca State, southern Mexico during the Middle Jurassic (abs.): EOS, American Geophysical Union Transactions, v. 67, p. 925.

Morán-Zenteno, D.J., J. Urrutia-Fucugauchi, H. Böhnel, and E. González-Torres, 1988, Paleomagnetismo de rocas Jurásicas del norte de Oaxaca y sus implicaciones tectónicas: Geofísica Internacional, v. 27, p. 485–518.

Morán-Zenteno, D.J., H. Koehler, V. Von Drach, and P. Schaaf, 1990, The geological evolution of Xolapa terrane, southern Mexico, as inferred from Rb-Sr and Sm-Nd isotopic data: Geowissenschaftliches Lateinamerika-Kolloquium, 21.11.90–23.11.90.

Morris, A.E.L., I. Taner, H.A. Meyerhoff, and A.A. Meyerhoff, 1990, Tectonic evolution of the Caribbean region; Alternative hypothesis: Geological Society of America, Geology of North America, v. H, p. 433–457.

Morris, L.K., S.P. Lund, and D.J. Bottjer, 1986, Paleolatitude drift history of displaced terranes in southern and Baja California: Nature, v. 321, p. 844–847.

Muehlberger, W.R., and P.R. Tanvers, 1985, Marathon fold-thrust belt, west Texas, *in* R.D. Hatcher, W.A. Thomas, and G.W. Viele, eds., The Geology of North America, v. F-2, The Appalachian and Ouachita Regions: U.S. Geological Society, p. 673–680.

Muehlberger, W.R., W.D. DeMis, and J.O. Leason, 1984, Geologic map and cross-sections, Marathon Region, Trans-Pecos Texas: Geological Society of America, Map and Chart Series, MC-28T.

Murillo-Muñetón, G., R. Torres-V., O.R. Navarrete, 1992, El macizo de la Mixtequita: redefinición: XI Convención Geológica Nacional, Sociedad Geológica Mexicana, Libro de Resúmenes, p. 131.

Murray, G.E., 1986a, Musings about some of the tectonics of the southwestern United States and northern Mexico: Geological Society of America Abstracts with Programs, v. 18, p. 701.

Murray, G.E., 1986b, Musings about some of the tectonics of the southwestern United States and northern Mexico: The Texas Journal of Science, v. 38, p. 301–326.

Murray, G.E., 1989, The California-Tamaulipas geosuture: A review of some facts, interpretations and speculations: West Texas Geological Society Publication 89-85, p. 211–221.

Nava, F., et al., 1988, Structure of the Middle America trench in Oaxaca, Mexico: Tectonophysics, v. 154, p. 241–251.

Nelson, T.H., 1991, Salt tectonics and listric-normal faulting, *in* A. Salvador, ed., The Gulf of Mexico Basin: Geological Society of America, v. J., p. 73–89.

Neuhaus, J.R., M. Cassidy, D. Kummenacher, and R.G. Gastil, 1988, Timing of protogulf extension and transtensional rifting through volcanic/sedimentary stratigraphy of SW Isla Tiburon, Gulf of California, Sonora, Mexico: Geological Society America Abstracts with Programs, v. 20, p. 218.

Nieto-Obregon, J., L. Delgado-Argote, and P.E. Damon, 1985, Geochronologic, petrologic and structural data related to large morphologic features between the Sierra Madre Occidental and the Mexican volcanic belt: Geofísica Internacional, v. 24, p. 623–663.

Nieto-Obregon, J., J. Urrutia-Fucugauchi, 1992, Listric faulting and continental rifting in western Mexico—A paleomagnetic and structural study: Tectonophysics, v. 208, p. 365–376.

Nieto-Samaniego, A.F., 1990 (1992) Fallamiento y estratigrafía Cenozoicos en la parte sudoriental de la Sierra de Guanajuato: Universidad Nacional Autónoma de México, Instituto de Geología, Revista, v. 9, p. 146–155.

Nieto-Samaniego, A.F., et al., 1992, Interpretación estructural de los rasgos geomorfológicos principales de la Sierra de Guanajuato: Universidad Nacional Autónoma de México, Instituto de Geología, Revista, v. 10, p. 1–5.

Nourse, J.A., 1990a, A thermal-gravitational driving mechanism for middle Tertiary extension in northern Sonora, Mexico: Geological Society of America Abstracts with Programs, v. 22, n. 3, p. 73.

Nourse, J.A., 1990b, Tectonostratigraphic and strain patterns in the Magdalena metamorphic core complex, northern Sonora, Mexico: Geological Society of America Abstracts with Programs, v. 22, n. 3, p. 73.

Nourse, J.A., 1990c, Tectonostratigraphic development and strain history of the Magdalena metamorphic core complex, northern Sonora, México, *in* G.E. Gehrels and J.E. Spencer, eds., Geologic excursions through the Sonoran Desert region, Arizona and Sonora: Arizona Geological Survey Special Paper 7, p. 155–164.

Nowicki, M.J., S.A. Hall, and I. Evans, 1990, Paleomagnetic evidence for local and regional post-Eocene rotations in northern Mexico: EOS, American Geophysical Union Transactions, v. 71, p. 491.

Omana, M.A.A., 1987, Gravity and crustal structure of the south-central Gulf of Mexico, the Yucatan Peninsula, and adjacent areas from 17º30′N to 26ºN and from 84ºW to 93ºW: Master's Thesis, Oregon State University, Corvallis, Oregon.

Ortega-Guerrero, B., and J. Urrutia-Fucugauchi, 1989, Paleogeography and tectonics of the Mixteca Terrane, southern Mexico during the interval of drifting between North and South America and Gulf of Mexico rifting (abs.): EOS, American Geophysical Union Transactions, v. 70, n. 43, p. 1314.

Ortega-Gutiérrez, F., 1986, Precambrian basement terranes of the southern Gulf of Mexico-northern Caribbean region: Geological Society of America Abstracts with Programs, v. 18, p. 712.

Ortega-Gutiérrez, F., 1987, New insights into the tectonic evolution of Mexico: Geological Society of America Abstracts with Programs, v. 19, p. 795.

Ortega-Gutiérrez, F., 1988, Essential aspects of the Phanerozoic tectonic history of Mexico: Geological Society of American Abstracts with Programs, v. 20, n. 7, p. A132.

Ortega-Gutiérrez, F., 1989, Suspect terranes of southern Mexico and nature of their boundaries: 28th International Geological Congress Abstracts, v. 2, p. 552.

Ortega-Gutiérrez, F., et al., 1990, Middle America Trench-Oaxaca-Gulf of Mexico: H-3: Geological Society of America Centennial Continent/Ocean Transect #14.

Ortiega-Gutiérrez, F., et al., eds., 1993, First Circum-Pacific and Circum-Atlantic Terrane Conference, Guanajuato, Mexico: Instituto de Geología, Universidad Nacional Autonoma Mexico, Book of Abstracts, 176 p. Forty-six papers on structural geology, tectonics, igneous and metamorphic petrology.

Ortiz-Hernández, et al., 1990 (1992), El arco intraoceánico alóctono (Cretácico Inferior) de Guanajuato—caracteristicas petrográficas, geoquímicas, estructurales e isotópicas del complejo filoniano y de las lavas basálticas asociadas; implicaciones geodinámicas: Universidad Nacional Autónoma de México, Instituto de Geología, Revista, v. 9, p. 126–145.

Ortiz-Ubilla, A., D. Santamaría-Orozco, and A. Riba-Ramírez, 1992, Consideraciones tectónicas sobre el basamento pre-Mesozóico de una porción del noreste de México: XI Convención Geológica Nacional, Sociedad Geológica Mexicana, Libro de Resúmenes, p. 143.

Ortuño-Arzate, S., and J. Delfaud, 1992, Evolución geodinámica de la Cuenca de Zongólica, situación en el contexto: Domínio Golfo de México/Sistema Cordillerano: XI Convención Geológica Nacional, Sociedad Geológica Mexicana, Libro de Resúmenes, p. 145–146.

Ortuño-Arzate, S., J.P. Xavier, and J. Delfaud, 1992, Análisis tectónico-estructural de la Cuenca de Zongólica a partir de imágenes de satolite LANDSAT MSS, Instituto Mexicano del Petróleo, v. 24, n. 1, p. 11–45.

Pacheco-Gutiérrez, C., 1992, Evolución tectónica de la porción nordoccidental de la Cuenca Mesozóica del centro de México: XI Convención Geológica Nacional, Sociedad Geológica Mexicana, Libro de Resúmenes, p. 148.

Pacheco-Gutiérrez, C., and M. Barba, 1986, El Precámbrico de Chiapas, un terreno estratotectónico: VIII Convención Nacional, Sociedad Geológica Mexicana, Libro de Resúmenes.

Pacheco-Gutiérrez, C., R. Castro-M., and M. A.-Gómez, 1984, Confluéncia de terrenos estratotectónicos en Santa María del Oro, Durango, México: Revista del Instituto Mexicano del Petróleo, v. 16, p. 7–20.

Padilla y Sánchez, R.J., 1985, Las estructuras de la Curvatura de Monterrey, estados de Coahuila, Nuevo León, Zacatecas y San Luis Potosí: Revista del Instituto de Geología, Universidad Nacional Autonóma México, v. 6, p. 1–20.

Padilla y Sánchez, R.J., 1986a, Post-Paleozoic tectonics of northeast Mexico and its role in the evolution of the Gulf of Mexico: Geofísica Internacional, v. 25, p. 157–206.

Padilla y Sánchez, R.J., 1986b, Post-Paleozoic tectonics of northeast Mexico and its role in the evolution of the Gulf of Mexico: Geological Society of America Abstracts with Programs, v. 18, p. 713.

Pardo-Casas, F., and P. Molnar, 1987, Relative motion of the Mazca (Farallon) and South American plates since Late Cretaceous time: Tectonics, v. 6, n. 3, p. 233–248.

Pasquare, G., and A. Zanchi, 1985, Cenozoic volcanism and tectonics in western-central Mexico: Atti della Accademia Nazionale dei Lincei, Rendiconti, Classe de Scienze Fisiche, Matematiche e Naturali, v. 78, p. 293–301.

Patterson, D.L., 1984, Paleomagnetism of the Valle Formation and the Late Cretaceous paleogeography of the Vizcaíno Basin, Baja California, Mexico, *in* V.A. Frizzel, ed., Geology of the Baja California Peninsula, Pacific Section SEPM, v. 39, p. 173–182.

Pattison, A. D., and F.E. Julian, 1990, Laramide deformation of San Francisco del Oro, Chihuahua, Mexico: Geological Society of America Abstracts with Programs, v. 22, n. 3, p. 75.

Pérez-Venzor, J.A., C. Neptali-Montoya, and R. Ibarra-González, 1992, Análisis geomorfológico estructural California, México: XI Convención Geológica Nacional en la región sur de la peninsula de Baja California: Sociedad Geológica Mexicana, Libro de Resúmenes, p. 156.

Pew, E., 1982, Seismic structural analysis of deformation in the southern Mexican Ridges: Master's Thesis, University of Texas at Austin, 102 p.

Pilger, R.H., Jr., 1981, The opening of the Gulf of Mexico: Implications for the tectonic evolution of the northern Gulf Coast: Gulf Coast Association of Geological Societies Transactions, v. 31, p. 377–381.

Pindell, J.L., 1985, Alleghenian reconstruction and subsequent evolution of the Gulf of Mexico, Bahamas, and Proto-Caribbean: Tectonics, v. 4, p. 1–39.

Pindell, J.L., 1988, Gulf of Mexico/Caribbean tectonostratigraphic evolution: Geological Society of America Abstracts with Programs, v. 20, p. A132.

Pindell, J. L., and S.F. Barrett, 1990, Geologic evolution of the Caribbean: a plate tectonic perspective, *in* J.E. Case and G. Dengo, eds., The Caribbean Region: Geological Society of America, The Geology of North America, v. H, p. 405–432.

Pindell, J.L., and J.F. Dewey, 1982, Permo-Triassic reconstruction of western Pangea and the evolution of the Gulf of Mexico/Caribbean region: Tectonics, v. 1, p. 179–211.

Pindell, J.L., and G.D. Karner, 1990, Regional analysis in/around Gulf of Mexico: A critical look at various scenarios for Mesozoic tectonic evolution: Geological Society of America Abstracts with Programs, v. 22, p. A186.

Pindell, J.L., S.F. Barrett, and J.F. Dewey, 1985, Tectonic evolution of the Gulf of Mexico and Caribbean: Geological Society of America Abstracts with Programs, v. 17, p. 690.

Pindell, J.L., et al., 1988, A plate-kinematic framework for models of Caribbean evolution: Tectonophysics, v. 155, p. 121–138.

Pindell, J.L., et al., 1989, Constraints on Mesozoic-Cenozoic motions of plates and terranes of Gulf of Mexico and Caribbean region: Mesozoic-Cenozoic paleogeographic analysis of Gulf of Mexico and Caribbean region: 28th International Geological Congress Abstracts, v. 2, p. 610–611.

Plauchut, B., 1989, Étude géologique des Cordilléres Nord-Americaines: Bull. Centres Research Exploration-Production, Elf-Aquitaine, v. 13, n. 2, p. 215–217.

Poole, F.G., 1990, Sonoran Orogen in the Barita De Sonora mine area, central Sonora, Mexico: Geological Society of America Abstracts with Programs, v. 22, p. 76.

Poole, F.G., R.J. Madrid, and J.M. Moráles-Ramírez, 1990, Sonoran orogen in the Barita de Sonora mine area, central Sonora, Mexico: Geological Society of America Abstracts with Programs, v. 22, p. 76.

Quintero-Legorreta, O., 1992, Geología de la región de Comanja, Estados de Guanajuato y Jalisco: Universidad Nacional Autónoma de México, Instituto de Geología, Revista, v. 10, p. 6–25.

Quintero-L., O., and M. Aranda-García, 1985, Relaciones estructurales entre el Anticlinario de Parras y el Anticlinario de Arteaga (Sierra Madre Oriental), en la región de Agua Nueva, Coahuila: Revista del Instituto de Geología, Universidad Nacional Autónoma México .v. 6, p. 21–36.

Radelli, L., 1990, The Mid-Cretaceous Olvidada nappe: A former Aptian–Albian basin between northern Baja California and Sonora, Mexico: Actas de la Facultad de Ciéncias de la Tierra de la Universidad Autónoma de Nuevo León, Linares, v. 4, p. 213–229.

Radelli, L., et al., 1986, Tectonic transpression in the Sierra Madre Oriental, northeastern Mexico; an alternative model; discussion and reply: Geology, v. 14, p. 808–810.

Radelli, L., et al., 1987, Allochthonous Paleozoic bodies of central Sonora: Departmento de Geología—UNISON (Univ. Sonora) Boletín, v. 4 (1 & 2), p. 1–15.

Ramírez, C.L.C., and J.G.F. Ronquillo, 1992, Tomografía sísmica, una alternativa para exploración petrolera, Revista del Instituto Mexicano del Petróleo, Universidad Nacional Autónoma México, v. 24, n. 4, p. 7–13.

Rangin, C., 1981 (1984), Aspectos geodinámicos de la región noroccidental de México: Universidad Nacional Autónoma de México, Instituto de Geología, Revista, v. 5, p. 186–194.

Ratschbacher, L., U. Riller, M. Meschede, U. Herrmann, and W. Frisch, 1991, Second look at suspect terranes in southern Mexico: Geology, v. 19, p. 1233–1236.

Riggs, N.R., and G.B. Haxel, 1990, The Early to Middle Jurassic magmatic arc in southern Arizona: Plutons to sand dunes, *in* G.E. Gehrels and J.E. Spencer, Geologic Excursions through the Sonoran Desert Region, Arizona and Sonora: Arizona Geological Survey Special Paper 7, p. 90–103.

Riggs, N.R., G.B. Haxel, and C.J. Busby-Spera, 1990, Paleogeography and tectonic setting of the Jurassic magmatic arc in southern Arizona: progress and problems: Geological Society of America Abstracts with Programs, v. 22, n. 3, p. 78.

Riggs, N.R., J.M. Mattinson, and C.J. Busby-Spera, 1991, Timing of Early-Middle Jurassic arc magmatism in southern Arizona and implications for paleogeographic setting of the magmatic arc: Geological Society of America Abstracts with Programs, v. 23, p. A250.

Robinson, K.L., 1991, U-Pb zircon geochronology of basement terranes and the tectonic evolution of southwestern mainland Mexico: Unpublished Master's Thesis, San Diego State University, 190 p.

Robinson, K.L., R.G. Gastil, M.F. Campa, and J. Ramírez, 1989, Geochronology of basement and metasedimentary rocks in southern Mexico and their relation to metasedimentary rocks in peninsular California: Geological Society of America Abstracts with Programs, v. 21, p. 135.

Robinson, K.L., R.G. Gastil, M.F. Campa, 1989, Early Tertiary extension in southwestern Mexico and the exhumation of the Xolapa metamorphic core complex: Geological Society of America Abstracts with Programs, v. 21, p. A92.

Robinson, K.L., R.G. Gastil, and M.S. Girty, 1990, Eocene intra-arc transtension: the detachment of the Chortis block from southwestern Mexico: Geological Society of America Abstracts with Programs, v. 22, p. 78.

Rodríguez-Castañeda, J.L., 1981 (1983), Notas sobre la geología del área de Hermosillo, Sonora: Universidad Nacional Autónoma de México, Instituto de Geologia, Revista, v. 5, p. 30–36.

Rodríguez-Castañeda, J.L., 1990a, Preliminary analysis of fault systems in the Tuape region, north-central Sonora, Mexico: Geological Society of America Abstracts with Programs, v. 22, n. 3, p. 79.

Rodriguez-Castañeda, J.L., 1990b, Relaciones estructurales en la parte centroseptentrional del Estado de Sonora: Universidad Nacional Autónoma de México, Instituto de Geología, Revista, v. 9, p. 51–61.

Roldán-Quintana, J., 1982, Evolución tectónica del Estado de Sonora: Revista del Instituto de Geológia, Universidad Nacional Autónoma México, v. 5, p. 178–185.

Roldán-Quintana, J., 1989, Geología de la Hoja Baviácora, Sonora: Revista del Instituto de Geología, Universidad Nacional Autónoma México, v. 8, p. 1–14.

Roldán-Quintana, J., 1990, Geology of the Tonibabi Quadrangle in northeastern Sonora, Mexico: Geological Society of America Abstracts with Programs, v. 22, n. 3, p. 79.

Rosas-Elguera, J., 1992, Análisis morfoestructural al sur de "Graben" de Tepic: rotación del Bloque Jalisco: XI Convención Geológica Nacional, Sociedad Geológica Mexicana, Libro de Resúmenes, p. 171.

Rosaz, T. M., 1989, Structural geology of Cerro de Cristo Rey, southern New Mexico and westernmost Texas, USA: A Paleocene–Eocene tectonic transcurrent zone along the Texas lineament: Geological Society of America Abstracts with Programs, v. 21, p. A90.

Rosenfeld, J.H., 1992, El margen suroeste de la plataforma de Yucatán en la Sierra de los Cuchamatanes de Guatemala: XI Convención Geológica Nacional, Sociedad Geológica Mexicana, Libro de Resúmenes, p. 171.

Ross, C.A., 1986, Paleozoic evolution of southern margin of Permian basin: Geological Society of America Bulletin, v. 97, p. 536–554.

Ross, M.I., and C.R. Scotese, 1988, A hierarchial tectonic model of the Gulf of Mexico and Caribbean region: Tectonophysics, v. 155, p. 139–168.

Rowley, D.B., J. Pindell, A.L. Lottes, and A.M. Ziegler, 1986, Phanerozoic reconstructions of northern South America, W. Africa, North America, and Caribbean region: Geological Society of America Abstracts with Programs, v. 18, p. 735.

Royden, L.H., B.C. Burchfiel, H. Ye, and M.S. Schuepbach, 1990, The Ouachita-Marathon thrust belt: Orogeny without collision: Geological Society of America Abstracts with Programs, v. 22, p. A112.

Rueda-Gaxiola, J., 1992, El Alogrupo Los San Pedros (= Alogrupo La Boca) del noreste de México y sus relaciones tectono-estratigráficas y paleogeográficas: XI Convención Geológica Nacional, Sociedad Geológica Mexicana, Libro de Resúmenes, p. 172.

Rueda-Gaxiola, J., E. López-Ocampo, M.A. Dueñas, and J.L. Rodríguez Benítez, 1991, Las fosas de Huizachal-Peregrina y de Huayacocotla; dos partes de un graben relacionado con el origen del Golfo de Mèxico: Memoria de la Convención sobre la Evolución Geológica de México, Primer Congreso Mexicano de Mineralogía, p. 189–192.

Salvador, A., and A.R. Green, 1980, Opening of the Caribbean Tethys: origin and development of the Caribbean and the Gulf of Mexico, *in* J. Aubouin, J. Debelmas, and M. Latreille, eds., Colloque C5-Geology of the Alpine Chains Born of the Tethys (26th International Geological Congress, Paris, 1980): Bureau de Recherches Géologiques et Minieres, Memoire 115, p. 224–229.

Santamaría, D., F. Ortuño, A. Artiz, A. Riba, and S. Franco, 1992, Estructura y evolución geodinámica Mesozóica de la Cuenca de Sabinas: XI Convención Geológica Nacional, Sociedad Geológica Mexicana, Libro de Resúmenes, p. 175–176.

Sawyer, D.S., 1984, Gulf of Mexico plate reconstruction by palinspastic restoration of extended continental crust (abs.): AAPG Bulletin, v. 68, p. 525.

Scheidegger, A.E., 1989, Estudios tectónicos de la parte sur de México: Revista de la Academia

Colombiana de Ciéncias Exactas, Físicas y Naturales, v. 17 (64), p. 125–132.

Schellhorn, R.W., 1987, Bouguer gravity anomalies and crustal structure of northern Mexico: Master's Thesis, University of Texas at Richardson, Texas.

Schellhorn, R.W., C.L.V. Aiken, and M.F. De La Fuente Duch, 1985, Regional gravity and airmagnetic studies in northern Mexico: Preliminary results (abs.): EOS, American Geophysical Union Transactions, v. 66, p. 379.

Schellhorn, R.W., S. Hall, C.L.V. Aiken, and M.F. de la Fuente, 1990, Bouguer gravity anomalies and crustal structure in northwestern Mexico: AAPG Memoir 47, p. 197–215.

Schlee, J.S., and K.D. Klitgord, 1986, Structure of the North American Atlantic continental margin: Journal of Geological Education, v. 34, p. 72–89.

Scholvink, D.V., 1990, Modelo transcurrente en la evolución tectónico-sedimentaria de México: Boletín Asociación Mexicana de Geólogos Petroleros, v. 40, n. 2, p. 1–35.

Scotese, C.R., L.M. Gahagan, and R.L. Larson, 1988, Plate tectonic reconstructuion of the Cretaceous and Cenozoic ocean basins: Tectonophysics, v. 155, p. 27–28.

Sedlock, R.L., 1988, Tectonic setting of blueschist and island-arc terranes of west central Baja California, Mexico: Geology, v. 16, p. 623–626.

Sedlock, R.L., and D.H. Hamilton, 1991, Late Cenozoic tectonic evolution of southwestern California: Journal of Geophysical Research, v. 96, p. 2325–2352.

Sedlock, R.L., F. Ortega-Gutiérrez, and R.C. Speed, 1990, Jurassic sinistral displacement in eastern Mexico: Geological Society of America Abstracts with Programs, v. 22, p. A186.

Sedlock, R.L., F. Ortega-Gutiérrez, and R.C. Speed, 1993, Tectonostratigraphic terranes and tectonic evolution of Mexico: Geological Society of America Special Paper 278, 153 p.

Servais, M., R. Rojo-Yaiz, and D. Colorado-Lievano, 1982, Estudio de las rocas básicas y ultrabásicas de Sinaloa y Guanajuato-postulación de un paleogolfo de Baja California y de una digitación Tethysiana en México central: Geomimet, v. 115, p. 53–71.

Servais, M., E. Cuevas-Pérez, and O. Monod, 1986, Une section de Sinaloa a San Luis Potosí: nouvelle approche de l'évolution du Mexique nord-occidentel: Societé Geolgique de France Bulletin, v. 8, n. 6, p. 1033–1047.

Shaub, F.J., 1982a, Growth faults on the southwestern margin of the Gulf of Mexico, in A.W. Bally, ed., Seismic Expression of Structural Styles; A Picture and Work Atlas: AAPG Studies in Geology 15, v. 2.

Shepherd, A., S. Hall, and K. Burke, 1987, Magnetic anomalies indicate Gulf of Mexico originated by counterclockwise rotation of Yucatan from Gulf Coast (abs.): AAPG Bulletin, v. 71, p. 614.

Shurbet, D.H., and S.E. Cebull, 1984, Tectonic interpretation of the Trans-Mexican volcanic belt: Tectonophysics, v. 101, p. 159–165.

Shurbet, D.H., and S.E. Cebull, 1987, Tectonic interpretation of the westernmost part of the Ouachita-Marathon (Hercynian) orogenic belt, west Texas-Mexico: Geology, v. 15, p. 458–461.

Silver, L.T., and T.H. Anderson, 1983, Further evidence and analysis of the role of the Mojave-Sonora Megashear(s) in Mesozoic cordilleran tectonics: Geological Society of America Abstracts with Programs, v. 15, p. 273.

Silver, L.T., and T.H. Anderson, 1988, Lithospheric reconstructions of southwestern North America in the Middle to Late Mesozoic: Geological Society of America Abstracts with Programs, v. 20, n. 7, p. A59.

Sivils, D.J., 1988, Post-Paleozoic structural styles in northern Sierra de Palomas: Chihuahua, Mexico (abs.): Southwest Section Meetings AAPG Bulletin, v. 72, n. 1, p. 103.

Smith, K.J., 1986, A gravity and tectonic study of the southwestern portion of the Ouachita system: Master's Thesis, University of Texas at El Paso, 98 p.

Sosson, M., T. Calmus, M. Tardy, and R. Blanchet, 1990, Nouvelles données sur le front tectonique Cenomano-Turonien dans le nord de l'Etat de Sonora (Mexique): Comptes Rendus Académie Science, Paris, v. 310, p. 417–423.

Soto-Guitérrez, M. A., 1992, Megaestructuras del occidente de México: XI Convención Geológica Nacional, Sociedad Geológica Mexicana, Libro de Resúmenes, p. 185.

Soto-Guitérrez, M.A., and J.M. Chávez-Aguirre, 1992, Geología de la Cuenca Media del Rio Mascota y sus estructuras regionales, Estado de Jalisco, México: XI Convención Geológica Nacional, Sociedad Geológica Mexicana, Libro de Resúmenes, p. 186.

Spencer, J.E., and S.J. Reynolds, 1986, Some aspects of the Middle Tertiary tectonics of Arizona and southeastern California: Arizona Geological Society Digest, v. 16, p. 102–107.

Stephens, W. E., and T.H. Anderson, 1986, La Lamina thrust sheet—a far traveled allochthon of crystalline basement, northwestern Mexico: Geological Society of America Abstracts with Programs, v. 18, p. 762.

Stevens, C.H., and P. Stone, 1988, Early Permian thrust faults in east-central California: Geological Society of America Bulletin, v. 100, p. 552–562.

Stevens, C.H., P. Stone, and R.W. Kistler, 1992, A speculative reconstruction of the middle Paleozoic continental margin of southwestern North America: Tectonics, v. 11, p. 405–419.

Stewart, J.H., 1988a, Latest Proterozoic and Paleozoic southern margin of North America and the accretion of Mexico: Geology, v. 16, p. 186–189.

Stewart, J.H., 1988b, Tectonic significance of late Proterozoic and Lower Cambrian Cordilleran miogeoclinal rocks in the Caborca region, Sonora, Mexico: Geological Society of America Abstracts with Programs, v. 20, p. A298.

Stewart, J.H., 1990, Position of Paleozoic continental margin in northwestern Mexico: Present knowledge and speculations: Geological Society of America Abstracts with Programs, v. 22, p. 86–87.

Stock, J.M., and K.V. Hodges, 1989, Pre-Pliocene extension around the Gulf of California, and the transfer of Baja California to the Pacific plate: Tectonics, v. 8, p. 99–116.

Stock, J., and P. Molnar, 1988, Uncertainties and implications of the Late Cretaceous and Tertiary position of North America relative to the Farallon, Kula, and Pacific plates: Tectonics, v. 7, p. 1339–1384.

Suárez, G., and S.K. Singh, 1985, Source mechanisms and tectonic deformation along the southwestern margin of the Gulf of Mexico (abs.), *in* G.A. Bollinger, chairperson, The Seismological Society of America, Annual Meeting, Seismological Research Letters, v. 55, p. 28.

Suárez, G., T. Monfret, G. Wittlinger, and C. David, 1990, Geometry of subduction and depth of the seismogenic zone in the Guerrero gap, Mexico: Nature, v. 345, p. 336–338.

Suter, M., 1987a, Structural section across the Sierra Madre Oriental fold-thrust belt in east-central Mexico: Geological Society of America Abstracts with Programs, v. 19, p. 860.

Suter, M., 1987b, Structural traverse across the Sierra Madre Oriental fold-thrust belt in east-central Mexico: Geological Society of America Bulletin, v. 98, p. 249–264.

Suter, M., 1987c, Orientational data on the state of stress in northeastern Mexico as inferred from stress-induced borehole elongations: Journal of Geophysical Research, v. 92, p. 2617–2626.

Suter, M., 1990, Reply to "Structural traverse across the Sierra Madre Oriental fold-thrust belt in east-central Mexico: Alternative interpretation: Geological Society of America Bulletin, v. 102, p. 264–266.

Tardy, M., J.C. Carfantan, and C. Rangin, 1986, Essai de synthese sur la structure du Mexique: Societé Geologique de France, Bulletin, v. 2 (6), p. 1025–1033.

Tardy, M., R. Blanchet, and M. Zimmermann, 1990, Les lineaments du Texas et Caltham entre Cordilleres Mexicaines: Nature, origine et évolution structurale: Bull. Centres Recherches Exploration-Production Elf-Aquitaine, v. 13, n. 2, p. 219–227.

Tauvers, P.R., 1985a, Decollement tectonics of the frontal zone, western domain, Marathon Basin, Texas, *in* P.W. Dickerson and W.R. Muehlberger, eds., Structure and Tectonics of Trans-Pecos, Texas: Field Conference: West Texas Geological Society Publication 85-81, p. 69–75.

Tauvers, P.R., 1985b, Reappraisal of the evolution of the Marathon fold-thrust belt in light of basement-control on sedimentation patterns and fault propagation: Geological Society of America Abstracts with Programs, v. 18, p. 769.

Tauvers, P. R., 1988, Basement-influenced deformation in the Marathon fold-thrust belt, West Texas: Journal of Geology, v. 96, p. 577–590.

Thomas, W.A., 1986, Early Paleozoic southern margin of North America: Geological Society of America Abstracts with Programs, v. 18, p. 772.

Thomas, W. A., 1990, The Paleozoic southern margin of North America: Geological Society of America Abstracts with Programs, v. 22, p. A111.

Tolson, G., 1990, Structural development and tectonic evolution of the Santa Rosa area, southwest State of Mexico, Mexico: Geological Society of America Abstracts with Programs, v. 22, p. A328.

Torres-Roldán, V., R. Vander Voo, and J. Urrutia-Fucugauchi, 1986, Remagnetization of Permian rocks from Las Delicias, Mexico (abs.): EOS, American Geophysical Union Transactions, v. 67, n. 16, p. 267.

Torres-Roldán, V., J. Ruiz, M. Grajales, and G. Murillo, 1992a, Permian magmatism in eastern and southern Mexico and its tectonic implications: Geological Society of America Abstracts with Programs, v. 24, p. A64.

Torres-Roldán, V., J. Ruiz, M. Grajales, and G. Murillo, 1992b, El plutonismo Pérmico en el oriente y sur de México y sus implicaciones tectónicas: XI Convención Geológica Nacional, Sociedad Geológica Mexicana, Libro de Resúmenes, p. 193–195.

Tosdal, R.M., G.B. Haxel, T.H. Anderson, C.D. Connors, D.J. May, and J.E. Wright, 1990, Highlights of Jurassic, Late Cretaceous to early Tertiary, and middle Tertiary tectonics, south-central Arizona and north-central Sonora, *in* G.E. Gehrels and J.E. Spencer, Geologic Excursions Through the Sonoran Deserts Region, Arizona and Sonora: Arizona Geological Survey Special Paper 7, p. 76–88.

Tristán-González, M., and J.R. Torres-Hernández, 1991, Geología de la Sierra de Charcas, Estado de San Luis Potosí: Memoria de la Convención sobre la Evolución Geológica de México: Primer Congreso Mexicano de Mineralogía, p. 221–223.

Urrutia-Fucugauchi, J., 1984, On the tectonic evolution of Mexico: paleomagnetic constraints, *in* R. Vander Voo, C.R. Scotese, and N. Bonhommet, eds., Plate Reconstructions from Paleozoic Paleomagnatism: American Geophysical Union, Geodynamics Series 12, p. 29–47.

Urrutia-Fucugauchi, J., 1986a, Crustal thickness, heat flow, arc magmatism, and tectonics of Mexico—preliminary report: Geofísica Internacional, v. 25, p. 559–573.

Urrutia-Fucugauchi, J., 1986b, Late Mesozoic–Cenozoic evolution of the northwestern Mexico magmatic arc zone: Geofísica Internacional, v. 25, p. 61–84.

Urrutia-Fucugauchi, J., 1988, Paleomagnetic study of the Cretaceous Morelos Formation, Guerrero State, southern Mexico: Tectonophysics, v. 147, p. 121–125.

Urrutia-Fucugauchi, J., and H. Böhnel, 1988, Tectonics along the Trans-Mexican volcanic belt according to palaeomagnetic data: Physics of Earth and Planetary Interiors, v. 52, p. 320–329.

Urrutia-Fucugauchi, J., and D.A. Valencio, 1986, Paleomagnetic study of Mesozoic rocks from Ixtapán de la Sal, México: Geofísica Internacional, v. 25, p. 485–502.

Urrutia-Fucugauchi, J., and R. Vander Voo, 1983, Reconnaissance paleomagnetic study of Cretaceous limestones from southern Mexico (abs.): American Geophysical Union Transactions, v. 64, p. 220.

Urrutia-Fucugauchi, J., E. Cabral-Cano, and D.J. Moran-Zenteno, 1985, Paleomagnetism and tectonostratigraphic terrains in Mexico (abs.): EOS, American Geophysical Union Transactions, v. 66, p. 866–867.

Urrutia-Fucugauchi, J., et al., 1987, Tectonic interpretation of the Trans-Mexican Volcanic Belt—Discussion: Tectonophysics, v. 138, p. 319–323.

Urrutia-Fucugauchi, J., D.J. Moran-Zenteno, and E. Cabral-Cano, 1987, Paleomagnetism and tectonics of Mexico: Geofísica Internacional, v. 26, n. 3, p. 429–458.

Urrutia-Fucugauchi, J., H. Böhnel, and E. Herrero-Bervera, 1988, What happened to terranes traveling along the western continental margin of Mexico (abs.): EOS, American Geophysical Union Transactions, v. 69, n. 44, p. 1170.

Valdes, C.M., et al., 1985, Crustal structure of Oaxaca, Mexico, from seismic refraction and gravity measurements (abs.): EOS, American Geophysical Union Transactions, v. 66, p. 309.

Valdes, C.M., et al., 1986, Crustal structure of Oaxaca, Mexico, from seismic refraction measurements: Bulletin of the Seismological Society of America, v. 76, p. 547–563.

Vander Voo, R., 1987, Precambrian and early Mesozoic paleomagnetism of the basement terranes of southern Mexico: evidence for displaced terranes: Geological Society of America Abstracts with Programs, v. 19, p. 876.

Vander Voo, R., J. Peinado, and C.R. Scotese, 1984, A paleomagnetic reevaluation of Pangea reconstructions, *in* R. Vander Voo, C.R. Scotese, and N. Bonhommet, eds., Plate Reconstructions from Paleozoic Paleomagnetism: American Geophysical Union, Geodynamics Series, v. 12, p. 11–26.

Van Siclen, D.C., 1984, Early opening of initially-closed Gulf of Mexico and Central North Atlantic Ocean: Gulf Coast Association of Geological Societies Transactions, v. 34, p. 265–275.

Van Siclen, D.C., 1989, Early Mesozoic rifting west of the Gulf of Mexico as it opened from a fully-closed condition: Transactions, Southwest Section, AAPG Convention, San Angelo, Texas, p. 127–131.

Van Siclen, D.C., 1990, Rifting west of Gulf of Mexico as it opened from a closed condition: Geological Society of America Abstracts with Programs, v. 22, p. 90.

Vázquez, M.M.E., et al., 1992, Neotectónica del sureste de México: Revista del Instituto Mexicano del Petróleo, v. 24, n. 3, p. 12–37.

Vega-Granillo, R., F. Pérez-Soto, and M. Chaparro-Meza, 1991, Overthrusting sheets with imbricated thrusts systems in central Sonora, Mexico: Geological Society of America Abstracts with Programs, v. 23, p. A128.

Vidal-Serratos, R., 1992, La estructura de Chilacachapa, en Guerrero, un duplex de orígen Laramídico: XI Convención Geológica Nacional, Sociedad Geológica Mexicana, Libro de Resúmenes, p. 211.

Viele, G.W., 1986, The subduction of Texas: Geological Society of America Abstracts with Programs, v. 18, p. 779.

Viele, G.W., 1988, Geologic history of the Ouachita orogenic belt: Geological Society of America Abstracts with Programs, v. 20, p. A132.

Viele, G.W., 1989, The Ouachita Orogenic Belt, *in* R.D. Hatcher, Jr., W.A. Thomas, and G.W. Viele, eds., The Appalachian-Ouachita Orogen in the United States: Geological Society of America, The Geology of North America, v. F-2, p. 555–561.

Viele, G.W., and W.A. Thomas, 1989, Tectonic synthesis of the Ouachita orogenic belt, *in* The Appalachian-Ouachita Orogen in the United States: Geological Society of America, The Geology of North America, v. F-2, p. 695–727.

Villaseñor-R., P.E., M.E. Vázquez-M., M.R. Sánchez-Q., M.A. Islas, 1992, Neotectónica del sureste de México: XI Convención Geológica Nacional, Sociedad Geológica Mexicana, Libro de Resúmenes, p. 213.

Villegas, C.C., 1988, Aplicación del perfil sísmico vertical para el delineámiento y localizacón de una falla estructural en el Domo Salino Tuzandépetl: Revista del Instituto Mexicano del Petróleo, v. 20, n. 1, p. 5–27.

Wadge, G., and K. Burke, 1983, Neogene Caribbean plate rotation and associated Central American tectonic evolution: Tectonics, v. 2, p. 633–643.

Walper, J.L., and R.E. Miller, 1985, Tectonic evolution of the Gulf Coast basins: Gulf Coast Section SEPM Foundation, Proceedings of the 4th Annual Research Conference, Austin, Texas, p. 25–42.

Watkins, J.S., 1990, Palinspastic restoration of Mesozoic extension in the northern Gulf of Mexico: Geological Society of America Abstracts with Programs, v. 22, p. A112.

Weber, B., M. Meschede, L. Ratschbacher, W. Frisch, U. Herman, U. Riller, 1992, La orogénesis Acadiana en el sur de México, etapa de deformación en el Complejo Acatlán: XI Convención Geológica Nacional, Sociedad Geológica Mexicana, Libro de Resúmenes, p. 216.

Winker, C.D., 1984a, Cross sections; Mexico to Florida, *in* R.T. Buffler et al., eds., Gulf of Mexico: Marine Science Institute, Woods Hole, Massachusetts, v. 6, p. 25.

Winker, C.D., 1984b, Initiation and evolution of a deep-water Gulf of Mexico, Jurassic to Middle Cretaceous (abs.): Geological Society of America Abstracts with Programs , v. 16, p. 117.

Winker, C.D., and R.T. Buffler, 1988, Paleogeographic evolution of early deep-water Gulf of Mexico and margins, Jurassic to Middle Cretaceous (Comanchean): AAPG Bulletin, v. 72, p. 318–346.

Woodcock, S.F., 1976, Crustal structure of the Tehuantepec Ridge and adjacent continental margins of southwestern Mexico and eastern Guatemala: Master's Thesis, Oregon State University, Corvallis, Oregon.

Yañez, P., P.J. Patchett, J. Ruiz, and F. Ortega-Gutiérrez, 1989, Nd isotopes and Cordilleran vs. Appalachian affinity for the Paleozoic of Mexico (abs.): EOS, American Geophysical Union, Transactions, v. 70, n. 43, p. 1404.

Yañez, P., P.J. Patchett, J. Ruiz, and G. Gehrels, 1990, The Paleozoic Acatlán terrane of southern Mexico: Implications for the tectonic history of North America: Geological Society of America Abstracts with Programs, v. 22, p. 96.

Yañez, P., J. Ruiz, P.J. Patchett, F. Ortega-Gutiérrez, and G. Gehrels, 1991, Isotopic studies of the Acatlán Complex, southern Mexico: Implications for Paleozoic North American tectonics: Geological Society of America Bulletin, v. 103, p. 817–828.

Zaldivar, R.J., and P.I. Gallo, 1992, Análisis estructural en el borde suroeste de la Peninsula de Coahuila; un intento en el uso de secciones balanceadas: XI Convención Geológica Nacional, Sociedad Geológica Mexicana, Libro de Resúmenes, p. 217.

3. Mineral Resources, Studies of Igneous Metamorphic Rocks, Neovolcanic Investigations, Basement Studies, and Age Dating

Aguirre-Diaz, G., and F.W. McDowell, 1987, Eocene and younger volcanism on the eastern flank of the Sierra Madre Occidental, Nazas, Durango, Mexico: Geological Society of America Abstracts with Programs, v. 19, p. 567.

Aguirre-Díaz, G., and F.W. McDowell, 1988, Nature and timing of faulting in the southern Basin and Range, central-eastern Durango, Mexico: EOS, American Geophysical Union Transactions, v. 69, p. 1412–1413.

Aguirre–Díaz, G., and F.W. McDowell, 1991, The volcanic section at Nazas, Durango, Mexico, and the possibility of widespread Eocene volcanism within the Sierra Madre Occidental: Journal of Geophysical Research, v. 96, p. 13,373–13,388.

Albrecht, A., and D.G. Brookins, 1989, Buenaventura magmatic complex (Chihuahua, Mex.): Mash followed by closed system fractional crystallization: Geological Society of America Abstracts with Programs, v. 21, p. A57.

Albrecht, A., and K.W. Sums, 1988, Geochemical variations across a major terrane boundary in northern Mexico: EOS, American Geophysical Union Transactions, v. 69, n. 44, p. 1493.

Alexandri, R., Jr., et al., 1985, The sedimentary manganese carbonate deposits of the Molango District, Mexico: Geological Society of America 98th annual meeting, Geological Society of America Abstracts with Programs, p. 511.

Allan, J.F., 1986, Geology of the Colima and Zacoalco grabens, SW Mexico: Late Cenozoic rifting in the Mexican volcanic belt: Geological Society of America Bulletin, v. 97, p. 473–485.

Allan, J.F., et al., 1990, Pliocene–Holocene rifting and associated vulcanism in southwest Mexico: an exotic terrane in the making, *in* J.P. Dauphin and B.R.T. Simoneit, eds., Gulf and Peninsular Provinces of the Californias: AAPG Memoir 47, p. 425–445.

Almazán-Vásquez, E., 1988, Geoquímica de las rocas volcánicas de la Formación Alisitos del Arroyo La Bocana en el Estado de Baja California Norte: Revista del Instituto de Geología, Universidad Nacional Autónoma México, v. 7, p. 78–88.

Anderson, T.H., L.T. Silver, and G.A. Salas, 1980, Distribution and U-Pb isotope ages of some lineated plutons, northwestern Mexico: Geological Society of America Memoir 153, p. 269–283.

Anthony, E.Y., and J.M. Hoffer, 1990, Petrology of Cenozoic mafic lavas from southern New Mexico: Geological Society of America Abstracts with Programs, v. 22, p. 3.

Aranda-Gómez, J.J., and J. Luhr, 1986, Spinel-herzolite-bearing basic volcanic centers of late-Quaternary age from San Luis Potosí, Mexico: Geological Society of America Abstracts with Programs, v. 18, p. 528.

Aranda-Gómez, J.J., and F. Ortega-Gutiérrez, 1987, Mantle xenoliths in Mexico, *in* P.H. Nixon, ed., Mantle Xenoliths: New York, John Wiley, p. 75–84.

Aranda-Gómez, J.J., and J.A. Pérez-Venzor, 1989, Estratigrafía del complejo cristalino de la región de Todos Santos, Estado de Baja California Sur: Revista del Instituto de Geología, Universidad Nacional Autónoma México, v. 8, p. 149–170.

Aranda-Gómez, J.J., J.F. Luhr, and J.G. Pier, 1989a, Geología de los volcanes cuaternarios portadores de xenolitos provenientes del manto y de la base de la corteza del Estado de San Luis Potosí: Volúmen Commemorativo del primer centenário del Instituto de Geología de la Universidad Nacional Autónoma de México.

Aranda-Gómez, J.J., J.F. Luhr, and J.G. Pier, 1989b, Petrology and geochemistry of basanitic rocks from the La Breña and El Jagüey Maar Complex, Durango, Mexico: Geological Society of America Abstracts with Programs, v. 21, n. 6, p. A201.

Aranda-Gómez, J.J., J.F. Luhr, and J.G. Pier, 1990 (1992), A new type of maar volcano from the State of Durango—the El Jaguey-La Breña complex reinterpreted; a discussion, and a reply by E.R. Swanson: Universidad Nacional Autónoma de México, Instituto de Geología, Revista, v. 9, p. 204–211.

Aranda-Gómez, J. J., J.F. Luhr, and J.G. Pier, 1992a, La influencia del agua subterranea en el desarrollo de los complejos volcanicos de la Joya Hondo, S.L.P. y la Breña-El Jagüey, Dgo: XI Convención Geológica

Nacional, Sociedad Geológica Mexicana, Libro de Resúmenes, p. 16.
Aranda-Gómez, J.J., J.F. Luhr, and J.G. Pier, 1992b, The La Breña—El Jagüey Maar Complex, Durango, Mexico: I. Geological evolution: Bulletin of Volcanology, v. 54, p. 393–404.
Arnold, M., and E. González-Partida, 1987, Un equilibrio químico e isotópico entre sulfatos y sulfuros de los fluidos hidrotermales del campo geotérmico Los Azufres, Michoacán: Implicaciones sobre el origen del azufre: Universidad Nacional Autónoma de México, Instituto de Geología, Revista, v. 7, p. 97–105.
Asmeron, Y., R.E. Zartman, P.E. Damon, and M. Shafiquillah, 1990, Zircon U-Th-Pb and whole-rock Rb-Sr age patterns of lower Mesozoic igneous rocks in the Santa Rita Mountains southeast, Arizona: Implications for Mesozoic magmatism and tectonics in the southern Cordillera: Geological Society of America Bulletin, v. 102, p. 961–968.
Baca-Carreón, C.J., and E. Flores-G., 1992, Los recursos minerales del Estado de Veracruz: XI Convención Geológica Mexicana, Libro de Resúmenes, p. 23.
Bazán-B., S., 1987, Génesis de las pegmatitas del arco insular de Telixtlahuaca: Geomimet, v. 149, p. 54–67.
Bazán-Perkins, S.D., 1992, Genesis del gráfito del complejo oaxaqueño y sus facies metamórficas: XI Convención Geológica Nacional, Sociedad Geológica Mexicana, Libro de Resúmenes, p. 27–29.
Bell, R.W., 1985, Geochemistry and petrology of Tertiary trachytic and alkali-basaltic rocks of Pico el Centinela, Northern Coahuila, Mexico: Master's Thesis, Sul Ross University, Alpine, Texas, 113 p.
Blount, J.G., 1990, Geochemistry of Late Proterozoic trondhjemites from Sierra el Carrizalillo, Chihuahua, Mexico: Geological Society of America Abstracts with Programs, v. 22, p. A174.
Blount, J.G., N.W. Walker, and W.D. Carlson, 1989, Geochemistry and U-Pb zircon ages of Mid-Proterozoic meta-igneous rocks from Chihuahua, Mexico: Geological Society of America Abstracts with Programs, v. 20, p. A205.
Bobier, C., and C. Robin, 1983, Paleomagnetisme de la Sierra Madre Occidentale dans les états de Durango et Sinaloa (Mexique): Variations du champ ou rotations de blocs au Paleocene et au Neogene?: Geofísica Internacional, v. 22, p. 57–86.
Bristol, D.A., and S. Mosher, 1989, Grenville-age, polyphase deformation of Mid-Proterozoic basement, NW Van Horn Mountains, Trans-Pecos, Texas: Journal of Geology, v. 97, p. 25–43.
Bustamante-García, J., and J.N. Espinal, 1992, Aspectos mineralógicos, geológicos y metalogenéticos del depósito aurifero Llanitos-El Real, municipio de Villa Madero, Mich: XI Convención Geológica Nacional, Sociedad Geológica Mexicana, Libro de Resúmenes, p. 40.
Calmus, Thierry, E. Pérez-Segura, and J. Roldán-Quintana, 1992, Los yacimientos de San Felipe (Sonora) un ejemplo de minas deslizadas en la provincia del Basin and Range: XI Convención Geológica Nacional, Sociedad Geológica Mexicana, Libro de Resumenes, p. 41.
Cameron, K.L., and M. Cameron, 1985, Rare earth element 87Sr/86Sr, and 143Nd/144Nd compositions of the Cenozoic orogenic dacites from Baja California, northwestern Mexico, and adjacent west Texas: evidence for the predominance of a subcrustal component: Contributions to Mineralogy and Petrology, v. 19, p. 1–11.
Cameron, K.L., and M. Cameron, 1986, The transition from high-K calc-alkalic to alkalic magmatism in West Texas. The quartz normative rocks: Geological Society of America Abstracts with Programs, v. 18, p. 555.
Cameron, K.L., and D. Kuentz, 1987, The mid-Tertiary basaltic andesite suite of northwestern Mexico and the southern U.S. Basin and Range: Backarc or arc magmatism? (abs.): EOS, American Geophysical Union Transactions, v. 68, p. 1529.
Cameron, K.L., and J.V. Robinson, 1988, Comments on "Nd-Sr isotope composition of lower crustal xenoliths—Evidence for the origin of mid-Tertiary felsic volcanics in Mexico" by J. Ruiz, P. J. Patchett, and R. J. Arculus: Contributions to Mineralogy and Petrology, v. 100.
Cameron, K.L., and R.L. Rudnick, 1990, Diverse ages in the lower crust of northern Mexico: EOS, American Geophysical Union Transactions, v. 71, p. 43.
Cameron, K.L., et al., 1980, Petrologic characteristics of mid-Tertiary volcanic suites, Chihuahua, Mexico: Geology, v. 8, p. 87–91.
Cameron, K.L., G.J. Hinz, S. Niemeyer, and J.V. Robinson, 1988, Sr-Nd-Pb isotopic composition of mafic granulites and metapyroxenitic xenoliths from Mexico: Implication for Cenozoic crustal growth (abs.): EOS, American Geophysical Union Transactions, v. 69, p. 1517.
Cameron, K.L., et al., 1989, Southern Cordillera basaltic andesite suite, southern Chihuahua, Mexico: A link between Tertiary continental arc and flood basalt magmatism in North America: Journal of Geophysical Research, v. 94, p. 7817–7840.
Cameron, K.L., J.V. Robinson, S. Niemeyer, and G.J. Nimz, 1990, Trace-element and isotope geochemistry of deep crustal xenoliths from La Olivina, central Chihuahua, Mexico: Geological Society of America Abstracts with Programs, v. 22, (3), p. 12.
Cameron, K.L., M. Cameron, R.S. Harmon, and B. Barriero, 1991, Role of crustal interaction in the genesis of ignimbrites from Batopilas, north central Sierra Madre Occidental (SMO), Mexico: Geological Society of America Abstracts with Programs, v. 23, n. 5, p. A332.
Cameron, K.L., J.V. Robinson., S. Niemeyer, and G.J. Nimz, 1991, Granulite-facies xenoliths from north-central Mexico: Evidence for a major pulse of mid-Cenozoic crustal growth: Journal of Geophysical Research, v. 96.
Cameron, K.L., R.D. Smith, D.C. Kuentz, and S.B. Tanner, 1992, Mafic xenoliths with orthocumulate

to granulite textures from northern Mexico, and their relationship to Tertiary volcanic rocks from the region (abs.): EOS American Geophysical Union Transactions, v. 73, p. 658.

Cameron, K.L., et al., 1992, Contrasting styles of pre-Cenozoic and mid-Tertiary crustal evolution in northern Mexico: Evidence from deep crustal xenoliths from La Olivina: Journal of Geophysical Research, v. 97, p. 17,353–17,376.

Cameron, M., K.L. Cameron, M. Sawlan, and R. Gunderson, 1983, Calc-alkalic volcanism in Chihuahua, Mexico: Its regional geochemical context, *in* Geology and Mineral Resources of North-Central Chihuahua: Guidebook for the 1983 Field Conference, El Paso Geological Society, p. 94–101.

Campa-Uranga, M.F., 1985, Metalogénesis y tectónica de placas: Ciéncias, v. 1, p. 22–29.

Campa-Uranga, M.F., and P.J. Coney, 1983, Tectonostratigraphic terranes and mineral resource distributions in Mexico: Canadian Journal of Earth Science, v. 20, p. 1040–1051.

Campos-Enríquez, J.O., M.A. Arroyo-Esquivel, and J. Urrutia-Fucugauchi, 1990, Basement Curie isotherm and shallow-crustal structure of the Trans-Mexican volcanic belt from aeromagnetic data: Tectonophysics, v. 172, p. 77–90.

Cantagrel, J.M., and C. Robin, 1979, K-Ar dating on eastern Mexican volcanic rocks; relations between the andesitic and the alkaline provinces: Journal of Volcanology and Geothermal Research, v. 5, p. 99–114.

Carballido-Sánchez, E.A., and L.A. Delgado-Argote, 1989, Geología del cuerpo serpentinítico de Tehuitzingo, Estado de Puebla—interpretación preliminar de su emplazamiento: Revista del Instituto de Geologia, Universidad Nacional Autónoma México, v. 8, p. 134–148.

Carrasco-Núñez, G., M. Milán, and S.P. Verma, 1989, Geología del Volcán Zamorano, Estado de Querétaro: Universidad Nacional Autónoma de México, Instituto de Geología, Revista, v. 8, p. 194–201.

Carrillo, A., H. Huyck, M. Sentilli, and A. Grist, 1992, Rocas encajonantes del yacimiento de oro diseminiado de Los Uvares, Baja California Sur: geología y edades radiométricas: XI Convención Geológica Nacional, Sociedad Geológica Mexicana, Libro de Resúmenes, p. 47.

Carrillo-Chavez, A., 1991, Las alteraciones como guías mineralogicas en yacimientos de oro tipo dike falla: Los Uvares, B.C.S., ejemplo característico: First International Meeting on Geology of the Baja California Peninsula, p. 14.

Carrillo-Chavez, A., 1992, Gneis milonitico de San Pedrito, sur de la peninsula de Baja California: el mas antiguo y meridional de los complejos metamórficos centrales (metamorphic core complex) de la Cordillera Americana? XI Convencion Geológica Nacional, Sociedad Geológica Mexicana, Libro de Resúmenes, p. 48.

Carrillo-Martinez, M., 1989, Tectonics and ore deposits in central Mexico: Geological Society of America Abstracts with Programs, v. 20, n. 7, p. A232.

Castillejos-Pastrana, R., 1992, Características principales de las rocas intrusivas que afloran en área de Tembabiche, B.C.S., México: XI Convención Geológica Nacional, Sociedad Geológica Mexicana, Libro de Resúmenes, p. 53.

Castillo-Rodriguez, H., 1988, Zur Geologie des kristallinen Grundgebirges der Sierra Madre Oriental—insbesondere des Granjeno-Schiefer-Komplexes—im Sudteil des Huizachal-Peregrina-Antiklinoriums (Raum Ciudad Victoria, Bundesstaat Tamaulipas, Mexiko): Unpublished Thesis, Universität Munster, Munster, Germany, 138 p.

Castillo-Rodriguez, H., Cossío-T., and H.J. Gursky, 1986, Rasgos litológicos principales del basamento cristalino de la Sierra Madre Oriental (área de Ciudad Victoria, Tamaulipas, México): Actas de la Facultad de Ciéncias de la Tierra, Universidad Autónoma de Nuevo León, Linares, v. 1, p. 1–10.

Clark, K.F., 1984, Lithologic framework and contained mineral resources in North-Central Chihuahua, *in* E.C. Kettenbrink, Jr., Geology and Petroleum Potential of Chihuahua, Mexico: West Texas Geological Society Publication 84-80, p. 126–148.

Clark, K.F., 1985a, Summary of the lithology, tectonic framework and metallic deposits in Sierra Madre Occidental, northwestern Mexico, *in* R.W. Nesbitt and I. Nichol, eds., Geology in the Real World—The Kingsley Dunham Volume, p. 31–50.

Clark, K.F., 1985b, Reconnaissance geological and mineral resource map of northern Sierra Madre Occidental, Mexico: Geological Society of America Abstracts with Programs, v. 17, p. 548.

Clark, K.F., 1992, Northern Sierra Madre Occidental—renewed interest in part of a well-known, mineral-rich province: XI Convención Geológica Nacional, Sociedad Geológica Mexicana, Libro de Resúmenes, p. 58.

Clark, K.F., C.T. Foster, and P.E. Damon, 1982, Cenozoic mineral deposits and subduction-related magmatic arcs in Mexico: Geological Society of America Bulletin, v. 93, p. 533–544.

Clark, K.F., P.K.M. Megaw, and J. Ruiz, eds., 1986, Lead-zinc-silver carbonate-hosted deposits of northern Mexico: University of Texas El Paso, Society of Economic Geologists Guidebook.

Córdoba, D.A., 1988, Estratigrafía de las rocas volcánicas de la región entre Sierra de Gamón y Laguna de Santaiguillo, Estado de Durango: Revista del Instituto de Geología, Universidad Nacional Autónoma México, v. 7, p. 136–147.

Corona-Esquivel, R.J., and J.R. Rodolfo, 1992, Geología y geoquímica del yacimiento de hierro Peña Colorada, Estado de Colima, México: XI Convención Geológica Nacional, Sociedad Geológica Mexicana, Libro de Resúmenes, p. 58.

Cossío-Torres, T., 1988, Zur Geologie des kristallinen Grundgebirges der Sierra Madre Oriental—insbesondere des Novillo-Gneis-Komplexes—im Südteil des Huizachal-Peregrina-Antiklinoriums (Raum Ciudad Victoria, Bundesstaat Tamaulipas, Mexiko): Unpublished Thesis, Universität Munster, Munster, Germany, 99 p.

Damon, P.E., 1984, Batholith-volcano coupling in the metalogeny of porphyry copper deposits: Symposium of Copper Hypabyssal Deposits, 27th International Geological Congress.

Damon, P.E., M. Shafiquillah, and K.F. Clark, 1984, Evolución de los arcos magmáticos en México y su relación con la metalogénesis: Revista del Instituto de Geología, Universidad Nacional Autónoma México, v. 5, p. 223–238.

Damon, P.E., M. Shafiquillah, K. DeJong, and J. Roldán-Quintana, 1991, Geochronology of Mesozoic magmatism in Sonora, northwestern Mexico: Geological Society of America Abstracts with Programs, v. 23, n. 5, p. A127.

Davis, G.H., A.F. Gardulkski, and T.H. Anderson, 1981, Structural and structural-petrologic characteristics of some metamorphic core complex terranes in southern Arizona and northern Sonora, *in* L. Ortlieb and J. Roldán-Quintana, eds., Geology of Northwestern Mexico and Southern Arizona: Instituto de Geología, Universidad Nacional Autónoma México, Estación Regional del Noroeste, p. 323–365.

Delgado-Argote, L.A., R. Rubinovich-Cogan, and A. Gasca-Durán, 1986, Descripción preliminar de la geología y mecánica de emplazamiento del complejo ultrabásico del Cretácico de Loma Baya, Guerrero, México: Geofísica Internacional, v. 25, p. 537–558.

Delgado-Argote, L.A., M. López-Martinez, D. York, and C.M. Hall, 1990, Geology and geochronology of ultramafic localities in the Cuicateco and Tierra Caliente complexes, southern Mexico: Geological Society of America Abstracts with Programs, v. 22, p. A326.

Demant, A., 1981 (1984), Interpretación geodinámica del volcanismo del Eje Neovolcánico Transmexicano: Universidad Nacional Autónoma de México, Instituto de Geología, Revista, v. 5, p. 217–222.

Demant, Alan, J.J. Cochemé, P. Delpretti, and P. Piquet, 1989, Geology and petrology of the Tertiary volcanics of the northwestern Sierra Madre Occidental, Mexico: Bulletin de la Societé Géologique de France, v. 5, p. 737–748.

Department of Geology and Geophysics, University of New Orleans and Department of Geology, Tulane University, 1990, The Tectonic, Geophysics, and Volcanism of Mexico, A Symposium, 36 abstracts, 47 p.

Deshler, R.M., J.D. Hoover, and R. Dyer, 1986, El Borracho Caldera: A 17 My rhyolitic edifice in the eastern Basin and Range Chihuahua Mexico: Geological Society of America Abstracts with Programs, v. 18, p. 584.

Díaz de León, J., 1981 (1984), Descubrimiento e identificación de la segunda localidad mundial del mineral bartonita en Michoacán: Universidad Nacional Autónoma de México, Instituto de Geología, Revista, v. 5, p. 261.

Diaz-U., R., 1986, Geology and mineralization of La Encantada silver-lead deposits, Coahuila, Mexico, *in* K.F. Clark, P.K.M. Megaw, and J. Ruiz, eds., Lead-zinc-silver carbonate-hosted deposits of northern Mexico: University of Texas at El Paso, Society of Economic Geologists Guidebook, p. 305–310.

Dreier, J., 1984, Regional tectonic control of epithermal veins in the western United States and Mexico: Arizona Geological Society Digest, v. 15, p. 228–50.

Duffield, W.A., R.I. Tlling, and R.F. Canul, 1985, Geology of the Chicon Volcano, Chiapas, Mexico: Journal of Vulcanology and Geothermal Research, v. 20, p. 117–132.

Elias-Herrera, M., 1987, Metamorphic geology of the Tierra Caliente Complex, Tejupilco region, State of Mexico: Geological Society of America Abstracts with Programs, v. 19, p. 654.

Elías-Herrera, M., and J.L. Sánchez-Zavala, 1990 (1992), Tectonic implications of a mylonitic granite in the lower structural levels of the Tierra Caliente complex (Guerrero terrane), southern Mexico: Universidad Nacional Autónoma de México, Instituto de Geología, Revista, v. 9, p. 113–125.

Ferriz, H., 1981 (1983), Geología de la Caldera de San Marcos, Chihuahua: Universidad Nacional Autónoma de México, Instituto de Geología, Revista, v. 5, p. 65–79.

Galicia, E.F., and J.L. Busto-Dias, 1989, Exploración del yacimiento de Azufre de Ojapa, Municipio de Oluta y Texistepec, Veracruz: Geomimet, 159, p. 16–30.

Garrison, J.R., Jr., C. Ramírez-Ramírez, and L.E. Long, 1980, Rb-Sr isotopic study of the ages and provenance of Precambrian granulite and Paleozoic greenschist near Ciudad Victoria, Mexico, *in* R.H. Pilger Jr., ed., The Origin of the Gulf of Mexico and the Early Opening of the Central North Atlantic Ocean: Baton Rouge, Louisiana State University, p. 37–49.

Gastil, R.G., 1983, Mesozoic and Cenozoic granitic rocks of southern California and western Mexico: Geological Society of America Memoir 159, p. 265–275.

Gastil, R.G., J. Diamond, and C. Knaack, 1986, The magnetite-ilmenite line in peninsular California: Geological Society of America Abstracts with Programs, v. 18, p. 109.

Gomberg, J., K.F. Priestly, T.G. Masters, and J.N. Brune, 1988, The structure of the crust and upper mantle of northern Mexico: Geophysical Journal of the Royal Astronomical Society, v. 94, p. 1–20.

Gomberg, J., K.F. Priestley, and J.N. Brune, 1989, The compressional velocity structure of the crust and upper mantle of northern Mexico and the border region: Seismological Society of America Bulletin, v. 79, p. 1496–1519.

Gómez-Morán, C., 1986, Geochemistry and petrology of crustal xenoliths from Xalapasco de la Joya, State of San Luis Potosí, Mexico: Master's Thesis, University of Houston, 324 p.

Gomez-Morán, C., and D. Elthon, 1986, Composition and equilibration conditions of the lower crust under Xalapasco de La Joya (State of San Luis Potosí, Mexico): Geological Society of America Abstracts with Programs, v. 18, p. 618.

González-Caver, E., and S.A. Nelson, 1989, Geology and K-Ar age dating of the Tuxtla volcanic field, Veracruz, Mexico: a Late Miocene to Recent alkaline volcanic field: Geological Society of America Abstracts with Programs, v. 21, p. A14.

González-Partida, E., and R. Martínez-Serrano, 1989, Geocronología, termomicrometría e isotopia de azufre y carbon de la brecha cuprífera La Sorpresa, Estado de Jalisco: Revista del Instituto de Geología, Universidad Nacional Autonóma México, v. 8, p. 202–210.

González-Partida, E., V. Torres-Rodríguez, and F. González-Sanchez, 1987, El Cretácico volcanosedimentario de la parte centro-occidental de México: implicaciones tectónicas y metalogenéticas: Actas de la Facultad de Ciéncias, Universidad Autónoma de Nuevo León, Linares, v. 2, p. 155–164.

González-Partida, E., I. Casar-Aldrete, P. Morales-Puente, and J. Nieto-Obregón, 1989, Fechas de Rb-Sr (Maastrichtiano y Oligoceno) de rocas volcánicas e intrusivas de la región de Zihuatanejo, Sierra Madre del Sur de México: Revista del Instituto de Geología, Universidad Nacional Autónoma México, v. 8, p. 248–249.

González-Partida, E., et al., 1992, Distribución y condiciones de formación de los minerales de arcilla en un campo geotérmico—el caso de Los Humeros, Estado de Puebla: Universidad Nacional Autónoma de México, Instituto de Geología, Revista, v. 10, p. 47–53.

González-Sandoval, J.R., and M. Morales Montana, 1992, Consideraciones sobre la génesis de los depósitos de zeolitas de Sonora, México: XI Convención Geológica Nacional, Sociedad Geológica Mexicana, Libro de Resúmenes, p. 80.

Goodell, P.C., 1984, The Chihuahua City uranium province, Chihuahua, Mexico, *in* E.C. Kettenbrink, ed., Geology and Petroleum Potential of Chihuahua: West Texas Geological Society Publication 84-80, p. 188–194.

Grajales-N., J.M., 1988, Geology, geochronology, geochemistry, and tectonic implications of the Juchatengo Green Rock Sequence, State of Oaxaca, Southern Mexico: Master's Thesis, University of Arizona, Tucson, 145 p.

Grajales-N., J.M., and M. López-Infanzón, 1983, Estudio petrogenético de las rocas ígneas y metamórficas en el prospecto Tomatlán-Guerrero-Jalisco: Instituto Mexicano del Petróleo, Internal Report, Proyecto C-1160, 69 p.

Grajales-N., J.M., and J. Záldívar-Ruiz, 1991, La franja magmática de Paleozóico Tardio en Oaxaca y sur de Puebla y sus implicaciones tectónicas (abs.): Convención sobre la Evolución Geológica de México Memoria, Primer Congreso Mexicano de Mineralogia, p. 61–62.

Grajales-N., J.M., R. Torres, and G. Murillo, 1986, Datos isotópicos potasio-argón para rocas ígneas y metamórficas en el Estado de Oaxaca: Sociedad Geológica Mexicana, VIII Convención Nacional, Libro de Resúmenes.

Gunderson, R., K. Cameron, and M. Cameron, 1986, Mid-Cenozoic high-K calc-alkalic and alkalic volcanism in eastern Chihuahua, Mexico: Geology and geochemistry of the Benavides-Pozos area: Geological Society of America Bulletin, v. 97, p. 737–753.

Gutmann, J.T., 1986, Origin of four-and five-phase ultramafic xenoliths from Sonora, Mexico: American Mineralogist, v. 71, p. 1076–1084.

Hayama, Y., K. Shibata, and H. Takeda, 1984, K-Ar ages of the low-grade metamorphic rocks in the Altar massif, northwest Sonora, México: Journal of the Geological Society of Japan, v. 90, p. 589–596.

Hayob, J.L., E.J. Essene, J. Ruiz, F. Ortega-Gutiérrez, and J.J. Aranda-Gómez, 1989, Young high-temperature granulites from the base of the crust in central Mexico: Nature, v. 342, p. 265–268.

Heinrich, W., and J.A. Ramírez-Fernández, 1987, Metamorfismo de contacto de formaciones Cretácicas en la aureola de contacto de la Bufa del Diente, Sierra de San Carlos, Tamaulipas, México: Actas de la Facultad de Ciéncias, Universidad Autónoma de Neuvo León, Linares, v. 2, p. 173–175.

Henry, C.D., and F.W. McDowell, 1986, Geochronology of magmatism in the Tertiary volcanic field, Trans-Pecos Texas: Bureau of Economic Geology Guidebook, v. 23, p. 91–122.

Henry, C.D., and J.G. Price, 1984, Variations in caldera development in the mid-Tertiary volcanic field of Trans-Pecos Texas: Journal of Geophysical Research, v. 89, p. 8765–8786.

Henry, C.D., and J.G. Price, 1986, Early Basin and Range development in Trans-Pecos Texas and adjacent Chihuahua: Magmatism and orientation, timing and style of extension: Journal of Geophysical Research, v. 91, p. 6213–6224.

Henry, C.D., and J.G. Price, 1987, The Van Horn Mountains caldera, Trans-Pecos Texas: Geology and development of a small (10-km^2) ash-flow caldera: The University of Texas at Austin, Bureau of Economic Geology, Report of Investigations 151, 46 p.

Hernández-Avila, L., 1992, Interpretación gravimétrica y magnetométrica de la porción oriental del Eje Neovolcánico: XI Convención Geológica Nacional, Sociedad Geológica Mexicana, Libro de Resúmenes, p. 87.

Herrera, M.E., 1984, Rocas alcalinas y mineralización de lantanidos en el área El Picacho, Sierra de Tamaulipas: Geomimet, v. 3, p. 61–75.

Hill, R.I., L.T. Silver, and H.P. Taylor, 1986, Coupled Sr-O isotope variations as an indicator of source heterogeneity for the northern Peninsular Ranges batholith: Contributions to Mineralogy and Petrology, v. 92, p. 351–361.

Hoffer, J.M., and R.L. Hoffer, 1984, A preliminary note on the Late Cenozoic basalts and associated volcanic rocks in southern New Mexico, northern Chihuahua, West Texas, and the Rio Grande Rift, *in* E.C Kettenbrink, ed., Geology and Petroleum Potential of Chihuahua: West Texas Geological Society Publication 84-80, p. 202–205.

Hubberten, H.W., 1986, The Sierra de San Carlos, Tamaulipas—an igneous complex of the eastern Mexican Alkaline Province: Zentralblatt für Geologie und Palaeontologie, Teil 1, v. 186, p. 1183–1191.

Hubberten, H.W., and K. Nick, 1986, La Sierra de San Carlos, Tamaulipas—un complejo ígneo de la Provincia Alcalina Mexicana Oriental: Actas de la Facultad de Ciéncias de la Tierra, Universidad Autónoma de Nuevo León, Linares, v. 1, p. 68–77.

Huspeni, J.R., et al., 1984, Petrology and geochemistry of rhyolites associated with tin mineralization in northern Mexico: Economic Geology, v. 79, p. 87–105.

Immitt, J.P., 1987, Mineralización tipo skarn y vetas epitermales en la región de la Caldera de San Carlos, noreste de Chihuahua: Boletín Instituto de Geología, Universidad Nacional Autónoma de Mexico, v. 103, 77 p.

Jacobo-A., J., M. Garcuno, F. Innocenti, G. Manetti, T. Pasquare-S., 1992, Datos sobre el vulcanismo Neogénico-Reciente del complejo volcánico de los Tuxtlas, Estado de Veracruz, México: Evolución petrológica y geo-vulcanológica: XI Convención Geológica Nacional, Sociedad Geológica Mexicana, Libro de Resúmenes, p. 97.

James, E.W., and C.D. Henry, 1991, Compositional changes in Trans-Pecos Texas magmatism coincident with Cenozoic stress realignment: Journal of Geophysical Research, v. 96, p. 13,561–13,575.

James, E.W., and N. Walker, 1992, Implications of initial Pb isotopic ratios for the source characteristics of Proterozoic rocks in the Llano uplift and Trans-Pecos Texas: Geological Society of America Abstracts with Programs, v. 24.

James, E.W., C.D. Henry, and J.G. Price, 1990, Geochemical variations in mid-Cenozoic magmatism at the eastern fringe of the Cordillera, Trans-Pecos Texas and adjacent Chihuahua, Mexico: Geological Society of America Abstracts with Programs, v. 22, p. 32.

Jones, N.W., and J.W. McKee, 1987, Pre-Cretaceous volcanic rocks at Sierra Diablo, Chihuahua: Universidad Autónoma de Chihuahua Gaceta Geológica, v. 1, n. 1, p. 82–96.

Jones, N.W., J.W. McKee, and T.H. Anderson, 1986, Pre-Cretaceous igneous rocks in north central Mexico: Geological Society of America Abstracts with Programs, v. 18, p. 649.

Kesler, S.E., and J. Ruiz, 1987, Tectonic controls on the metallogenesis of Mexico: Geological Society of America Abstracts with Programs, v. 19, p. 725.

Kesler, S.E., J. Ruiz, and L.M. Jones, 1983, Strontiuim-isotopic geochemistry of fluorite mineralization (Coahuila, Mexico): Chemical Geology, 41, p. 65–75.

Kesler, S. E., L.M. Jones, and J. Ruiz, 1988, Strontium and sulfur isotope geochemistry of the Galeana barite district, Nuevo León, Mexico: Economic Geology, v. 83, p. 1907–1917.

Kimbrough, D.L., J.J. Hickey, and R.M. Tosal, 1987, U-Pb ages of granitoid clasts in upper Mesozoic arc-derived strata of the Vizcaíno Peninsula, Baja California, Mexico: Geology, v. 15, p. 26–29.

Kohler, H., et al., 1990, Geochronological and geochemical investigations on plutonic rocks from the complex of Puerto Vallarta, Sierra Madre del Sur (Mexico): Geofísica Internacional.

Kuentz, D.C., et al., 1988, The Southern Cordilleran basaltic andesite suite, Mexico and the U.S.A.: a link between Tertiary continental arc and flood basalt magmatism in North America: Geological Society of America Abstracts with Programs, v. 20, p. A194.

Kuentz, D.C., et al., 1990, Mid-Cenozoic transition from orogenic to intraplate-like magmatism in West Texas: A perspective from Mexico: Geological Society of America Abstracts with Programs, v. 22, n. 3, p. 35.

Kuentz, D.C., et al., 1991, Mid-Cenozoic basaltic rocks from West Texas: Trace element and isotopic evidence for mixing between orogenic and intraplate sources: Journal of Geological Research.

Kuentz, D., K.L. Cameron, and S. Niemeyer, 1992, Isotopic provinces of southeastern Chihuahua and adjacent West Texas defined by mid-Cenozoic volcanic rocks, *in* K.F. Clark, J. Roldán-Quintana, and R.H. Schmidt, eds., Geology and Mineral Resources of Northern Sierra Madre Occidental, Mexico: El Paso Geological Society Guidebook, p. 111–118.

López-Infanzón, M., 1986a, Petrología y radiometría de rocas ígneas y metamórficas de México: Boletín Asociación Mexicana de Geólogos Petroleros, v. 38, n. 2, p. 59–98.

López-Infanzón, M., 1986b, Estudio petrogenético de las rocas ígneas en las formaciones Huizachal y Nazas: Boletín Sociedad Geológica Mexicana, v. 47, n. 2, p. 1–42.

López-Infanzón, M., and S.A. Nelson, 1990, Geology and K-Ar dating of the Sierra de Chiconquiaco-Palma Sola Volcanics, central Veracruz, Mexico: Geological Society of America Abstracts with Programs, v. 22, p. A165.

López-Infanzón, M., and S.A. Nelson, 1991, Geochemistry and geochronology of subduction related magmatism in the Sierra de Chiconquiaco-Palma Sola area, Veracruz State. Memoria de la Convención sobre la Evolución Geológica de México: Primer Congreso Mexicano de Mineralogía, p. 86.

López-Infanzón, M., et al., 1984, Estudio petrográfico y tectónico de las rocas metamórficas del área Capoas-Rodeo, Zacatecas (abs.): Sociedad Geológica Mexicana, VII Convención Nacional, Resúmenes, p. 18.

Lozano-Santa Cruz, R., E. Herrera-Mariano, and L. Reyes-Ortega, 1991, Estudio por rayos X de la eudialita del intrusivo sienítico de Cerro del Diente, Sierra de San Carlos, Estado de Tamaulipas: Memoria de la Convención sobre la Evolución Geológica de México: Primer Congreso Mexicano de Mineralogía, p. 89–91.

Luhr, J.F., et al., 1989, Primitive calc-alkaline and alkaline rock types from the western Mexican Volcanic

Belt: Journal of Geophysical Research, v. 94, p. 4515–4530.
Luhr, J.F., J.J. Aranda-Gómez, and J.G. Pier, 1989, Spinel-herzolite-bearing Quaternary volcanic centers in San Luis Potosí, Mexico, 1, Geology, mineralogy and petrology: Journal of Geophysical Research, v. 94, p. 7916–7940.
Luhr, J.F., J.G. Pier, and J.J. Aranda-Gómez, 1990, Geology and petrology of the Late-Neogene los Encinos volcanic field of north-central Mexico: Geological Society of America Abstracts with Programs, v. 22, p. A165.
Luhr, J.F., J.J. Aranda-Gómez, and J.G. Pier, 1991, Crustal contamination in early basin and range alkalic basalts: Petrology of the Los Encinos volcanic field, Mexico (abs.): EOS, American Geophysical Union Transactions, v. 72, n. 44, p. 560.
MacDonald, A.J., M.J. Kretzmer, and S.E. Kesler, 1986, Vein, manto and chimney mineralization at Fresnillo silver-lead-zinc mine, Mexico: Canadian Journal of Earth Sciences, v. 23, p. 1603–1614.
Magonthier, M.C., 1988, Distinctive rhyolite suites in the mid-Tertiary ignimbritic complex of the Sierra Madre Occidental western Mexico: Bulletin de la Societé Géologique de France, v. 4, n. 8, p. 57–68.
Marvin, R.F., H.H. Mehnert, and R.E. Zartman, 1988, Radiometric ages: "C." Part One, Mexico, Wisconsin, Michigan, and North and South Carolina: Isochron/West, U. S. Geological Survey, v. 51, p. 3–4.
Mauger, R.L., 1983, The geology and volcanic stratigraphy of the Sierra Sacramento block near Chihuahua City, Chihuahua, Mexico, *in* Geology and Mineral Resources of North-central Chihuahua: Guidebook for the 1983 Field Conference, El Paso Geological Society, p. 137–156.
Mauger, R.L., 1987, Tertiary volcanic history of the Tinaja Lisa block (TLB) north-central Chihuahua, Mexico: Geological Society of America Abstracts with Programs, v. 19, p. 762.
Mauger, R.L., 1990, Tertiary volcanic history of the Tinaja Lisa-Nido area, north-central Chihuahua, Mexico: Geological Society of America Abstracts with Programs, v. 22, p. 65.
Mauger, R.L., F.W. McDowell, and J.G. Blount, 1983, Grenville-age Precambrian rocks of Los Filtros area, near Aldama, Chihuahua, Mexico, *in* K.F. Clark and P.C. Goodell, eds., Geology and Mineral Resources of North-Central Chihuaha, Mexico: El Paso Geological Society Guidebook, p. 165–168.
McDowell, F.W., 1987, The magmatic record of western Mexico and its mismatch to tectonic models: Geological Society of America Abstracts with Programs, v. 19, p. 765.
McDowell, F.W., and S. Clabaugh, 1979, Ignimbrites of the Sierra Madre Occidental and their relation to the tectonic history of western Mexico: Geological Society of America Special Paper, 180, p. 113–124.
McDowell, F.W., D.A. Wark, and G. Aguirre-Díaz, 1990, The Tertiary ignimbrite flare-up in western Mexico: Geological Society of America Abstracts with Programs, v. 22, p. 66.
Megaw, P.K.M., J. Ruiz, and S.R. Titley, 1988, High-temperature, carbonate-hosted Ag-Pb-Zn (Cu) deposits of northern Mexico: Economic Geology, v. 83, p. 1856–1885.
Milán-Valdez, M., 1992, Características petrológicas y estratigráficas de la ignimbrita San Francisco, Caldera de Huichapan, Estado de Hidalgo: XI, Convención Geológica Nacional, Sociedad Geológica Mexicana, Libro de Resúmenes, p. 123.
Miranda-Avilés, R., 1992, Generalización del modelo genético de una veta auro-argentífera en el cuerpo dioritico de San Joaquin, San Antonio, Baja California Sur: XI Convención Geológica Nacional, Sociedad Geológica Mexicana, Libro de Resúmenes, p. 125.
Monod, O., and J. de J. Parga-Pérez, 1991, Una nueva interpretación estructural del distrito minero de Fresnillo, Estado de Zacatecas: Memoria de la Convención sobre la Evolución Geológica de México, Primer Congreso Mexicano de Mineralogía, p. 118–120.
Monod, O., et al., 1989, Complete oceanic island arc of Lower Cretaceous age in central Mexico: Guanajuato Magmatic Series: 28th International Geological Congress Abstracts, v. 2, p. 452.
Monod, O., et al., 1990, Reconstitution d'un arc insulaire intra-oceanique au Mexique central: La sequence volcano-plutonique de Guanajuato (Crétacé Inferieur): Comptes Rendus Académie Science, Paris, v. 310, p. 45–51.
Montigny, R. A., P. Demant, P. Delpretti, P. Piguet, and J.J. Cocheme, 1987, Chronologie K/A des séquences volcaniques du Nord de la Sierra Madre (Mexique): Comptes Rendus Acad. Sci., v. 304, p. 987–992.
Mora, C.I., and J.W. Valley, 1985, Ternary feldspar thermometry in granulites from the Oaxacan Complex, Mexico: Contributions to Mineralogy and Petrology, v. 89, p. 215–225.
Mora, C.I., J.W. Valley, and F. Ortega-Gutiérrez, 1986, The temperature and pressure conditions of Grenville-age granulite-facies metamorphism of the Oaxacan complex, southern Mexico: Revista del Instituto de Geología, Universidad Nacional Autónoma México, v. 7, p. 222–242.
Morales-Montaño, M., C. Bartolini, P. Damon, and M. Shafiquillah, 1990, K-Ar age dating, stratigraphy and extensional deformation of Sierra Lista Blanca, central Sonora, Mexico: Geological Society of America Abstracts with Programs, v. 22, p. A364.
Morán-Zenteno, D.J., J. Urrutia-Fucugauchi, and H. Koehler, 1990, Nuevos fechamientos de Rb-Sr en rocas cristalinas del Complejo Xolapa en el Estado de Guerrero: Sociedad Geológica Mexicana, X Convención Nacional, Libro de Resúmenes, p. 92–93.
Murillo-Muñetón, G., 1991, Análisis petrológico y edades K-Ar de las rocas metamórficas e ígneas pre-Cenozóicas de la región de La Paz-Los Cabos, B.C.S., México: First International Meeting on Geology of the Baja California Peninsula, p. 55–56.
Negendank, J.F.W., 1987, The granitoids of the Sierra Madre del Sur, Mexico: Actas de la Facultad de

Ciéncias, Universidad Autónoma de Nuevo León, Linares, v. 2, p. 165–172.

Nelson, D.O., and J.H. Schieffer, 1990, Ultramafic inclusions and megacrysts in mafic rocks of the Terlingua district, West Texas, *in* P. Dickerson and J.B. Stevens, eds., Geology of the Big Bend and Trans-Pecos Region, Field Trip Guidebook: San Antonio, Texas, South Texas Geological Society, p. 211–249.

Nelson, S.A., 1986, Geología del Volcán Ceboruco, Nayarit, con una estimación de riesgos de erupciones futuras: Universidad Nacional Autónoma de México, Instituto de Geología, Revista, v. 9, p. 243-258.

Nelson, S.A., 1990, Volcanic hazards in Mexico—a summary: Universidad Nacional Autónoma de México, Instituto de Geología, Revista, v. 9, p. 71–81.

Nelson, S.A., 1991, Pliocene to Recent basic magmatism in the Mexican volcanic belt: Memoria de la Convención sobre la Evolución Geológica de México, Primer Congreso Mexicano de Mineralogía, p. 131–134.

Nelson, S.A., and E. González-Caver, 1989, Geochemistry of Late Miocene to Recent alkaline and calc-alkaline volcanic rocks of the Tuxtla volcanic field, Veracruz, Mexico: Geological Society of America Abstracts with Programs, v. 21, p. A14.

Nelson, S.A., and E. González-Caver, 1992, Geology and K-Ar dating of the Tuxtla volcanic Field, Veracruz, Mexico: Bulletin of Volcanology, v. 55, p. 85–96.

Nelson, S.A., and J. Hegre, 1990, Volcán Las Navajas, a Pliocene-Pleistocene trachyte/peralkaline rhyolite volcano in the northwestern Mexican volcanic belt: Bulletin of Volcanology, v. 52, p. 186–204.

Nimz, G.J., 1989, The geochemistry of the mantle xenolith suite from La Olivina, Chihuahua, Mexico: Ph.D. Thesis, University of California at Santa Cruz, 306 p.

Nimz, G.J., and K.L. Cameron, 1987, Rare earth element (REE) evidence for the petrogensis of mantle pyroxenites from La Olivina, Chihuahua, Mexico: Geological Society of America Abstracts with Programs, v. 19, p. 789.

Nimz, G.J., K.L. Cameron, M. Cameron, and S.L. Morris, 1986, The petrology of the lower crust and upper mantle beneath southeastern Chihuahua, Mexico: A progress report: Geofísica Internacional, Union Geofísica Mexicana, v. 25, p. 85–116.

Nimz, G.J., K.L. Cameron, and S. Niemeyer, 1988, Pb/Sr/Nd isotopic evidence from mantle xenoliths and related basalts for mid-Cenozoic North American arc petrogenesis and mantle Pb systematics: Geological Society of America Abstracts with Programs, v. 20, n. 7, p. A74.

Nixon, G.T., 1982, The relationship between Quaternary volcanism in central Mexico and the seismicity and structure of subducted oceanic lithosphere: Geological Society of America Bulletin, v. 93, p. 514–523.

Nixon, G.T., 1989, The geology of Iztaccíhuatl volcano and adjacent areas of the Sierra Nevada and Valley of Mexico: Geological Society of America Special Paper 219, 45 p.

Nixon, G.T., A. Demant, R.L. Armstrong, and J.E. Harakal, 1987, K-Ar and geologic data bearing on the age and evolution of the Trans-Mexican Volcanic Belt: Geofísica Internacional, v. 26, p. 109–158.

Nuñez-B., J., and G. Arriaga-G., 1992, El distrito minero El Limón, Municipio de Cocula, Guerrero: XI Convención Geológica Nacional, Sociedad Geológica Mexicana, Libro de Resúmenes, p. 134–135.

Nuñez-Espinal, J., A. Márquez-Ruiz, and J. Bustamante-Garcia, 1992, Origen sintectónico del depósito aurífero de "El Zapatillo" Municipio de Lazaro Cardenas, Michoacán: XI Convención Geológica Nacional, Sociedad Geológica Mexicana, Libro de Resúmenes, p. 136.

Okita, P.M., J.B. Maynard, and A. Martinez-Vera, 1986, Molango; giant sedimentary manganese deposit in Mexico (abs.): AAPG Bulletin, v. 70, p. 627.

Orozco, M.T., 1991, Geotermobarometría de granulitas Precámbricas del basamento de la Sierra Madre Oriental: Memoria de la Convención sobre la Evolución Geológica de México, Primer Congreso Mexicano de Mineralogía, p. 138–141.

Ortega-Gutiérrez, F., and G. Sánchez-Rubio, 1985, Xenolitos plutónicos de Isla Socorro, Archipiélago Revillagigedo: Universidad Nacional Autónoma de México, Instituto de Geología, Revista, v. 6, p. 37–47.

Ortigoza-Cruz, F., A. Changkakoti, R.D. Morton, and J. Gray, 1992, Strontium isotope geochemistry of barite mineralization at La Minita, SW Mexico: XI Convención Geológica Nacional, Sociedad Geológica Mexicana, Libro de Resúmenes, p. 138.

Ortiz-Hernández, L.E., H. Lapierre, and M. Yta, 1991, Late Jurassic-Early Cretaceous tholeitic and calc-alkaline arc series in Mexico; implications for the magmatic evolution of the Mexican Cordillera: Memoria de la Convención sobre la Evolución Geológica de México, Primer Congreso Mexicano de Mineralogía, p. 147–149.

Ortiz-Hernández, L.E., H. Lapierre, and M. Yta, 1992, Evolution magmatica del Terreno Guerrero en Mexico centro-meridional: XI Convencion Geologica Nacional, Sociedad Geológica Mexicana, Libro de Resúmenes, p. 141–142.

Ortiz-Hernández, L.E., H. Lapierre, G. Ayala-Rojas, and C. Yañez-Mondragon, 1992, Petrogénesis y significado tectónico de las rodingitas del Cretácico Inferior de la Sierra de Guanajuato: XI Convención Geólogica Nacional, Sociedad Geológica Mexicana, Libro de Resúmenes, p. 139.

Ortiz-Hernández, L.E., J. Romo, and G. Mercado, 1992, Estructura paleovolcánica asociada con mineralización polymetálica en los estados de Jalisco, Michoacán, Guerrero y México: XI Convención Geológica Nacional, Sociedad Geológica Mexicano, Libro de Resúmenes, p. 140.

Pantoja-Alor, J., 1983, Geocronométría de magmatismo Cretácico-Terciário de la Sierra Madre del

Sur: XIV Convención Nacional, Sociedad Geológica Mexicana, Libro de Resúmenes, p. 29.

Pantoja-Alor, J., 1988, Petrología e implicaciones tectónicas del evento magmático Balsas (Paleogene) de la Sierra Madre del Sur: Unión Geofísica Mexicana, Reunion Anual, Resúmenes.

Parga-Pérez, J. de J., 1992, El significado del Cretácico Inferior volcanosedimentario en la metalogénesis del centro de México: XI Convención Geológica Mexicana, Libro de Resúmenes, p. 152–163.

Pasquare, G., L. Ferraro, V. Perazzoli, M. Tiberi, and F. Turchetti, 1987, Morphological and structural analysis of the central sector of the Transmexican volcanic belt: Geofísica Internacional, v. 26, p. 177–193.

Pasquare, G., V.H. Garduno, A. Tibaldi, and M. Ferrari, 1988, Stress pattern evolution in the central sector of the Mexican Volcanic Belt: Tectonophysics, v. 146, p. 353–364.

Patchett, P.J., and J. Ruiz, 1987, Nd isotopic ages of crust formation and metamorphism in the Precambrian of eastern and southern Mexico: Contributions to Mineralogy and Petrology, v. 96, p. 523–528.

Patchett, P.J., and J. Ruiz, 1989, Nd isotopes and the origin of Grenville-age rocks in Texas: Implications for Proterozoic evolution of the United States midcontinent region: Journal of Geology, v. 97, p. 685–695.

Patchett, P.J., J. Ruiz, and J.G. Price, 1989, Nd isotopes, Grenville crust and 1.5–1.3 Ga igneous activity of the United States and Mexico (abs.): EOS, American Geophysical Union Transactions, v. 70, n. 15, p. 485.

Pérez-Segura, E., 1990, A new Au-Te paragenesis in Sonora, Mexico: The San Francisco ore deposit: Geological Society of America Abstracts with Programs, v. 22, n. 3, p. 75.

Pier, J.G., 1989, Isotopic and trace element systematics in a spinel herzolite-bearing, alkali basalt suite, San Luis, Potosí, Mexico: Ph.D. Thesis, Washington University, St. Louis, Missouri.

Pier, J.G., F.A. Podosek, J.F. Luhr, and J.C. Brannon, 1989, Spinel-herzolite-bearing Quaternary volcanic centers in San Luis Potosí, Mexico, 2, Sr and Nd isotopic systematics: Journal of Geophysical Research, v. 94, p. 7941–7951.

Pier, J.G., F.A. Podosek, J.F. Luhr, and J.J. Aranda-Gómez, 1991, Crustal contamination in alkalic basalts from Los Encinos volcanic field, Mexico: Modification of isotopic and elemental compositions (abs.): EOS, American Geophysical Union Transactions, v. 72, n. 44, p. 560.

Pier, J.G., J.F. Luhr, F.A. Podosek, and J.J. Aranda-Gómez, 1992, The La Breña-El Jagüey Maar complex, Durango, Mexico: II. Petrology and geochemistry: Bulletin of Volcanology, v. 54, p. 405–428.

Ponce, B.F., and K.F. Clark, 1988, The Zacatecas Mining District: A Tertiary caldera complex associated with precious and base metal mineralization: Economic Geology, v. 83, p. 1668–1682.

Poole, F.G., B.L. Murchey, and D.L. Jones, 1983, Bedded barite deposits of Middle and Late Paleozoic age in central Sonora, Mexico: Geological Society of America Abstracts with Programs, v. 15, p. 299.

Poole, G.B., 1990, Petrology, geochemistry and geochronology of lower-crustal xenoliths, central Mexico: Master's Thesis, Washington University, St. Louis, Missouri, 116 p.

Poole, G.B., J.F. Luhr, and J.J. Aranda-Gómez, 1990, Petrology and geochemistry of granulite-grade, lower-crustal xenoliths from central Mexico: Geological Society of America Abstracts with Programs, v. 22, p. A257.

Poole, G.B., J.F. Luhr, and S.A. Bowring, 1990, Trace element geochemistry and U-Pb zircon geochronology from lower-crustal xenoliths, central Mexico (abs.): EOS, American Geophysical Union Transactions, v. 71, p. 1689.

Prakash, G.O., et al., 1991, A rare wollastonite-quartz-graphite assemblage from a high-grade regional metamorphic terrain of Late Precambrian age in Oaxaca, Mexico: Revista del Instituto Mexicano del Petróleo, v. 23, n. 3, p. 5–13.

Prol-Ledesma, R.M., Mediciones y estimaciones de flujo térmico en México; un análisis comparativo: Geotérmia, v. 5 (1), p. 19–32.

Prol-Ledesma, R.M., and M.G. Juarez, 1986, Geothermal map of Mexico: Journal of Volcanology and Geothermal Research, v. 28, p. 251–362.

Puy-Alquiza, M.I., R. Miranda-Áviles, and J.A. Pérez-Vanzor, 1992, Evidencias de las rocas graniticas a lo largo de la zona de falla San Juan de los Planes, Baja California Sur, Mexico: XI Convención Geológica Nacional, Sociedad Geológica Mexicana, Libro de Resúmenes, p. 159.

Ramos-Salinas, A., 1992a, Plutonismo del Neógeno al noreste de Tepic, Nayarit: XI Convención Geológica Nacional, Sociedad Geológica Mexicana, Libro de Resúmenes, p. 163.

Ramos-Salinas, A., 1992b, Volcanismo andesítico basáltico del Neógeno, en la porción suroriental de la Sierra de Guanajuato, Salamanca, Gto: XI Convención Geológica Nacional, Sociedad Geológica Mexicana, Libro de Resúmenes, p. 164.

Ramos-Salinas, A., and D.J. Terrell, 1992, Fechamiento K-Ar de rocas volcánicas al norte de Salamanca, Gto: XI Convención Geológica Nacional, Sociedad Geológica Mexicana, Libro de Resúmenes, p. 162.

Reinhardt, B.K., and S.A. Nelson, 1989, Volcanic hazards study of the Tuxtla volcanic field, Veracruz, Mexico: Geological Society of America Abstracts with Programs, v. 21, n. 6, p. A14.

Ritter, S.P., and J.C. Cepeda, 1991, The Hechiceros Caldera: a recently identified mid-Tertiary caldera in eastern Chihuahua, Mexico: Journal of Geophysical Research, v. 96, n. B10, p. 16,241–16,250.

Roberts, S.J., and J. Ruiz, 1989, Geochemistry of exposed granulite facies terrains and lower crustal xenoliths in Mexico: Journal of Geophysical Research, v. 94, p. 7961–7974.

Robin, C., 1982, Mexico, *in* R.S. Thorpe, ed., Andesites, Orogenic Andesites and Related Rocks: New York, Wiley, p. 137–147.

Robinson, J. V., 1988, The geochemistry and petrography of crustal xenoliths from La Olivina, Mexico: Unpublished Master's Thesis, University of California at Santa Cruz, 106 p.

Robinson, J.V., K.L. Cameron, and M.H. Ort, 1987, Orthogneiss xenoliths from central Chihuahua Mexico: restite or cumulate?: Geological Society of America Abstracts with Programs, v. 19, p. 822.

Rodríguez-Elizarras, S., 1992, La Caldera de las Cumbres, estados de Veracruz y Puebla. Una interpretación geológica preliminar: XI Convención Geológica Nacional, Sociedad Geológica Mexicana, Libro de Resúmenes, p. 168.

Roldán-Quíntana, et al., 1989, Pegmatitas de la Sierra El Jaralito al suroeste de Vaviácora, Sonora: Universidad Nacional Autónoma de México, Revista, v. 8, p. 15–21.

Rosas-Solís, A., and P. Samano-Tirado, 1990, Tectonics and mineralization in NW Mexico: Geological Society of America Abstracts with Programs, v. 22, p. 79.

Ross, M.L., and D.D. Lambert, 1986, K, Rb, Ba, REE and Sr isotope geochemistry of alkalic rocks from the Trans-Pecos magmatic province, Texas: Evidence for multiple source regions: Geological Society of America Abstracts with Programs, v. 18, p. 734.

Rubalcaba-Ruiz, D.C., 1987, Geology of Galeana barite district, Nuevo León, Mexico (abs.): AAPG Bulletin, v. 71 (5), p. 609.

Rubalcaba-Ruiz, D.C., and T.B. Thompson, 1988, Ore deposits at the Fresnillo mine, Zacatecas, Mexico: Economic Geology, v. 83, p. 1583–1596.

Rudnick, R.L., and K.L. Cameron, 1990, Ages of granulite-facies xenoliths from northern Mexico (abs.): EOS, American Geophysical Union Transactions, v. 71, n. 43, p. 1689.

Rudnick, R.L., and K.L. Cameron, 1991, Age diversity of the deep crust in northern Mexico: Geology, v. 19, p. 1197–1200.

Ruiz, J., 1988, Geology and isotope geochemistry of fluorine-rich rocks from the Sierra Madre Occidental, Mexico: Canadian Institute of Mining.

Ruiz, J., and P. Coney, 1985, Correlation between lead isotopes in Mexican ore deposits and tectono-stratigraphic terranes: Geological Society of America Abstracts with Programs, v. 17, p. 704.

Ruiz, J., and P.J. Patchett, 1988, Nd-Sr isotopes and the origin of Grenville-age crust in Mexico (abs.): Geological Association of Canada; Mineralogical Association of Canada; Canadian Geophysical Union, v. 13, p. A106.

Ruiz, J., and S.J. Roberts, 1988, U-Th-Pb geochemistry of the lower crust: Geological Society of America Abstracts with Programs, v. 20, p. A303.

Ruiz, J., F. Ortega-Gutiérrez, and E.J. Essene, 1983, Geochemical and petrographic characteristics of inclusions in Cenozoic alkalic basalts from central Mexico (abs.): EOS, American Geophysical Union Transactions, v. 64, (18), p. 343.

Ruiz, J., L.M. Jones, and W.C. Kelly, 1984, Rubidium-strontium dating of ore deposits hosted by Rb-rich rocks, using calcite and other common Sr-bearing minerals: Geology, v. 12, p. 259–262.

Ruiz, J., W.A. Duffield, D.M. Burt, and C. Reece, 1986, Sm-Nd and Rb-Sr isotopic compository of Mid-Tertiary, high silica rhyolites from the Black Range, New Mexico, and the Sierra Madre Occidental, Mexico: Geological Society of America Abstracts with Programs, v. 18, p. 736.

Ruiz, J., P.J. Patchett, and F. Ortega-Gutiérrez, 1987, Proterozoic and Phanerozoic basement terranes of Mexico: Geological Society of America Abstracts with Programs, v. 19, p. 826.

Ruiz, J., P.J. Patchett, and R.J. Arculus, 1988a, Nd-Sr isotope composition of lower crustal xenoliths—Evidence for the origin of mid-Tertiary volcanics in Mexico: Contributions to Mineralogy and Petrology, v. 99, p. 36–43.

Ruiz, J., P.J. Patchett, and R.J. Arculus, 1988b, Reply to "Comments on Nd-Sr isotopic composition of lower crustal xenoliths—evidence for the origin of mid-Tertiary felsic volcanics in México" by K.L. Cameron and J.V. Robinson: Contributions to Mineralogy and Petrology, v. 104, p. 615–618.

Ruiz, J., P.J. Patchett, and F. Ortega-Gutiérrez, 1988, Proterozoic and Phanerozoic basement terranes of Mexico from Nd isotopic studies: Geological Society of America Bulletin, v. 100, p. 274–281.

Ruiz, J., P.J. Patchett, and F. Ortega-Gutiérrez, 1990, Proterozoic and Phanerozoic terranes of Mexico based on Nd, Sr and Pb isotopes: Geological Society of America Abstracts with Programs, v. 22, p. A113–A114.

Sánchez-Rubio, G., 1985, La erupción del 28-29 de marzo (1982) del Volcán Chichonal; un estudio breve de su tefra: Universidad Nacional Autónoma de México, Instituto de Geología, Revista, v. 6, p. 48–51.

Sawlan, M.G., and J.G. Smith, 1984, Petrologic characteristics, age, and tectonic setting of Neogene volcanic rocks in northern Baja California Sur, Mexico, *in* V.A. Frizzel, ed., Geology of the Baja California Peninsula, Pacific Section SEPM, v. 39, p. 237–251.

Sedlock, R.L., 1988a, Metamorphic petrology of a high-pressure, low-temperature subduction complex in west-central Baja California, Mexico: Journal of Metamorphic Geology, v. 5, p. 205–253.

Sedlock, R.L., 1988b, Lithology, petrology, structure, and tectonics of blueschists and associated rocks in west-central Baja California, Mexico: Unpublished Ph.D. Dissertation, Stanford University, 223 p.

Shafiquillah, M., P.E. Damon, and K.F. Clark, 1983, K-Ar chronology of Mesozoic-Cenozoic continental magmatic arcs and related mineralization in Chihuahua, *in* K.F. Clark and P.C. Goodell, eds., Geology and Mineral Resources of North-Central Chihuahu: El Paso Geological Society Guidebook, p. 303–315.

Shearer, C.D., 1985, Geology of the igneous intrusions of northern Valle Las Norias, Coahuila, Mexico (abs.): Geological Society of America Abstracts with Programs, v. 17, p. 191.

Siem, M.E., and R.G. Gastil, 1990, Active development of the Sierra Mayor metamorphic core complex, northeastern Baja California: Geological Society of America Abstracts with Programs, v. 22, p. A228.

Silberman, M.L., A. Moore-Hall, and B.M. Smith, 1991, Gold-bearing quartz veins along the Mojave-Sonora megashear zone, northern Sonora, Mexico: Seventh Annual V.E. McKelvey Forum on Mineral and Energy Resources, p. 70–72.

Silva-Mora, L., 1987, Algunos aspectos de los basaltos y andesitas cuaternarios de Michoacán oriental: Universidad Nacional Autónoma de México, Instituto de Geología, Revista, v. 7, p. 89–96.

Silva-Mora, L., and D.A. Cordoba, 1992, Vulcanismo y fácies de la Caldera de Huichapán, Estado de Hidalgo: XI Convención Gelógica Nacional, Sociedad Geológica Mexicana, Libro de Resúmenes, p. 181.

Simmons, W.B., Jr., 1985, Authigenic mineralization from Sierra del Fraile, near Monterrey, Nuevo León, Mexico; an example of a sedimentary skarn? (abs.): Geological Society of America Abstracts with Programs, v. 17 (7), p. 717.

Smith, D.P., and C.J. Busby-Spera, 1989, Evidence for the early remagnetization of Cretaceous strata on Cedros Island, Baja California Norte (Mexico): EOS, American Geophysical Union Transactions, v. 70, p. 1067–1068.

Smith, D.P., and C.J. Busby-Spera, 1991, Shallow magnetic inclinations in Cretaceous Valle Fm., Baja California: remagnetization, compaction or 18 degrees northward drift compared to North America?: Geological Society of America Abstracts with Programs, v. 23, p. 99.

Smith, J.A., J.F. Luhr, J.G. Pier, and J.J. Aranda-Gómez, 1989, Extension-related magmatism of the Durango Volcanic Field, Durango, Mexico: Geological Society of America Abstracts with Programs, v. 21, n. 6, p. A201.

Smith, R.D., K.L. Cameron, and S. Niemeyer, 1991, Mid-Cenozoic volcanic rocks and related deep crustal xenoliths from La Olivina, southeastern Chihuahua, Mexico: Geological Society of America Abstracts with Programs, v. 23, n. 5, p. A332.

Stoiber, R.E., and M.J. Carr, 1973, Quaternary volcanic and tectonic segmentation of Central America: Bulletin Volcanologique, v. 37, p. 304–325.

Swanson, E.R., 1990, A new type of maar volcano from the State of Durango—the El Jaguey-La Breña complex reinterpreted: Universidad Nacional Autónoma de México, Instituto de Geología, Revista, v. 10, p. 243–247.

Swanson, E.R., and F.W. McDowell, 1984, Calderas of the Sierra Madre Occidental volcanic field, western Mexico: Journal of Geophysical Research, v. 89, p. 8787–8799.

Swanson, E., and D. Wark, 1988, Mid-Tertiary silicic volcanism in Chihuahua, Mexico, *in* K.F. Clark, P.C. Goodell, and J.M. Hoffer, eds., Stratigraphy, Tectonics and Resources of Parts of Sierra Madre Occidental Province, Mexico, Guidebook for the 1988 Field Conference: El Paso Geological Society, p. 229–240.

Terrell, D.J., 1992a, Geocronologia K-Ar en campos geotérmicos alteración del patron de gases nobles en rocas: XI Convención Geológica Nacional, Sociedad Geológica Mexicana, Libro de Resúmenes, p. 192.

Terrell, D.J., 1992b, Fechamiento radiométrico del volcanismo en la Isla Clarión, Archipiélago Revillagigedo: XI Convención Geológica Nacional, Sociedad Geológica Mexicana, Libro de Resúmenes, p. 191.

Terrell, D.J., and B.M. Escudero, 1992, La termocronométria $^{40}Ar/^{39}Ar$ y su aplicación a rocas sedimentarias: Revista del Instituto Mexicano del Petróleo, v. 24, n. 2, p. 16–24.

Terrell, D.J., M. Escudero-B., M. Balcazar-G., and E. Volbert-R., 1992, Fechamiento $^{40}Ar/^{39}Ar$ de rocas basálticas de la región de Tres Vírgenes, Baja California: XI Convención Geológica Nacional, Sociedad Geológica Mexicana, Libro de Resúmenes, p. 192.

Thorpe, R.S., 1977, Tectonic significance of alkaline volcanism eastern México: Tectonophysics, v. 40, p. 19–26.

Torres-Roldán, V., G. Murillo, and J. Ruiz, 1992, Geoquímica y geocronométría de las rocas ígneas del Paleozóico y Mesozóico en el Paleogolfo de Huamuxtitlán, estados de Guerrero y Oaxaca: XI Convención Geológica Nacional, Sociedad Geológica Mexicana, Libro de Resúmenes, p. 195–196.

Torres-Vargas, R., G. Murillo, and M. Grajales, 1986, Estudio petrográfico y radiométrico de la portión norte del límiteentre Los complejos Acatlán y Oaxaca: Sociedad Geológica Mexicana, VIII Convención Nacional, Libro de Resúmenes.

Vassallo, L.F., 1988, Características de la composición mineralógica de las minas de la Veta Madre de Guanajuato: Universidad Nacional Autónoma de México, Instituto de Geología, Revista, v. 7, p. 232–243.

Vassallo, L.F., 1992, Criterios energéticos sobre la génesis de la Veta Madre de Guanajuato, GTO, México: XI Convención Geológica Nacional, Sociedad Geológica Mexicana Libro de Resúmenes, p. 204.

Vassallo, L.F., and M. Reyes-Salas, 1992, Mineralogía y zoneamiento composicional de los minerales argentíferos de la parte norte de la Veta Madre, Guanajuato, GTO, México: XI Convención Geológica Nacional, Sociedad Geológica Mexicana Libro de Resúmenes, p. 205.

Vassallo, L.F., et al., 1989, Alteración hidrotermal de las rocas encajonantes de la parte central de la Veta Madre de Guanajuato, Estado de Guanajuato—características petrofísicas y químicas: Universidad Nacional Autónoma de México, Instituto de Geología, Revista, v. 8, p. 211–222.

Vassallo, L.F., P. Girón-García, and A. Lozano-Cobo, 1992, Cambios químicos durante la alteración hidrotermal en domos riodacíticos contenedores de mineralización aurífera Guanajuato, GTO, Mexico:

XI Convención Geológica Nacional, Sociedad Geológica Mexicana Libro de Resúmenes, p. 206.

Verma, S.P., 1984, Sr and Nd isotopic evidence for petrogenesis of mid-Tertiary felsic volcanism in the mineral district of Zacatecas, Zacatecas (Sierra Madre Occidental), Mexico: Isotope Geology, v. 2, p. 37–53.

Verma, S.P., 1987, Mexican volcanic belt: present state of knowledge and unsolved problems: Geofísica International, v. 26, p. 309–340.

Verma, S.P., 1989, Isotopic and trace element constraints on the origin and evolution of alkaline and calc-alkaline magmas in the northwestern Mexican Volcanic Belt: Journal of Geophysical Research, v. 94, p. 4531–4544.

Victoria-Morales, A., 1992, Geología del yacimiento de grafito cristalino de Telixtlahuaca, Oax: XI Convención Geológica Nacional, Sociedad Geológica Mexicana, Libro de Resúmenes, p. 209.

Walker, N.W., 1988, U-Pb zircon evidence for 1305-1231 Ma crust in the Llano uplift, central Texas: Geological Society of America Abstracts with Programs, v. 20, p. A205.

Wark, D.A., 1991, Oligocene ash flow volcanism, northern Sierra Madre Occidental: Role of mafic and intermediate-compositon magmas in rhyolitic genesis: Journal of Geophysical Research, v. 96, p. 13,389–14,411.

Wark, D.A., and G.L. Farmer, 1988, Isotopic and geochemical constraints on the genesis of rhyolite ash flows and related magmas, Sierra Madre Occidental (abs): EOS, American Geophysical Union Transactions, v. 69, n. 16, p. 519.

Wark, D.A., K.A. Kempter, and F.W. McDowell, 1990, Evolution of waning, subduction-related magmatism, northern Sierra Madre Occidental, México: Geological Society of America Bulletin, v. 102, p. 1555–1564.

Ziagos, J.P., D.D. Blackwell, and F. Mooser, 1985, Heat flow in southern Mexico and the thermal effects of subduction: Journal of Geophysical Research, v. 90, p. 5410–5420.

4. Petroleum and Gas, Structure and Mapping, Engineering and Environmental Geology, Hydrology, and Remote Sensing

Aguilera-I., R., 1983, Evaluación probabilistica de las provincias geológicas de México: Ingenieria Petrolera, v. 23, p. 5–11.

A.I.P.M., 1990, Integración geofísica-geológica en los proyectos prioritarios de la Sierra de Chiapas y su importancia económica-petrolera: Asociación de Ingenieros Petroleros de México, Villahermosa, Tabasco, 35 p.

Albrecht, A., et al., 1987, The petrochemistry of the Tertiary volcanic rocks of the northern Sierra Madre Occidental; tracer of changes in basement composition (abs.): EOS, American Geophysical Union Transactions, v. 68, p. 1516.

Alcántara-García, J.R., 1990, Integración y evaluación petrolera regional de una porción de las provincias geológicas Sierra de Chiapas y Cuencas del Sureste, estados de Veracruz, Oaxaca, Tabasco y Chiapas: Unpublished Master's Thesis, Universidad Nacional Autónoma México, 86 p.

Alfaro-Montoya, J.R., 1992, Proyecto geotécnico para la construcción de la Presa Cueva Amarilla, municipio de Tepozotlán, México: XI Convención Geológica Nacional, Sociedad Geológica Mexicana, Libro de Resúmenes, p. 11–13.

Anderson, B.D., and V.M. Aguilera, 1986, Push faults; a conceptual model for groundwater exploration in the Sierra Madre Oriental foreland, Mexico, *in* H. Miller, et al., eds., Symposium on Latin-America: Geosciences, p. 1149–1160.

Angeles-Aquino, F.J., J.R. Nuñez, and J.M. Quezada-Muñetón, 1992, Evolución tectónica de la sonda de Campeche, estilos estructurales resultantes y su implicación en la generación y acumulación de hidrocarburos (abs.): II Symposio Asociación Mexicana de Geológos Petroleros, Instituto Mexicano del Petróleo, México, D.F.

Anonymous, 1983, Primera sesión foro de geología y exploración; estratégia y exploración nacional con aplicación de nuevas tecnicas de exploración en México: Geomimet, v. 3, n. 126, p. 55–57.

Aranda-Gómez, J., and J.A. Pérez-Venzor, 1986, Reconocimiento geológico de las islas Espíritu Santo y La Partida, Baja California Sur: Universidad Nacional Autónoma de México, Instituto de Geología, v. 6, p. 103–116.

Aranda-Gomez, J.J., and J.J. Pérez-Venzor, 1988, Estudio geólogico de Punta Coyotes, Baja California Sur: Revista del Instituto de Geología, Universidad Nacional Autónoma México, v. 8, p. 149–170.

Arellano-Gil, J., C. Mendoza-Rosales, and G. Silva-Romo, 1992, Características estructurales de la Sierra de Guadalupe, Zacatecas: XI Convención Geológica Nacional, Sociedad Geológica Mexicana, Libro de Resúmenes, p. 18.

Arenal, Rodolfo del, 1985, Estudio hidrogeoquímico de la porción centro-oriental del Valle del Mezquital, Hidalgo: Universidad Nacional Autónoma de México, Instituto de Geología, Revista, v. 6, p. 86–97.

Arévalo-Mendoza, V., 1992, Excavación de galerias de inyección, drenaje, e inspección del prospecto hidroeléctrico Zimapán: XI Convención Geológica Nacional, Sociedad Geológica Mexicana, Libro de Resúmenes, p. 19.

Arias-Cortes, J., 1992, Asentamiento y urbanización, investigación, y monitoréo en Torreón, Coahuila, México: XI Convención Geológica Nacional, Sociedad Geológica Mexicana, Libro de Resúmenes, p. 20–21.

Arizabalo, R.D., and D.J. Terrell, 1991, Aplicación de los isótopos estables del hidrögeno y carbono en la exploración petrolera: Revista del Instituto Mexicano del Petróleo, v. 23, n. 4, p. 7–15.

Avenius, C.G., 1982, Tectonics of the Monterrey Salient, Sierra Madre Oriental, northeast Mexico: Master's Thesis, University of New Orleans, 83 p.

Ayala, S.R., J.L. Granados-G., C.A. Piñeda, and L. M. Villalobos, 1984, Explotación de petróleo en México *in* Schlumberger Evaluación de Formaciones en México: Petróleo Internacional, v. 43, p. 1–21.

Ayala, S.R., J.L. Granados, C. Piñeda, and M. Villalobos, 1985, Yacimientos supergigantes de México, *in* Schlumberger, Evaluación de Formaciónes en México: Petróleo Internacional, v. 43, p. 30–31.

Back, W., 1984, Role of groundwater in shaping the eastern coastline of the Yucatan Peninsula, Mexico, *in* R.G. LaFleur, ed., Groundwater as a Geomorphic Agent: 13th Annual Geomorphology Symposium, Boston, Allen & Unwin, p. 281–293.

Back, W., and B. Hanshaw, 1984, Karst processes in the ground-water mixing zone of coastal aquifers, *in* Friends of the Karst: GEO, 11, Puerto Rico meeting abstracts, v. 11, n. 3, p. 44.

Back, W., et al., 1981, Chemical characterization of cave, cove, caleta, and karst creation in Quintana Roo: Geological Society of America Abstracts with Programs, v. 13, p. 400.

Back, W., B. Hanshaw, W.C. Ward, et al., 1982, Geology and hydrogeology of carbonate rocks of the northeastern Yucatan Peninsula; road log and supplement to 1978 Guidebook, Geology and hydrogeology of northeastern Yucatan: New Orleans Geological Society.

Back, W., B. Hanshaw, and J.N. Van Driel, 1984, Role of groundwater in shaping the eastern coastline of the Yucatan Peninsula, Mexico, *in* R.G. LaFleur, ed., Groundwater as a Geomorphic Agent: The Binghamton Symposia in Geomorphology, International Series, 23: Boston, Allen & Unwin, p. 281–293.

Back, W., B. Hanshaw, et al., 1986, Differential dissolution of a Pleistocene reef in the ground-water mixing zone of coastal Yucatan, Mexico: Geology, v. 14, p. 137–140.

Baldwin, S.L., and T.M. Harrison, 1989, Geochronology of blueschists from west-central Baja California and the timing of uplift of subduction complexes: Journal of Geology, v. 97, p. 149.

Baldwin, S.L., T.M. Harrison, and R.L. Sedlock, 1987, $^{40}Ar/^{39}Ar$ geochronology of high-pressure blocks in melánge, Cedros Island, Baja California: Geological Society of America Abstracts with Programs, v. 19.

Baltazar-Chongo, O., and F. Díaz-Zamora, 1993, Trampas: Asociación Mexicana Geólogos Petroleros, I Simposio de Geología de Subsuelo, Ciuadad del Carmen, Resúmenes, p. 116–117.

Belew, D., 1982, México Petrolero 1983; Chiapas/Tabasco/Golfo de Campeche: Petróleo Internacional, v. 40, p. 49.

Brandi-Purata, J.M., V. Martínez-Ramírez, and E. Hernández-Flores, 1992, Análisis de la anomalia gravimétrica medida en Chilicote Sur, área Aldama, Chihuahua: XI Convención Geológica Nacional, Sociedad Geológica Mexicana, Libro de Resúmenes, p. 37.

Brown, M.L., 1985, Geology of Sierra de los Chinos-Cerro La Cueva, northwest Chihuahua, Mexico: Master's Thesis, University of Texas at El Paso, 163 p.

Camacho-Angulo, F., and O.J. Bermúdez-Villegas, 1992, Geología general de la franja costera entre Tampico y Soto La Marina para la localización de sitios para plantas generadoras de electricidad: XI Convención Geológica Nacional, Sociedad Geológica Mexicana, Libro de Resúmenes, p. 42.

Camargo-Zanoguera, A., and J.M. Quezada-Muñetón, 1991, Análisis geológico-económico de las áreas del Golfo de México con posibilidades petroleras: Boletín, Asociación Mexicana de Geólogos Petroleros, v. 41, p. 1–32.

Campbell, M. P., 1984, The geology of the southern Sierra Santa Rita, Chihuahua, Mexico: Master's Thesis, University of Texas at El Paso, 110 p.

Carlson, S., 1984, Mexico's oil plans; problems and potentials of the 1980's: Platt's Oilgrams/McGraw-Hill, Washington, DC, 369 p.

Carrillo-Martínez, M., 1992, El sistema de fallas imbricadas El Angel. Cinemática de secuéncias de deformación cordillerana; resultados preliminares: XI Convención Geológica Nacional, Sociedad Geológica Mexicana, Libro de Resúmenes, p. 49.

Castro-García, Arturo, J.J. Aranda, and J.T. Vázquez-Ramírez, 1992, Geología de la región de San Antonio de las Minas, Sierra de Guanajuato: XI Convención Geológica Nacional, Sociedad Geológica Mexicana, Libro de Resúmenes, p. 53.

Chapa-Guerrero, J.R., K. Schetelig, and P. Meiburg, 1992, Riesgos geológicos en zonas urbanizadas en el noreste de la Sierra Madre Oriental en el área metropolitana de Monterrey, N.L.: XI Convención Geológica Nacional, Sociedad Geológica Mexicana, Libro de Resúmenes, p. 60.

Collins, Edward W., and J.A. Raney, 1990, Tertiary and Quaternary faulting of the Hueco basin, Trans-Pecos Texas and Chihuahua, Mexico: Geological Society of America Abstracts with Programs: v. 22, p. A329–A330.

Córdoba, D.A., and L. Silva-Mora, 1989, Marco geológico del área de Revolución-Puerta de Cabrera, Estado de Durango: Revista del Instituto de Geología, Unversidad Nacional Autónoma México, v. 8, p. 111–122.

Corona-Esquivel, R., F. Ortega-Gutiérrez, J. Martínez-Reyes, and E. Centeno-García, 1988, Evidencias de levantamiento tectónico asociado con el sismo del 19 de Septiembre de 1985, en la región de Caleta de Campos, Estado de Michoacán: Revista del Instituto de Geologia, Universidad Nacional Autónoma México, v. 7, p. 106–111.

Correa-Pérez, I., and J. Gutiérrez y Acosta, 1983, Interpretación gravimétrica y magnetométrica del occidente de la Cuenca Salina del Istmo: Revista del Instituto Mexicano del Petróleo, v. 15, p. 5–25.

Cortes-Peña, E.A., and G. González-Pech, 1993, Caracterización geológica inicial y evaluación de reservas del Campo Secadero en el sureste de México: Asociación Mexicana Geólogos Petroleros, I Simposio de Geologia de Subsuelo, Ciudad del Carmen, Resúmenes, p. 118–119.

Coustere, G., et al., 1983, El DCA, un nuevo metodo para el análisis de fracturamiento atravesado en pozos: Ingeniería Petrolera, v. 23, p. 13–20.

Dahl, J.E., et al., 1993, Sistemas petroleros del sureste de México: Asociación Mexicana Geólogos Petroleros, I Simposio de Geología de Subsuelo, Ciudad del Carmen, Resúmenes, p. 139–145.

Davis, G.H., 1983, Hoja Viesca 14Q-g(9), con resúmen de la geología de la Hoja Tejupilco, estados de Guerrero, México y Michoacán: Carta Geológica de México, série de 1:100,000.

DeJong, K.A., and M.J. Westerfield, 1987, Backsliding in the Caborca area, northwest Mexico: Geological Society of America Abstracts with Programs, v. 19, p. 639.

DeJong, K.A., J.A. Escarcega-Escarcega, and P.E. Damon, 1988, Eastward thrusting southwestward folding, and westward backsliding in the Sierra La Víbora, Sonora, Mexico, Geology, v. 16, p. 904–907.

DeLeón-Gómez, H., K. Schetelig, and P. Meiburg, 1992, Problemas de ingenieria geológica y geohidrología durante la operación de la Presa Cerro Prieto/Linares, N.L. Mexico: XI Convención Geológica Nacional, Sociedad Geológica Mexicana, Libro de Resúmenes, p. 64.

Delgado-Argote, L.A., and J.G. Abdeslem, 1992, Reconocimiento geológico-geofísico en la Cuenca de Bahia de Los Angeles, Golfo de California: XI Convención Geológica Nacional, Sociedad Geológica Mexicana, Libro de Resúmenes, p. 67.

Del Vallo, G.R., and C.L.C. Ramírez, 1990, Nuevas perspectivas de interpretación sismoestratigráfica, por medio de atributos sismicos no convencionales: Revista del Instituto Mexicano del Petróleo, v. 22, n. 4, p. 6–13.

DeMis, W.D., 1985, Geology of Hell's Half Acre and its significance to the structural evolution of southern Marathon Basin West Texas, *in* P.W. Dickerson and W.R. Muehlberger, eds., Structure and Tectonics of Trans-Pecos Texas: West Texas Geological Society Publication 85-81, p. 27–37.

Dietzman, W.D., et al., 1983, The petroleum resources of Mexico: U.S. Department of Energy, Energy Information Administration, Washington, DC, 107 p.

Dillman, G.D., and J.F. Casey, 1985, Structural geology and tectonic history of the east-central Parras Basin Coahuila, Mexico: Gulf Coast Association of Geological Societies Transactions, v. 35, p. 45.

Drewes, H.D., and R. Dyer, 1993, Geologic map and structure sections of the Sierra Juárez, Chihuahua, Mexico: U.S. Geological Survey Miscellaneous Investigations Series Map I-2287, scale 1:12,500.

Dyer, R., 1988, Bibliography of geologic studies in northwestern Chihuahua Mexico, *in* K.F. Clark, P.C. Goodell, and J.M. Hoffer, eds., Stratigraphy, Tectonics and Resources of Parts of Sierra Madre Occidental Province, Mexico: Guidebook for the 1988 Field Conference, El Paso Geological Society, p. 173–181.

Enciso-Estrada, G., and D.P. Griffith, 1992, Prospecting with pressures, temperatures and velocities in the northwest Gulf of Mexico: XI Convención Geológica Nacional, Sociedad Geológica Mexicana, Libro de Resúmenes, p. 73.

Figueroa-Correa, J.L., et al., 1985, Estúdio geológico-geofísico del Prospecto Sama, área Hércules, Estado de Coahuila, México: Revista del Instituto Mexicano del Petróleo, v. 17, p. 6–26.

Flatt, D., 1984, Landsat reconnaissance of Chihuahua area, *in* E.C. Kettenbrink, ed., Geology and Petroleum Potential of Chihuahua: West Texas Geological Society Publication 84-80, p. 149–160.

Ford, B.H., 1985, Geochemistry of water in a coastal mixing zone, northeastern Yucatan Peninsula: Master's Thesis, University of New Orleans, 90 p.

Ford, B.H., J.D. Schuffert, R.K. Stoessell, and W.C. Ward, 1985, Fluid geochemistry in a coastal mixing zone on the Yucatan Peninsula: Geological Society of America Abstracts with Programs, v. 17, p. 585.

Gallo-Padilla, I., and A. Ortíz-Ubilla, 1992, Interpretación estructural en el sector Concepción del Oro, Zac., Aramberri, N.L.: XI Convención Geológica Nacional, Sociedad Geológica Mexicana, Libro de Resúmenes, p. 78.

Gamboa, V.J., S.C. Lazcano, E.C. Perry, and P.M. Villasuso, 1986, A self-sealing carbonate aquifer, Yucatan Peninsula, Mexico: Abstracts, International Symposium on Water-Rock Interaction 5, p. 222–224.

García, C., and M. Fernández Bollo, 1983, Definition du fond rocheux du lit du Rio Grijalva a Chicoasen: Bulletin of the International Association of Engineering Geology, v. 26, p. 243–246.

Garduño-M., V.H., 1984, Estúdio de accidentes del basamento en base a teledetección y análísis microstructural on el área de Matehuala, S.L.P., con un módelo experimental: Sociedad Geológica Mexicana, VII Convención Nacional, Memorias, p. 75–82.

Geological Map of Mexico, 1981: Secretaria de Programación y Presupuesto, Dirección General de Geografia del Territorio Nacional, scale 1:1,000,000.

Goodwin, L.B., and G.B. Haxel, 1990, Structural evolution of the southern Baboquivari Mountains, south-central Arizona and north-central Sonora: Tectonics, v. 9, p. 1077–1095.

Guatemala Ministry of Energy and Mines, 1988, Guatemala—a new petroleum exploration opportunity: Editor y Dirección General de Hidrocarburos, 12 p.

Gutiérrez-Acosta, J., 1992, Determinación de la anomalía de Bouguer con corrección isostática en la Plataforma de Córdoba y Cuenca de Veracruz: XI Convención Geológica Nacional, Sociedad Geologica Mexicana, Libro de Resúmenes, p. 86.

Gutiérrez-Nuñez, J.J., 1981, Procesos e interpretación de secciones de pseudovelocidades de intervalo en el sureste de México: Revista del Instituto Mexicano del Petróleo, v. 13, p. 22–38.

Guzmán-Vega, M., 1992, La pirólosis rock-eval como instrumento de interpretación geológica, GACETA, v. 1, n. 7, p. 1–2.

Hallman, S.E., et al., 1984, Geologic interpretation of geothermal movements in the Cerro Prieto field, Baja California, Mexico: AAPG Bulletin, v. 68, p. 18–30.

Harkey, D.A., and R. Dyer, 1985, Structural geology of Sierra San Ignacio, Chihuahua, Mexico: Field Trip Guidebook, West Texas Geological Society Publication 85-81, p. 195–204.

Henry, C.D., and G. Frederickson, 1987, Geology of part of southern Sinaloa, Mexico adjacent to the Gulf of California: Geological Society America Map and Chart Series, MCG063, scale 1:250,000, 14 p.

Hernández-Estevez, S., 1982, Indicadores de hidrocarburos en el Estado de Colima y su interpretación geológico petrolera: Ingeniería Petrolera, v. 22, p. 5–14.

Hernández-García, G., and J.M. Berlanga-Gutiérrez,

1983, Evaluación de volumenes originales de hidrocarburos empleando métodos geoestadísticos; aplicación al Campo Miguel Alemán-Chicontepec: Ingeniería Petrolera, v. 23, p. 18–40.

Hernández-García, J.F., 1981, Determinación de residuales gravimétricos en el Proyecto Cerro Minerva, Coahuilla, con el método de aproximación polinomial por mínimos cuadrados: Geos (Unión Geofísica Mexicana), v. 1 (4, A-B), p. C3.

Hernández-M., M., J.L. Gárrido-U., and S.G. Diáz-Velarde, 1992, Estudio de Geología y geohidrología aplicados a la selección de un sitio para una central hidroeléctrica en terranos cársticos; XI Convención Geológica Nacional, Sociedad Geológica Mexicana, Libro de Resúmenes, p. 92.

Hernández-Ramírez, J., and M. Salazar-Bustamante, 1993, Desarrollo de los yacimientos del Campo Bellota: Asociación Mexicana Geólogos Petroleros, I Simposio de Geología de Subsuelo, Ciudad del Carmen, Resúmenes, p. 120–135.

Hernández-Sánchez, C., 1984, Definición de zonas con presiones anormalmente altas en el área Huimanguillo-Reforma: Revista del Instituto Mexicano del Petróleo, v. 16, p. 73–79.

Hoffelt, J., 1985, Karst hydrology expedition to the Sierra Madre Oriental of Mexico: Geo, v. 12, p. 26–27.

Huízaar-Alvarez, R., and O. Oropeza-Orozco, 1989, El karst en la región de Galeana, Nuevo León: Universidad Nacional Autónoma de México, Instituto de Geología, Revista, v. 8, p. 71–83.

Hurtado-Carador, M., 1981, Aplicación de registros geofíscos de pozo a la exploración del carbón: Resúmenes de la reunión, Geos, 1 (4,A-B), p. C6–C7.

INEGI, 1980, Carta geológica de México: Instituto Nacional de Estadística Geográfia e Informática, Scale 1:1 million, 8 sheets.

Iparrea, V., and M. Pintado, 1992, Geología del acuifero Huazuntlán, Ver: XI Convención Geológica Nacional, Sociedad Geológica Mexicana, Libro de Resúmenes, p. 95.

Jiménez-Nava, F.J., 1992. Estratigrafía del Valle de Aguascalientes. Aportes de la exploración geohidrológica a profundidad: XI Convencíon Geológica Nacional, Sociedad Geológica Mexicana, Libro de Resúmenes, p. 99.

Jiménez-Salas, O.H., 1984, Application of the Shuttle imaging radar-A (Sir-A) imagery to stratigraphic and tectonic studies in the Sierra Madre Oriental, northeastern Mexico: Master's Thesis, University of Texas at Dallas, Richardson, Texas, 120 p.

Johnson, C.A., J.A. Barros, and C.G.A. Harrison, 1989, Tectonics and volcanism in central Mexico: a landsat thematic mapper prespective: Remote Sensing of the Environment, v. 28, p. 273–286.

Lancin, M., 1985, Geomorfología génesis de las flechas litorales del Canal del Infiernillo, Estado de Sonora: Universidad Nacional Autónoma de México, Instituto de Geología, Revista, v. 6, p. 52–72.

Lesser, J.M., 1988, Quantification and prediction of evolution of a limestone aquifer in the north part of Mexico, *in* Juan Daxian, ed., Karst Hydrogeology and Karst Environment Protection: Proceedings IAHS-AISH Publication, 176, v. 1, p. 421.

Lesser, J.M. and G. Lesser, 1988, Region 9, Sierra Madre Oriental, *in* W. Back et al., eds., The Geology of North America: Geological Society of America, v. 02, p. 89–92.

Lesser, J.M., and A.E. Weidie, 1988, Region 25, Yucatan Peninsula, *in* W. Back et al., eds., The Geology of North America: Geological Society of America, v. 2, p. 237–241.

Lomnitz, C., 1983, Seismicity and earthquake risk at the NPP site of Laguna Verde, Veracruz: Geofísica Internacional, v. 22, p. 113–135.

Longoria, J.F., and O.H. Jimenez, 1985, Spaceborne radar imagery in regional geologic mapping of the Sierra Madre Oriental, northeastern Mexico: International Symposium on Remote Sensing of Environment: ERIM, Ann Arbor, Michigan, April 1985, p. 437–446.

Lopez-Vega, J., 1980, Evaluación económica-petrolera del área Comitán-Pedregal, Estado de Chiapas: Boletín de la Asociación Mexicana de Geólogos Petroleros, v. 32, n. 1, p. 57–77.

Lovejoy, E.M.P, ed., 1980, Sierra de Juarez, Chihuahua, Mexico: Structure and stratigraphy: El Paso Geological Society, 59 p.

Lugo-Hubp, J., 1990, Él relieve de la República Mexicana: Universidad Nacional Autónoma de México, Instituto de Geología, Revista, v. 9, p. 82.

Marín, L.E., E. Perry, C. Booth, and M. Villasuso, 1987, Hydrogeology of the northwestern peninsula of Yucatan, Mexico (abs.): EOS, Transactions, American Geophysical Union, v. 68, p. 1292.

Marín, L.E., J.M. Quezada, V.L. Sharpton, G. Ryder, and B. Schuraytz, 1992, La estructura de impacto de Chicxulub: XI Convención Geológica Nacional, Sociedad Geológica Mexicana, Libro de Resúmenes, p. 104–105.

Márquez, G., and R. Piñeda, 1983, Uso de geotextil en la construcción de carreteras: Proceedings of the Panamerican Conference on Soil Mechanics and Foundation Engineering, v. 7, p. 381–388.

Martínez, R.D., J.D. Schroeder, and G.A. King, 1987, Formation pressure prediction using seismic data from the Gulf of Mexico. Volume 3: Proceedings, Offshore Technology Conference, v. 19, Houston, Texas, p. 259–268.

Martínez-Flotte, L., 1983, Descripción y utilización del tonstein en la Cuenca Fuentes Rio Escondido, Coahuila: Geomimet, v. 3, p. 69–76.

Martínez-Goves, M., and J.O. Ortiz-Moreno, 1992, Factibilidad geológica del prospecto hidroeléctrico Arroyo Hondo, Jalisco: XI Convención Geológica Nacional, Sociedad Geológica Mexicana, Libro de Resúmenes, p. 110.

Martínez-Rodríguez, E., 1992a, Exploración geológica para la modernización de la central hidroeléctrica Tuxpango, Estado de Veracruz: XI Convención Geológica Nacional, Sociedad Geológica Mexicana, Libro de Resúmenes, p. 112.

Martínez-Rodríguez, E., 1992b, Exploración en casa de máquinas del proyecto hidroeléctrico Xuchiles, en el Rio Blanco, Estado de Veracruz: XI Convención Geológica Nacional, Sociedad Geológica Mexicana, Libro de Resúmenes, p. 113.

Maya-Cortes, V., 1992, Estudio de las propiedades hidrodinámicas de una porción del acuitardo de Tlahuac, D.F., Chalco, México: XI Convención Geológica Nacional, Sociedad Geológica Mexicana, Libro de Resúmenes, p. 115–116.

McLean, H., 1988, Reconnaissance geologic map of the Loreto and part of the San Javier quadrangles, Baja California Sur, Mexico: United States Geologic Map MF-2000, scale 1:50,000.

Méndez, V.H., 1993, Proyecto de inversión Ocosingo incorporación de reservas: Asociación Mexicana Geólogos Petroleros, I Simposio de Geología de Subsuelo, Ciudad del Carmen, Resúmenes, p. 20–21.

Mendoza-Antillón, S., 1981, Anomalia Zeuz, Coahuila: Resúmenes de la reunión, Geos, Union Geofísica Mexicana, v. 1 (4, A-B), p. C9.

Mendoza-Romero, G., and C. Pérez-Rosales, 1984, Determinación de propiedades petrofísicas a traves de un nuevo procedimiento, su aplicación en la Sonda de Campeche y en el área de Tabasco-Chiapas (abs.): Ingeniería Petrolera, v. 24, n. 5, p. 22.

Miranda, M.A., 1984, Reconocomiento geológico de la Cuenca del Rio Santa Clara, Estado de Chihuahua: SARH-ECOTERRA (internal report), 39 p.

Molina, G.G., 1983, Aplicacion del procesado por ondicula en la interpretación sismico-estratigráfica del área Piedras Negras, Coahuila: Ingeniería Petrolera, v. 23, p. 5–12.

Montiel Hernández, D., et al., 1984, Evaluación de pozos intermedios en yacimientos de gas de baja permeabilidad parcialmente drenados; Distrito Frontera Noreste: Ingeniería Petrolera, v. 24, p. 5–14.

Morales-Recinos, L., and H. Macias-González, 1992, Marco estructural y su influéncia en las aguas subterraneas en la zona de Lara Grajales-Perote, Estados de Puebla y Veracruz: XI Convención Geológica Nacional, Sociedad Geológica Mexicana, Libro de Resúmenes, p. 130.

Moreno-Castillo, R., 1982, Método de interpetación cualitativa de mapas gravimétricos y su aplicación en la Zona Sur de Pétroleos Mexicanos: Boletín, Asociación Mexicana de Geólogos Petroleros, v. 34.

Nehring, R., 1991, Oil and gas resources, *in* A. Salvador, ed., The Gulf of Mexico Basin: Geological Society of America, v. J, p. 445–494.

Nick, K., 1988, Mineralogische, geochemische und petrographische Untersuchungen in der Sierra de San Carlos (Mexiko): Unpublished Ph.D. Dissertation, Universitaet Karlsruhe, Germany, 167 p.

Nieto-Samaniego, A.F., J.J. Aranda-Gómez, R. Machorro-Sagastume, 1992, Fases y estilos Cenozóicos de fallamiento en el límite septentrional del Bajío: XI Convención Geológica Nacional, Sociedad Geológica Mexicana, Libro de Resúmenes, p. 132.

Ortega-Gutiérrez, F., and J.C. Guerrero-García, 1983, The geologic regions of Mexico, *in* A.R. Palmer, ed., Perspectives in Regional Geological Synthesis: Planning for the Geology of North America: Geological Society of America, p. 99–104.

Ortlieb, L., and C. Pierre, 1981 (1983), Génesis evaporítica en tres áreas supralitorales de Baja California: Contextos sedimentarios y procesos actuales: Universidad Nacional Autónoma de México, Instituto de Geología, Revista, v. 5, p. 94–116.

Padilla y Sánchez, R.J., 1983, Resúmen de la geología de la Hoja Ventura, estados de Coahuila, Nuevo León, Zacatecas y San Luis Potosí: Carta Geológica Mexicana, 1:100,000 14R-J8, 9 p.

Paris, J.P., 1992, Percepción remota e investigación geológica: XI Convención Geológica Nacional, Sociedad Geológica Mexicana, Libro de Resúmenes, p. 154–155.

Pérez-Venzor, J. A., R. Miranda-Aviles, and M. de J. Puy-Alquíza, 1992, Características de los afloramientos del sistema de falla de La Paz entre la Ciudad de La Paz y San Juan de los Planes B.C.S., México: XI Convención Geológica Nacional, Sociedad Geológica Mexicana, Libro de Resúmenes, p. 157.

Peterson, J.A., 1983, Petroleum geology and resources of southeastern Mexico, northern Guatemala, and Belize: U.S. Geological Survey Circular 760, 44 p.

Peterson, J.A., 1985, Petroleum geology and resources of northeastern Mexico: U. S. Geological Survey Circular 943, 30 p.

Petroconsultants, S.A., 1990, Mexico Oil and Gas Map, Foreign Scouting Service.

Puerto-Zapata, C., and F. Díaz-Zamora, 1984, Estudio sismológico-estructural determinado con el método tridimensional en el área marina de Campeche: Ingeniería Petrolera, v. 24, n. 5, p. 25.

Ramos-Aracén, R., 1992, Exploración petrolera en fallas y fracturas en la región de Tampico: XI Convención Geológica Nacional, Sociedad Geológica Mexicana, Libro de Resúmenes, p. 160.

Ramos-Rodríguez, J.M., 1992, Geomorfología del Valle de Poanas-Vicente Guerrero, Dgo: XI Convención Geológica Nacional, Sociedad Geológica Mexicana, Libro de Resúmenes, p. 161.

Rodríguez de Barbarín, C., M. Hoffmann, and M.M. Rangél-Rodríguez, 1992, Investigaciones hidrogeológicas en el campus Universitario, Linares, Nuevo León: XI Convención Geológica Nacional, Sociedad Geológica Mexicana, Libro de Resúmenes, p. 167.

Ronquillo, G., and G. López, 1986, Avance en la elaboración de la carta sismotectónica de la Republica Mexicana 1:2,000,000 (abs.): Geologie und Paläontologie, Sonderband, p. 179–180.

Sánchez-Pérez, J., J. Azara-Z., S. Flores-C., and M.A. Soto-G., 1992, Estudios geológico-geohidrológicos, para la construcción de plantas termoeléctricas en Baja California: XI Convención Geológica Nacional, Sociedad Geológica Mexicana, Libro de Resúmenes, p. 173.

Santamaría-Guevara, N.E., and P.M.A. Hernández, 1989, Desarrollo de correlaciones para determinar las propiedades físicas de los hidrocarburos, producidos de los principales campos petroleros de México: Revista del Instituto Mexicano del Petróleo, v. 21 (1), p. 63–79.

Santana-Palomino, C., 1992, Exploración geohidrológica en la Meseta Tarasca: XI Convención Geológica Nacional, Sociedad Geológica Mexicana, Libro de Resúmenes, p. 176.

Santiago-Valencia, R., 1984, Método de inversión geofísica aplicado a sondeos eléctricos verticales: Revista del Instituto Mexicano del Petróleo, v. 16 (3), p. 7–17.

Saucedo-Quiñones, D., P. Barriga-Lanos, and J. Alcalá-Amaro, 1992, Investigación de la intrusión salina por medio de sondeos eléctricos verticales en el área de Petacalco, Gro: XI Convención Geológica Nacional, Sociedad Geológica Mexicana, Libro de Resúmenes, p. 177.

Solís-Estrada, R.A., and J.L. Hernández-Velázquez, 1992, Geología aplicado en el proyecto Boca del Cerro, Tab., Sitio Santa Margarita: XI Convención Geológica Nacional, Sociedad Gelógica Mexicana, Libro de Resúmenes, p. 184.

Stewart-Gordon, 1991, Mexico to boost exports in the 1990's: Petroleum Engineer International, November 1991, p. 32–34.

Stoessell, R.K., et al., 1989, Water chemistry and $CaCO^3$ dissolution in the saline part of an open-flow-mixing zone, coastal Yucatan Peninsula, Mexico: Geological Society of America Bulletin, v. 101, p. 159–169.

Tomaszewski, J., et al., 1989, Estudio cronológico en la parte meridional de la Cuenca de México: Universidad Nacional Autónoma de México, Instituto de Geología, Revista, v. 8, p. 223.

Uresti-Ariceaga, J.C., 1992, Estudio geohidrológico para el abastecimiento de agua subterranea a la C.T. Pdte. a. López Mateos: XI Convención Geológica Nacional, Sociedad Geológica Mexicana, Libro de Resúmenes, p. 200.

U.S. Department of Energy, Office of Oil and Gas, 1983, The Petroleum Resources of Mexico: Energy Information Administration, 109 p.

Valladares-Lagunas, E., J. Casique-Vázquez, and J. Avalos-Salazar, 1992, Geología de una porción de la Sierra de Juárez para un proyecto hidroeléctrico: XI Convención Geológica Nacional, Sociedad Geológica Mexicana, Libro de Resúmenes, p. 202.

Vargas-Badillo, J., 1992, Factibilidad geológica para el proyecto hidroeléctrico de rebombeo en Monterrey, Nuevo León: XI Convención Geológica Nacional, Sociedad Geológica Mexicana, Libro de Resúmenes, p. 203.

Vázquez-Izar, M., and S. Moreno-Gárnica, 1992, La importancia de la geología aplicado a la construcción de obras de contención en el P.H. Aguamilpa: XI Convención Geológica Nacional, Sociedad Geológica Mexicana, Libro de Resúmenes, p. 207.

Weidie, A.E., 1988, Fracture control of groundwater flow in folded and horizontal carbonate rocks in Mexico (abs.), *in* Y. Daoxian, ed., Karst Hydrogeology and Karst Environment Protection: IAHS-AISH Publication, 176, v. 1, p. 422.

West, R.C., P.C. Psuty, and B.G. Thom, 1985, Las tierras bajas de Tabasco: Gobierno del Estado de Tabasco, Biblioteca Básica Tabasqueña, v. 8, 490 p.

Yañez-M., M., 1984, Planeación de la perforación exploratoria a partir de la sismología: Ingeniería Petrolera, v. 24, p. 5–17.

Zenteno-B., M.A., 1984, Exploración y explotación petrolera en México y en el Mundo: Ingeniería Petrolera, v. 24, p. 5–12.

Index